Fluid Power and the Mechanics of Fluids

Fluid Power and the Mechanics of Fluids

Second Edition

Larry Villasmil and Robert Garrick

Rochester Institute of Technology

cognella

SAN DIEGO

Bassim Hamadeh, CEO and Publisher
Carrie Baarns, Senior Manager, Revisions and Author Care
Kaela Martin, Senior Project Editor
Rachel Kahn, Production Editor
Alexa Lucido, Licensing Manager
Emely Villavicencio, Senior Graphic Designer
Alia Bales, Director, Project Editorial and Production
Natalie Piccotti, Director of Marketing
Kassie Graves, Senior Vice President, Editorial

Printed in the United States of America.

cognella® | ACADEMIC PUBLISHING
320 South Cedros Ave., Ste. 400, Solana Beach, CA 92075

Contents

Preface

This book was conceived and intended to address the needs of instructors and students desiring to integrate fluid mechanics and fluid power studies. Traditionally, fluid power is taught as a prerequisite to fluid mechanics/fluid dynamics classes. Some engineering technology educators prefer a more effective and efficient treatment of fluid power and fluid mechanics in one integrated text. The authors hope that this book will appeal to these educators. The objective of this book is to provide an important structure for the beginning learner. The objective is to help the learner organize basic principles in their mind, since this type of learners tends to treat concepts as an undifferentiated collection of unrelated ideas. The authors' objective is to provide an organizing hierarchy that will help the beginning learner organize fluid power and fluid mechanic concepts and aid with retention.

The specific structure of the book is divided into three basic sections: Chapters 1 through 4 provide the basic concepts and introduction to fluid power. Chapter 5 introduces fluid in motion and the conservation principles of conservation of mass, energy, and momentum as a natural integrating device for the study of fluid power, fluid sciences, and thermal sciences. Chapters 5 through 9 then cover fluid power components, fluid power theory, fluid power circuit design, and applications. Chapter 10 reviews special topics in the study of fluid power, specifically virtual experiments using Automation Studio and a review of the state-of-the-art in fluid power and fluid mechanic analysis. In addition to this organizing structure of the book, we have also attempted to use significant color figures and images to illustrate important applications and concepts. Each chapter also begins with learning objectives and a chapter overview to provide students with a summary of the chapter's objectives. Each chapter also contains example problems that follow the standard problem-solving format outlined in the text. Key equations in the text are denoted in each chapter. All these features were incorporated into the text to assist student learning and for ease of use by the instructor. The figures help the students connect the subject reviewed to real-world applications. This connection of theory to practical applications is key for many engineering technology students. The incorporation of learning objectives at the beginning of each chapter is also useful to students as they desire to understand what could be incorporated into a formative or summative test or assessment. These learning objectives should also help the instructor in the selection of review problems and during the creation of assessments and tests.

Computer-Assisted Problem Solving and Design

Practicing engineers use computer animations and simulations to solve engineering problems. Therefore, the text will introduce computer animations, simulations, and smartphone apps to assist students in mastering fluid power concepts. In this text, we will use computer-assisted problem solving and computer fluid power design programs to allow students to explore multiple scenarios and simulate circuit designs.

Acknowledgments

We would like to acknowledge the support of our wives and families while we dedicated time to writing this text. We also appreciate the support and guidance given by the Cognella publishing team.

About the Authors

Larry Villasmil, MSc, PhD, is an associate professor in the Department of Manufacturing and Mechanical Engineering Technology (MMET) of the College of Applied Science and Technology (CAST). Before joining the Rochester Institute of Technology (RIT) faculty, Dr. Villasmil accumulated five years of research experience, was a graduate teaching assistant, and was a fellow of the Graduate Teaching Academy sponsored by the Center of Teaching Excellence at Texas A&M University. Prior to his enrollment at Texas A&M, he worked in the Exploration and Production division of one of the largest state-owned oil corporations for 12 years as a conceptual engineering leader and rotating equipment and machinery adviser while serving as an instructor for the international educational branch and content specialist to periodically revise and update the Standards and Design Engineering Manual. Dr. Villasmil's current areas of research include computational fluid dynamics and turbulence modeling and its application to multiphase flow and internal combustion engines; wind power as alternative energy; fluid power; and the use of technology in the classroom.

Robert D. Garrick, PhD, PE, is an associate professor in the Department of MMET of CAST at RIT. Dr. Garrick worked for 25 years in automotive engineering research and holds eight patents.

CHAPTER 1

Introduction

Overview and Learning Objectives

This chapter will cover the definition of fluid power and the two major fields of fluid power, including hydraulics and pneumatics. We will also review the fundamentals of fluid mechanics and fluid dynamics. This chapter will also review the applications of fluid power systems, including the basic components, and review why the study of fluid power systems is a key aspect for mechanical engineers.

The learning objectives of this chapter should allow students to do the following:

- Explain in their own words the terms *fluid power*, *hydraulics*, *pneumatics*, *fluid transport*, *gas*, and a *liquid* (i.e., what are they?).
- Sketch a fluid power system and label the major components.
- Explain the differences between a pneumatic system and a hydraulic system.
- Explain how power is transmitted and converted in a fluid power application.
- Explain in their own words the difference between fluid power and fluid transport systems.

Fluids: Liquids and Gases

We will use the term *fluid* for substances that flow under an applied force or stress. In engineering applications, the term *fluids* is used to describe a large set of matter that includes liquids, gases, and plasmas. For fluid power applications, we will only concern ourselves with liquids and gases. As mentioned previously, the term *fluid* includes both gases and liquids in engineering terminology. This can be confusing to a student because in common usage, *fluid* is often a synonym for a liquid, with no implied reference to a gas. For example, when you go to your doctor with a cold and they say to take plenty of fluids, they are referring obviously to liquids, with no intention for you to ingest gases. In the engineering study of fluid power and fluid mechanics, the term *fluid* is used for both liquids and gases. Fluids display two key properties. These are their ability to flow in a system (inability to resist a deforming force) and to take the shape of their container. The key difference between gases and liquids is that liquids form a free surface that is not created by the container by which they are held. Liquids differ from gases in that liquids do not always mix readily with other liquids. An example of this is water and oil; they do not mix easily with each other, and if they are left sitting together, they will separate.

Figure 1-1 Two phases (water and oil) in the same state of aggregation (liquid).

Gases, on the other hand, readily mix with each other and will always fill every space in the container in which they are held. Very importantly, liquids will not compress significantly except under extremely high pressure. The ability of liquids to flow and distribute applied pressure evenly along with their incompressibility (i.e., they do not compress) makes liquids very suitable for fluid power hydraulic applications. Gases do not form a free surface and expand to take up all of a container's volume.

It is interesting to note that most known matter in the universe is in a gaseous form, with liquids being a relatively rare form of a state of matter in the universe. This is very different from our experience on Earth, where water is by far the most common state of matter (i.e., a liquid). Liquids exist in a relatively narrow temperature and pressure range. If the temperature is increased, the liquid changes to a gas; if the temperature decreases, the liquid changes to a solid.

What Is Fluid Power?

Fluid power is the energy transmitted over time by a pressurized fluid. This fluid can be either a liquid or a gas.

Figure 1-2a Hollow empty glass and wineglass with blue liquid

Figure 1-2b

When a liquid is put into a container, it takes the shape of it (Figure 1-2). The liquid covers the bottom and sides of the container up to the level to which it is filled. The top surface of the liquid is in contact with the atmosphere above it and maintains a near-flat top level. As the container holding the liquid is tipped, the top of the fluid also tips with it. In contrast, when a gas is held in a closed container, it expands and completely fills the container, touching all sides (as shown in Figure 1-3).

If the container is moved or tipped, the fluid—in this case, a gas—will not change inside the container. If the container is opened, allowing the top of the gas to be exposed to the atmosphere, the gas will expand and escape from the container into the atmosphere. A significant difference between these two fluids is that if a gas is contained in a container or cylinder, you can compress it fairly easily by pushing on the sides of the container. By reducing the container's volume, the gas will be compressed and increase in pressure. An example of this is seen when using a hand-operated bicycle pump to pump up a tire or basketball (Figure 1-4).

As the piston in the bicycle pump is pushed down, the volume of the gas is reduced, increasing the pressure inside the pump. This air at increased pressure is then put through a hose to the tire or ball being inflated. If the fluid inside of the pump is a liquid instead of a gas, you could apply a very large force to it, which would increase the pressure of the liquid, but its volume would change very little. This key difference between liquids and gases leads to the following generalizations that we will use in this text:

Figure 1-3 Gas bottle for nitrogen gas.

- Generalization 1: Gases are easily compressible.
- Generalization 2: Liquids are only slightly compressible, and in many problems, we will consider them incompressible—in other words, their volume remains the same under pressure; it does not change.

There are two different types of fluid systems. The first type is fluid power; the second is fluid transport.

Fluid power systems are designed to exert forces over a distance, pushing, pulling, or rotating parts or elements. These forces over a distance generate work. The work in the fluid power system is accomplished by the pressurized fluid operating on a cylinder or fluid power motor.

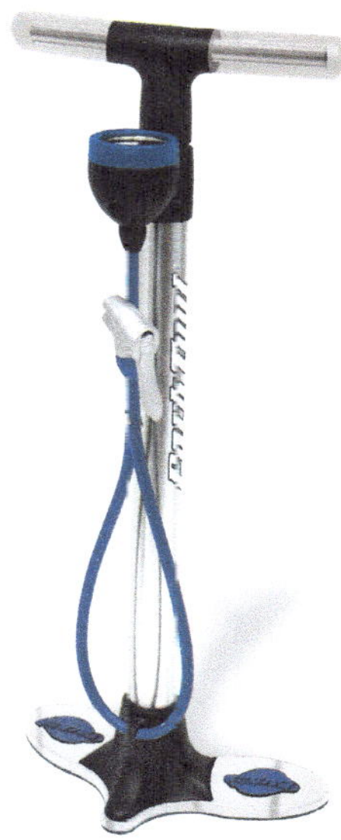

Figure 1-4 Bicycle floor pump.

Figure 1-5 A hydraulic cylinder.

A fluid power cylinder produces linear force, resulting in linear motion. A fluid power motor produces rotary force, or torque, resulting in rotary motion. Fluid power cylinders and fluid power motors are called *actuators*, as they provide the force over a distance to accomplish the work desired.

A fluid power system transmits energy and controls a system. Fluid power systems that use liquids are called hydraulic systems. These may use liquids such as oil or water. Fluid power systems that use gases are called **pneumatic systems**; they use compressed air or other inert gases, such as nitrogen.

Fluid power systems can be combined with other mechanical or electrical technologies to produce larger systems. The components of these larger systems can include sensors, microprocessors, programmed logic, linear or rotary acting devices, and human user interface devices. Fluid power systems are used for linear motion, rotary motion, and control of many modern devices in industrial and commercial applications.

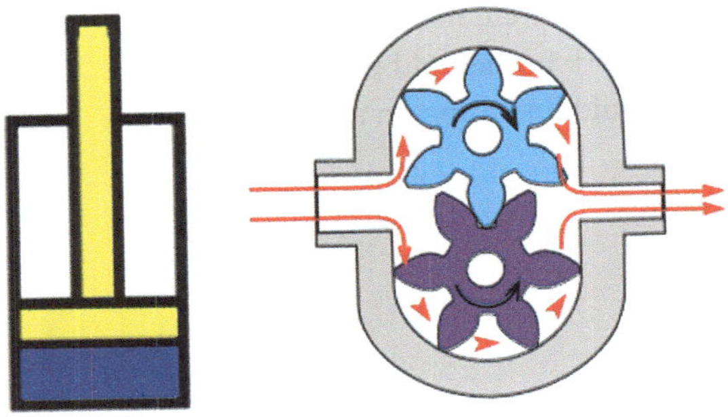

Figure 1-6 Hydraulic piston cylinder (colored) and gear pump.

Fluid Transport Introduction

The second type of fluid system is **fluid transport**. Fluid transport systems move fluids across distances, such as across containers or conduits. For example, in your vehicle, gasoline is stored in the tank and delivered through a pump and fuel lines to the engine. Other fluids are also stored and transported in your vehicle, such as windshield washer fluid, engine coolant, engine oil, power steering fluid, transmission fluid, intake air, and the airflow in the heating and ventilation system. In all these cases, fluid in both liquid and gas states is pumped and transmitted through tubing, hoses, and valves to other components. Also in each of these cases, energy is introduced into the system and dissipated away as the fluid travels through the system. In fluid transport systems, we are concerned with the flow of fluid and the conservation and transformation of energy in the fluid.

Figure 1-7 Excavator.

An Application of Fluid Power

Hydraulic Hybrid

We will now look at an application of fluid power systems to understand the work accomplished by the pressurized fluid and the forces generated by it when under pressure.

Fluid power can be used to power hybrid vehicles; in that case, pressurized fluid is used in the place of electric power to move the vehicle. An additional internal combustion engine can also be used to extend the range of the hydraulic hybrid. Similar to an electric hybrid, a hydraulic hybrid vehicle is employed to achieve greater fuel economy for the driver. In a hydraulic hybrid vehicle, the hydraulic fluid is stored in a reservoir and pumped from a low-pressure reservoir to a high-pressure container. When the driver wants to move the vehicle, the high-pressure fluid is released from the high-pressure container and moves through a hybrid hydraulic motor to turn the drive wheels of the vehicle. As the fluid exits the hydraulic motor, the fluid returns to the low-pressure fluid reservoir. In this instance, the energy contained in the high-pressure fluid has been removed to do work and turn the vehicle wheels, propelling the vehicle forward.

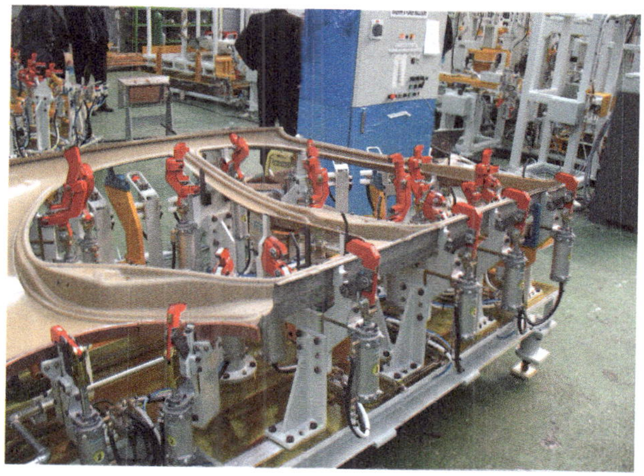

Figure 1-8 Manufacturing equipment.

There are four main components in a hydraulic hybrid vehicle. The first component is the hydraulic fluid; the second is the low-pressure reservoir; the third is the pump or motor used to rotate the vehicle wheels; and the last component is the high-pressure container, also called an *accumulator*. Hydraulic hybrids can also recover the kinetic energy of the vehicle during braking and deceleration to recover the vehicle's forward-motion kinetic energy and store it in the high-pressure reservoir for future use. The regenerative braking of hydraulic hybrids is significantly more efficient than that of electric systems, as hydraulic hybrids can recover more than 70% of the vehicle's kinetic energy during braking. Hydraulic hybrids are used currently for trucks and buses, as the kinetic energy in a truck or bus can be very significant, given their size and weight, and recovered as the truck or bus brakes. Hydraulic hybrid systems can be less expensive than electric systems because currently, in electric hybrid vehicles, batteries alone make up a substantial chunk of their cost.

http://www.epa.gov/otaq/technology/research/research-hhvs.htm

Components of a Fluid Power System

Hydraulic Fluid Power System

A schematic of a typical hydraulic system is shown in Figure 1-9.

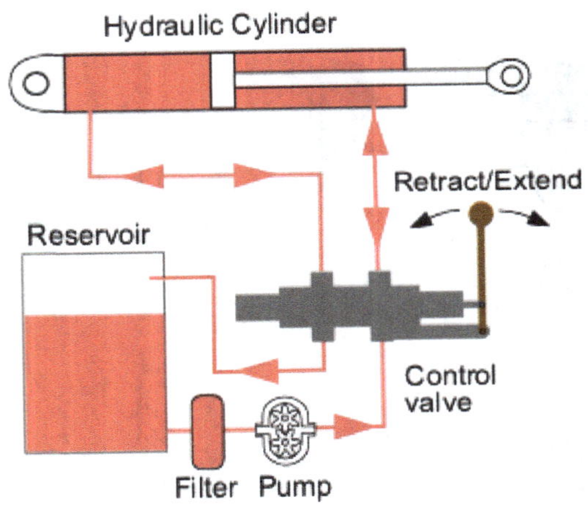

Figure 1-9 Typical hydraulic system.

The main components of a hydraulic system are the following:

1. Tank or reservoir
2. Actuator/pump
3. External power source to drive the hydraulic pump (prime mover)
4. Control valves
5. Hydraulic actuators (cylinders or hydraulic motors)
6. Tubes/piping/hoses

The first component is a **tank or reservoir**. The hydraulic fluid power reservoir holds the excess hydraulic fluid used in the system. This is used to accommodate the changes in the system's volume as cylinders extend or retract or motors rotate. The tank also provides extra fluid for the system to account for temperature-driven changes in volume and system leakage. The tank also provides a function to separate any air in the system from the fluid—in this case, the liquid/hydraulic oil. The tank provides additional features such as dirt separation, as foreign material typically settles at the bottom of the tank/reservoir, and the tank can act as a cooling source as heat in the fluid can dissipate in the tank and surroundings.

The second component of a hydraulic fluid power system is the **pump**. The hydraulic pump supplies the fluid at the pressure and flow rate required for the system. The pressure developed by the pump allows the actuator to generate the force required by the system. Hydraulic pumps generally have a much greater ability to generate power than electric motors by volume. Common hydraulic pumps include gear pumps, vane pumps, and both axial and radial piston pumps. These are driven by the third component of the fluid power system, which is the external power source. Throughout this text, we will call this source the *prime mover*. The prime mover can be an electric motor, a gas or diesel engine, or any other device connected through belts or gears or directly coupled to the pump.

The fourth component of the hydraulic fluid power system is the **directional control valve**. This valve controls the movement of the fluid in the fluid power circuit. The control valve typically uses a spool inside of a housing. As the spool slides to different locations, the fluid is able to flow between different inlet and outlet port locations based on the position of the valve spool. Control valves can accomplish various functions, such as relieving pressure relief, checking valves, and regulating pressure regulation, cartridge valves, hydraulic fuses, and sequence valves.

The fifth component of the hydraulic fluid power system is the **actuator**. This component transfers the power in the fluid to accomplish linear or rotary motion. Linear motion is accomplished using a hydraulic cylinder. Rotary motion is accomplished using a hydraulic motor. A hydraulic motor can be thought of as a hydraulic pump operating in a reverse condition. Here, fluid power is provided to the motor and rotary motion is output.

The last component of the fluid power hydraulic system is the tubing, pipes, or hoses needed to transport the fluid to and from all other components. These connection systems provide flexibility in the location of the fluid power components. The connection systems need to be carefully designed to withstand the high fluid-power hydraulic pressures and avoid causing excessive pressure losses during operation.

While the complexity of a fluid power hydraulic system can vary depending upon the function desired, most hydraulic systems will have the six components listed above. This text will study in detail the entire system's individual components and design in subsequent chapters. We will review the design and options of each system component and the proper integration of the components into a system to deliver the functionality required.

Pneumatic Fluid Power System

A schematic of a typical pneumatic system is shown in Figure 1-10.

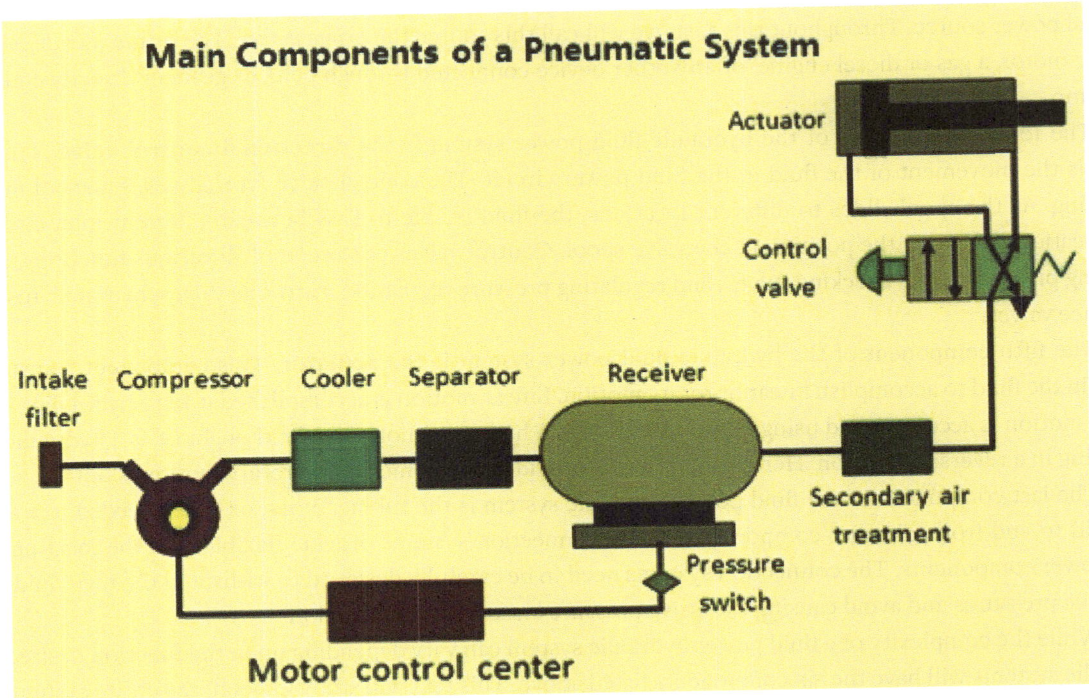

Figure 1-10 Typical pneumatic system.

The main components of a pneumatic system are the following:

1. Air tank
2. Air compressor
3. External power source to drive the air compressor (prime mover)
4. Control valves
5. Pneumatic actuators (cylinders or air motors)
6. Tubes/piping/hoses

All components in a pneumatic system perform the same function as the hydraulic counterparts. The only (and major) difference between pneumatic and hydraulic systems is the fluid used by the systems to operate. In a pneumatic system, the fluid is air that is drawn from the atmosphere and compressed through the air compressor. In a hydraulic system, the fluids are liquids, typically oils, that are drawn from the hydraulic reservoir. Similarly, in a pneumatic system, the air or gas is exhausted after use back into the atmosphere. In a hydraulic system, the fluid or liquid is returned to the reservoir after its use.

Motivation to Study Fluid Power Systems

We study fluid power systems because of their wide uses in everything from household components to large industrial systems. Fluid power allows us to move large forces or control movement of systems precisely. These systems also allow us to operate components under special circumstances or in certain environments. The fluid power industry operates globally or beyond in our space systems. Our transportation systems, material moving systems, and manufacturing systems that produce goods globally for our use rely on fluid power systems. They are also prevalent in agriculture and processing systems that produce and deliver our food. Therefore, the study of systems and their design are a critical part of many engineers' jobs. Fluid power engineers design and develop new fluid power systems or components to achieve various functions.

Figure 1-11 A small air compressor.

Advantages and Disadvantages of Hydraulic and Pneumatic Systems

In discussing the advantages and disadvantages of fluid power systems, it is important to understand that no system is better than another. The specific system design features are selected based on their suitability for a specific purpose or function. Power systems are divided into three basic types, which are electrical power systems, mechanical power systems, and fluid power systems. A typical manufacturing or product system will use a combination of these three basic power systems: electrical, mechanical, and fluid power.

Consequently, it is very important to know the features of each of these systems. In ranking the systems' ability to transmit power economically over distances, we can say that mechanical systems are generally restricted to transmitting power over short distances. For example, a mechanical gear system can convey power over distances of inches or feet, but it would be very challenging to build a gear system that could transmit power over hundreds or thousands of feet.

Figure 1-12 Farm tractor.

At the other extreme, electrical power systems can transmit electrical power more economically over greater distances in the order of thousands of feet, as compared with mechanical or fluid power systems.

In terms of transmitting power over distance, fluid power systems operate in the range between mechanical power systems and electrical power systems in transmitting power in small- to medium-distance ranges. As for transmitting power over various geometries, we can again review our simple mechanical gear system and see that power typically has to be in line or at fixed angles by a gear system or strictly in line by a belt or chain mechanical power system.

Fluid or electrical power systems are not restricted by geometry. We can design an electrical or fluid power system with power sources and actuators at different heights and locations, as the wiring or fluid power tubing can transmit the system's power to various locations without the restriction of being in line like the gears, chains, or belts of the mechanical power system. In terms of absolute power, transmitted fluid power systems generally can absorb and transmit power at higher levels/rates than mechanical or electrical power systems. Electrical power systems typically are limited by the battery's physical materials and ability to quickly absorb and disburse electrical energy. We all have had the experience that charging a cell phone battery occurs over a long period of time in comparison with the almost instantaneous ability of a hydraulic system to charge a pressurized accumulator. Therefore, in a typical real-world application, a combination of electrical, mechanical, and fluid power systems is normally used. In developing a system, a design engineer must consider the limitations of distance, geometry, and power transmission and the magnitude of power needed.

Figure 1-13 Gear train.

Fluid power systems are well suited for applications requiring ease of control. Through the use of simple user interfaces such as levers and push buttons, the user can easily begin, stop, or control the system's speed.

These small user interface devices can control very large structures and direct very large forces due to the force multiplication achievable in fluid power. As an example, the movement of a lever by an operator of heavy-duty equipment can generate thousands of pounds of lifting force to move gravel or large rocks with little input force required by the operator. This example leads to the second area well suited for fluid power systems. This is where multiplication of forces is required without the use of typical mechanical gears, chains, or linkages. Fluid power systems can multiply forces efficiently and easily in the scale of ounces to thousands of tons of output. This typically allows fluid power systems to excel in applications where mechanical or electrical power systems would be challenged.

Fluid power systems can also produce constant force or torque under different changes in output speed. The ability to produce a constant force, regardless of whether the actuator is moving very slowly or very quickly, is an important feature in many systems. A final advantage of fluid power systems is their relative simplicity as compared with mechanical or electrical power systems. In general, a fluid power system can contain fewer moving parts or sensitive connections than mechanical or electrical power systems.

Figure 1-14 Transmission lines of an electrical power station.

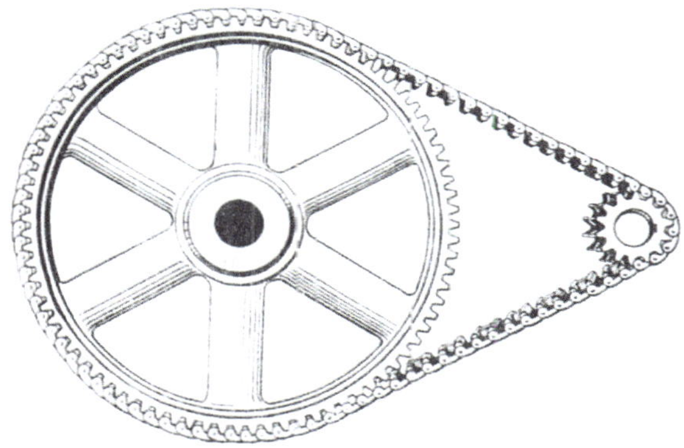

Figure 1-15 Silent chain drive.

Exercises

Figure 1-16 Push button; start/stop power switch.

1. Define the term *fluid power* and explain why hydraulic power is especially used for applications requiring high forces/loads.
2. Name the basic components in a hydraulic and a pneumatic fluid power circuit. Explain the differences between the two.
3. Name and very briefly describe (two to four sentences per application) three hydraulic applications.

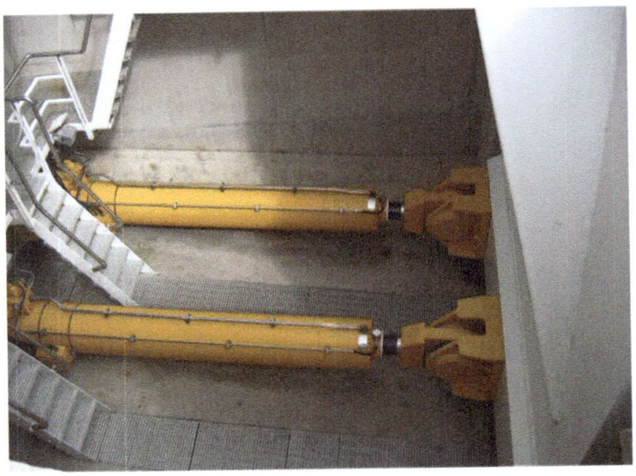

Figure 1-17 London's Millennium Bridge hydraulic rams. South-side pier hydraulic rams.

4. Describe the difference between fluid power and fluid transport in your own words.
5. Review the three power systems described in the text. Discuss which system or combination of power systems you would use to design a brick splitter to allow you to split 500 bricks a day *and why*. Draw a simple sketch of your design and describe how it works in terms of power systems used. (A brick splitter, as shown below, is able to cut/break bricks.) *Cost is not a factor.*

Figure 1-18 Log splitter.

Figure Credits

Fig. 1-1: Copyright © by Victor Blacus (CC BY-SA 4.0) at http://commons.wikimedia.org/wiki/File:Water_and_oil.jpg.
Fig. 1-2a: Paolo Neo, http://commons.wikimedia.org/wiki/File:Hollow_empty_glass.jpg, 2013.
Fig. 1-2b: Copyright © 2007 by Acdx. Reprinted with permission.
Fig. 1-3: Ratel, http://commons.wikimedia.org/wiki/File:Gas_bottle.png, 2009.
Fig. 1-4: U.S. Consumer Product Safety Commission, http://commons.wikimedia.org/wiki/File:Bike_pump.jpg, 2006.
Fig. 1-5: Copyright © by Markus Schweiss (CC BY 4.0) at http://commons.wikimedia.org/wiki/File:Hydraulikzylinder01.jpg.
Fig. 1-6a: Daniele Pugliesi, http://commons.wikimedia.org/wiki/File:Hydraulic_piston_cylinder_coloured.png, 2009.
Fig. 1-6b: Copyright © by Duk (CC BY-SA 3.0) at http://commons.wikimedia.org/wiki/File:Gear_pump.png.
Fig. 1-7: U.S. Bureau of Land Management, http://commons.wikimedia.org/wiki/File:Blmexcavsm.jpg, 2004.
Fig. 1-8: Copyright © by Mixabest (CC BY-SA 3.0) at http://commons.wikimedia.org/wiki/File:Manufacturing_equipment_087.jpg.
Fig. 1-9: Copyright © by Duk (CC BY-SA 3.0) at https://commons.wikimedia.org/wiki/File:Hydraulic_circuit_directional_control.png.
Fig. 1-10: "Typical Pneumatic System," https://www.ispatguru.com/wp-content/uploads/2015/11/Components-of-a-pneumatic-system.jpg. Copyright © by Ispat Guru.
Fig. 1-11: Yalaman2002, http://commons.wikimedia.org/wiki/File:Air_compressor_small_2009.6.15.JPG, 2006.
Fig. 1-12: Copyright © by Stanley Howe (CC BY-SA 2.0) at http://www.geograph.org.uk/photo/1324610.
Fig. 1-13: Copyright © by Robbie Sproule (CC BY 2.0) at http://www.flickr.com/photos/85278812@N00/4644315.
Fig. 1-14: Arthur Tress / U.S. National Archives and Records Administration, http://commons.wikimedia.org/wiki/File:TRANSMISSION_LINES_OF_ELECTRICAL_POWER_STATION_BUILT_ON_STATEN_ISLAND_MARSHLAND._OPENING_OF_VERRAZANO_BRIDGE_BROUGHT..._-_NARA_-_547936.tif, 1973.
Fig. 1-15: Andy Dingley, http://commons.wikimedia.org/wiki/File:Silent_chain_drive_%28Army_Service_Corps_Training,_Mechanical_Transport,_1911%29.jpg, 1911.
Fig. 1-16: Michael Holley, http://commons.wikimedia.org/wiki/File:Start_Stop_Power_Switch.jpg, 2012.
Fig. 1-17: Copyright © by Stephen Horncastle (CC BY-SA 2.0) at http://www.geograph.org.uk/photo/477276.

CHAPTER 2

Basic Concepts

Overview and Learning Objectives

In this chapter, we will review the major unit systems as well as the terms energy, power, and linear/rotary mechanics. Additionally, we will distinguish between a liquid and a gas and the desired properties of various fluids. Lastly, we will define a problem-solving method.

The learning objectives of this chapter should allow students to do the following:

- Convert within the U.S. customary unit system and the SI unit system (Le Système Internationale d'Unités, or International System of Units).
- Convert between the U.S. and SI unit systems.
- Calculate the force, work, and power in a linear motion system and an angular motion system.
- Explain in your own words the terms fluid, gas, and liquid (i.e., what they are and how they differ).
- Solve problems using the problem-solving method.

Dimensions and Units

The primary dimensions that we will use in this book are mass, length, and time. The other dimensions such as force, pressure, area, work, and power are derived from these primary dimensions. We will employ the U.S. customary unit system and the international system of units, or Le Système Internationale d'Unités (SI). The SI unit system is also known as the metric system.

The U.S. customary unit system is based on the English unit system from the British Empire (imperial unit system). In 1988, the U.S. government passed the Omnibus Foreign Trade and Competitiveness Act, which mandated that the metric system be "the preferred system of weights and measures of U.S. trade and commerce." Even though this law was passed more than 35 years ago, the U.S. customary unit system is widely used in the engineering of products and in manufacturing systems. The basic units from the U.S. system that we will employ are the Fahrenheit and Rankine scales for temperature; the second for time; the inch and foot for length; the gallon for volume; and the pound-mass for mass. It is worth noting that some students are confused when we refer to the pound-mass as the unit of mass, as you may have used the slug in a previous class as the unit of mass. The unit slug is from the imperial unit system and is not typically used in engineering work in the United States. The use of pound-mass allows us to easily move from mass to weight, assuming that we are on the surface of the Earth, as 1 pound-mass equals 1 pound-force.

The SI or metric system is built around seven basic units and a set of prefixes that act as the multipliers of the basic units, such as milli, kilo, and giga. The basic units from the SI system that we will work with in this text are Celsius and Kelvin for temperature; the second for time; the meter for length; and the kilogram for mass. The

SI unit system has been globally adopted. The only major countries not following the SI system are the United States, Liberia, and Burma.

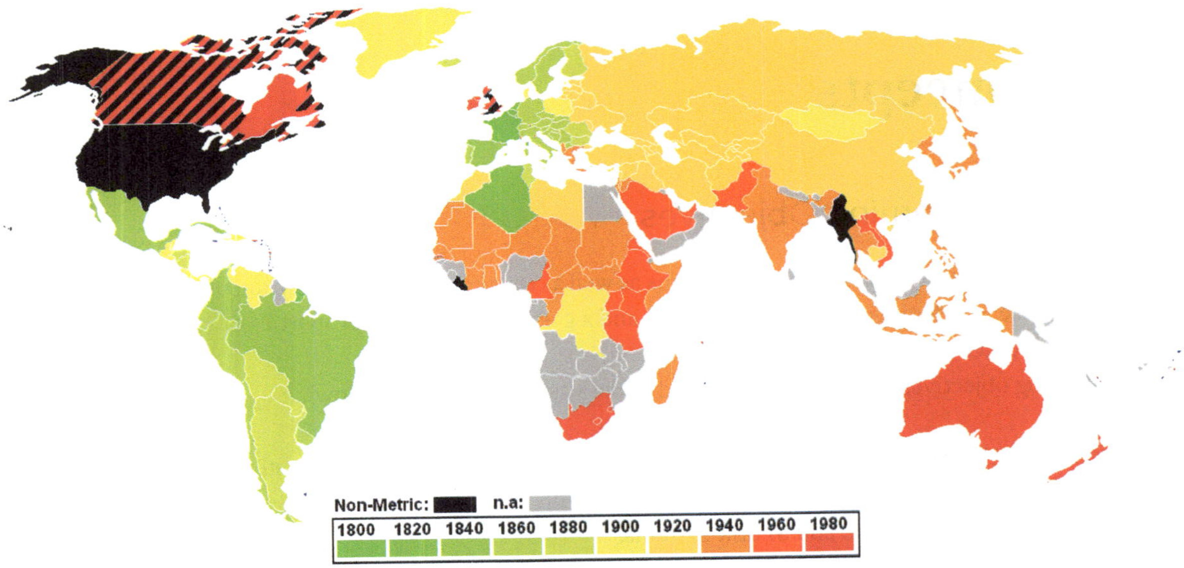

Figure 2-1 World map showing metrication, color-coded by year of conversion, which uses data from the U.S. Metric Association.

As a practicing engineer working with U.S. and global suppliers/customers, you will be required to move easily between the U.S. customary unit system and the SI (metric) system.

The units most frequently used in this book are based on the basic units from either the U.S. customary unit system or the SI unit system. The derived dimensions we will use are force, work, and power. Conversion factors from the U.S. customary unit system and the SI system are presented in the appendix.

Review Angular and Linear Motion, Mechanics

We will use linear and angular motion in our analysis of fluid power systems. From your previous work in physics, you are familiar with the terms linear velocity, linear acceleration, and linear displacement. Uniform linear velocity is a function of the change in displacement over the change in time, as seen in the following equation:

$$\mathrm{v} = \frac{distance}{time} = \frac{\Delta d}{\Delta t}$$

Uniform linear acceleration is a function of the change in velocity over the change in time, as shown in the following equation:

$$a = \frac{velocity}{time} = \frac{\Delta v}{\Delta t}$$

Linear displacement can be expressed as the initial velocity multiplied by time plus half the constant acceleration multiplied by time squared.

$$\Delta d = Vo * t + \left(\frac{1}{2}\right) * a * t^2$$

To change the velocity of a body (increase or decrease velocity), we would have to exert a force on the body in accordance with Newton's first law of motion. Newton's first law states that a force is required to change the velocity of a body, either in magnitude or direction.

$$Force\,(F) = Mass \times acceleration = m \times a$$

The application of forces brings us to the concepts of energy and work. Energy is defined as the ability to perform work. If a force is applied to a body and the body moves through a distance, then energy has been applied and work has been accomplished. The magnitude of the work completed is the product of the force applied and the distance moved. Note that the direction of the force applied and the distance the body moved are the same.

$$Work\,(W) = Force * Distance = F * S$$

Power is the rate of accomplishing work or using energy per unit time. This yields the following formula for linear motion power:

$$Power = Force * \frac{distance}{time} = Force * velocity = F * V$$

Note that with this definition, the units of power in the U.S. customary unit system would be pound(foot/second). In the U.S. customary system, we historically have used horsepower (hp) as the unit of power. One horsepower is defined as 550 pound(foot/second) [1 hp=550 lbf(ft/s)]. In the SI unit system, the units from our power equation would be newton(meter/second). In the SI system, 1 newton(meter/second) [1 Nm/s] is equal to 1 watt (in honor of James Watt).

With the basic linear motion equations stated above, we can think of angular displacement as being similar to the displacement in linear motion. We will call theta (θ) this angular displacement, which is measured in radians.

This allows us to have an angular component for many of the linear motion equations that we are both familiar with and have defined. The variable substitution necessary to get the angular motion is angular velocity, with omega (ω) meaning the number of radians traveled per unit time. Therefore, the angular (uniform) velocity equation is the change in angular displacement over the change in time, as shown in the following equation:

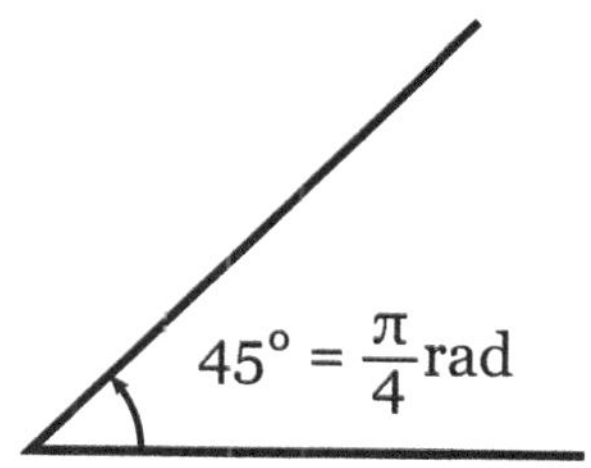

Figure 2-2 Radians and degrees figure.

$$\omega = \Delta\theta / \Delta t$$

We will use angular (uniform) acceleration as the unit of the change in radians per second squared. This gives us angular acceleration equals the change in angular velocity over the change in time.

$$\alpha = \frac{\Delta\omega}{\Delta t}$$

For example, if you have a ball tied to the end of a string and you want to determine the angular velocity of the ball as you spin it around in a complete circle in 0.5 seconds, you would use our relationship for angular velocity. We have to keep in mind for this problem that angular displacement is in units of radians. There are two pi (π) radians in one complete circle. So, our angular velocity is two pi radians divided by 0.5 seconds, or an angular velocity of four pi radians per second.

Similar to linear motion, when we want to change or create an angular velocity, we must apply a force. A classic example of this is the application of a force to a wrench to remove a bolt. The force acts at a distance from the center of the bolt, thus creating a torque on the bolt. The torque is what causes the bolt to rotate through a defined angle until it is removed. The magnitude of the applied torque is a product of the force times the distance. It is important to note that the distance or moment arm is the perpendicular distance between the bolt center and the perpendicular line of action of the force.

$$Torque = Force * Distance(radius) = F * r$$

We can use our definitions of torque and angular velocity to define angular motion power.

$$AngularPower = Torque * AngularVelocity = T * \omega$$

From our earlier discussion, we know the units of torque in the U.S. customary system foot*pound (ft*lbf), and the units of angular velocity are per second (1/sec). Therefore, our unit for angular power is foot*pound/sec or pound*foot/second. In the U.S. customary system, we historically have used hp as the unit of power. One hp is defined as 550 pound*(foot/second) [1 hp = 550 lbf*(ft/s)]. In the SI unit system, the units from our angular power equation would be newton*(meter/second) or watts. In fluid power applications, we will use these definitions of work and power to calculate the power delivered to the system from the prime mover and from the fluid power system to the external load.

Efficiency

In all real-world systems, the output power delivered by a system is always less than the input power to the system due to losses (e.g., friction, leakage). We will use efficiency as the measure for this transfer of power (energy/time).

$$Efficiency = output\_power/input\_power$$

As mentioned, the output power is always less than the input power; thus, the efficiency is numerically always less than 1, meaning less than 100%.

Problem-Solving Method

To solve engineering fluid power problems consistently, you will need to develop a process that helps you to visualize problems, determine solution paths, and identify errors as you proceed to solutions. The following problem-solving method produces these outputs and is recommended when you are solving both textbook and real-world problems:

1. State what is known or given in the problem, along with the symbol for the item, the value given, and—most importantly—the unit of the value.
2. State what quantities you are asked to determine; it is very important at this step to understand what units are being asked for in the solution (U.S. customary or SI).
3. Draw and label a sketch or diagram of the system where possible. If present, the sketch should contain all pressures, forces, and components to fully describe the system.
4. List any initial valid assumptions that allow you to simplify the problem.
5. Analyze the problem and identify the key definitions and principles that you would like to apply toward the solution. Additional reasonable assumptions might be needed.
6. Write the symbolic or algebraic solution to your problem. It is important to delay substitution of numerical values as late as possible in the process to allow you to make sure you are not mixing unit systems.
7. Substitute the numerical given values as indicated for all of the physical data given. Please note that as you write numerical values, you must always include the units associated with these values. While this appears to add extra time to your solution, you will find that this helps you not mix or handle units inappropriately.
8. Check the units associated with each calculation in your final answer.
9. Examine your answer critically. Does this answer appear to be reasonable and consistent with your expectations? If you calculate a number that is extremely small or extremely large, this may indicate that you have made an error.

Key Equations

$$V = \frac{distance}{time} = \frac{\Delta d}{\Delta t}$$

$$a = \frac{velocity}{time} = \frac{\Delta V}{\Delta t}$$

$$\Delta d = Vo * t + \left(\frac{1}{2}\right) * a * t^2$$

$$Force(F) = mass * acceleration = m * a$$

$$Work(W) = Force * Distance = F * S$$

$$Power = Force * \frac{distance}{time} = Force * velocity = F * v$$

$$\omega = \Delta\theta/\Delta t$$

$$Torque = Force * Distance(radius) = F * r$$

$$AngularPower = Torque * Angular\ Velocity = T * \omega$$

$$Efficiency = output\_power/input\_power$$

$$\frac{W}{\Delta t} = power = T \cdot \omega$$

Units Table

Velocity	ft/s	m/s
Acceleration	ft/s^2	m/s^2
Displacement	ft	m
Force	lbf	N
Work	ft*lb	J
Power	hp	Watt
Angular acceleration	rad/s^2	rad/s^2
Torque	ft*lb	N*m
Angular power	hp	Watt
Efficiency	Percentage (%)	Percentage (%)

Exercises

1. Explain what the U.S. customary units and pound-force and pound-mass are.
2. What is efficiency? Can any process ever achieve efficiency over 1.0 (100%)? Why or why not?
3. A typical vehicle's engine is approximately 40% efficient in producing power to move the vehicle. Where is the "other" 60% of the power (energy) going from the fuel?
4. Describe a common misuse of the term fluid in common language. (Remember: A fluid in engineering is a liquid or a gas.)
5. Look up the power of any large ship engine, a motorcycle engine, and a truck engine. Calculate or look up the operating revolutions per minute and torque of these specific engines and prove that for these three examples of engines: $rotarypower = torque * Omega$. Comment on the values you found.
6. A 60-foot-long pipe is inclined at a 30-degree angle with the horizontal. The pipe is filled with hydraulic oil with a specific gravity of 0.92. What is the pressure at the base of the pipe if the top of is vented to the atmosphere in pounds per square inch?

7. A cylinder has a diameter of 0.5 meters (m) and a height of 2 m. If it is filled with a liquid that has the specific weight of 2000 newton/cubic meters, how many kilograms (kg; mass) of this liquid must be in the cylinder?

Figure Credits

Fig. 2-1: Copyright © by Maphobbyist (CC BY-SA 3.0) at http://commons.wikimedia.org/wiki/File:SI-metrication-world.png.
Fig. 2-2: Gustavb, http://commons.wikimedia.org/wiki/File:Angulo_positivo.svg, 2008.

CHAPTER 3

Physical Properties of Fluids and Their Measurement

Overview and Learning Objectives

We will define various fluid terms, including *mass, force, weight, volume, specific weight, density, specific gravity, pressure, head,* and *force*. This chapter will also review the terms *kinematic viscosity* and *absolute viscosity* and convert viscosity from one set of units to another.

The learning objectives of this chapter should allow students to do the following:

- Define the terms *specific weight, density,* and *specific gravity.*
- Understand the terms *pressure, head,* and *force.*
- Determine the difference between gauge pressure and absolute pressure.
- Understand the relationship between the terms *vacuum* and *absolute pressure.*
- Define the terms *kinematic viscosity* and *absolute viscosity.*

Mass, Force, Weight, and Volume

As we discussed in the units section, mass is one of our fundamental dimensions. As such, it cannot be defined in terms of other dimensions. This is similar to the fundamental dimension of time. We use the term *mass* in its relationship to force from Newton's second law of motion:

$$F_{\text{net}} = ma$$

From this relationship, we can see that the force required to produce a given acceleration is proportional to the mass of the object being moved. Please note that in the U.S. customary system, the unit for mass is pound-mass.

Weight is defined as the force of an object due to gravity. Weight is the product of the mass of the object times the local gravitational constant, or gravitational acceleration. In this case, weight equals mass times g, or W = mg. The terms *mass* and *weight* can be confused in everyday language, but to engineers, it is *very important* to understand that weight and mass are two different, distinct quantities. The unit of measurement for weight is force; in the U.S. customary system, this is pound—or, more exactly, we will use the term *pound-force*. In the International System of Units (SI) system, the unit of weight is in newtons (N). As an example, an object with a mass of 1 kilogram (kg) has a weight of approximately 9.8 N on the surface of Earth. In the U.S. customary system, 1 pound-mass equals 1 pound-force. A body can be weightless or have zero weight only if it is far away from any gravitational mass.

Volume is a familiar property that is defined as the space occupied in three dimensions of an object. In the U.S. customary system, the unit of volume is typically the cubic inch or the cubic foot. In the SI unit system, the unit of volume is the cubic meter.

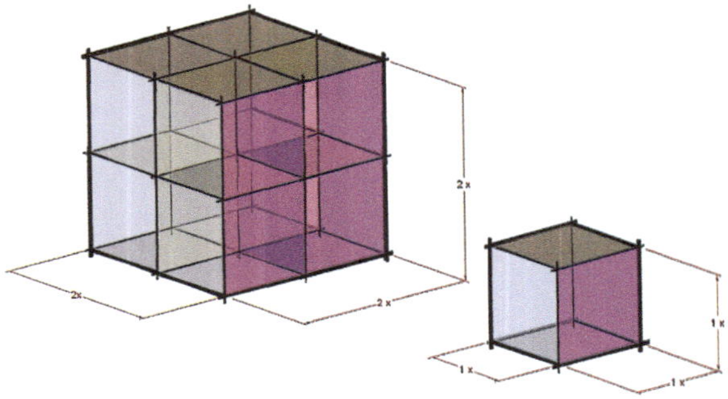

Figure 3-1 The volume of two cubes.

Density and Specific Weight

Density is the ratio of the mass of an element to its volume. If we think of a substance such as lead, it has a very high mass to its volume. In comparison, a gas has a relatively low mass to its volume. We will use the Greek letter *rho* (ρ) to represent density.

$$\rho = \frac{m}{V}$$

It is important to note that increasing the amount of a substance does *not* increase its density because as the mass of the object increases, so does the volume that the object occupies. Thus, for a given standard condition of pressure and temperature of a substance, the density is constant.

Specific weight is very closely related to density. **Specific weight** is the ratio of an object's weight to its volume.

$$\gamma = \frac{W}{V}$$

Specific weight and density are related by the local gravitational acceleration. On the surface of Earth, the gravitational acceleration will be represented by the term g.

Specific Gravity

Density or specific weight can also be expressed in a dimensionless quantity called *specific gravity* or *relative density*. **Specific gravity** or *relative density* is the ratio of the density or specific weight of the material to that of a standard material. In other words, relative density is the comparison of the density of a substance relative to the standard. The standard for specific gravity or relative density is usually water at 4 degrees Celsius (°C) for liquids and air at standard temperature and pressure for gases. Specific gravity is used to convey the relative difference between the substance of interest and the reference material without having to remember the density of the reference material. For example, if a liquid substance has a specific gravity greater than 1, we know that it has a larger density than water and is also heavier.

In this example, by definition, water has a specific gravity of 1. It is important to not confuse the term *specific gravity* with the term *specific weight*.

Figure 3-2 Layers of glycerin, propylene glycol, ethylene glycol, and water demonstrating their increasing densities from top to bottom.

Pressure and Temperature

Pressure is defined as the normal force exerted by a fluid on a container surface or on a contacting fluid surface per unit area.

$$P = \frac{F}{A}$$

Pressure is a scalar quantity having a nominal value but not a specific direction associated with it. In other words, pressure at a point is independent of orientation. The pressure force is always directed normal or 90 degrees to the boundary surface. As mentioned previously, a fluid by definition cannot sustain any force that is not normal to its surface (not 90 degrees to the surface) without exhibiting flow movement. Pressure forces have a very important role in understanding fluid power and the flow of fluids. We can think of pressure as a measure of the rate at which the momentum of the fluid molecules carry when colliding with the container walls.

The U.S. customary unit for pressure is pounds per square inch (psi). The SI unit for pressure is the pascal (Pa), which is defined as 1 newton per meter squared. One pascal is a very small unit in fluid power applications; thus, we typically use kilopascals, or 1,000 Pa. For larger systems, we will use megapascals, or 1 million Pa. Another common term for pressure is the *term bar*. One bar equals 100,000 Pa. One bar is approximately equal to one standard atmosphere of pressure on the surface of the Earth. One standard atmosphere is the pressure of 101,325 Pa.

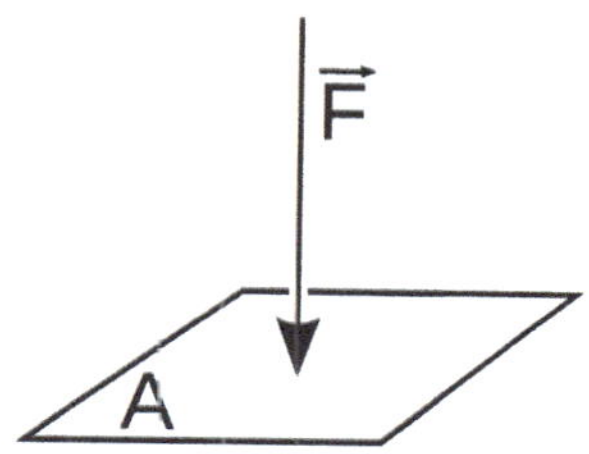

Figure 3-3 Definition of pressure.

We need to distinguish between the terms *gauge (gage) pressure* and *absolute pressure*. As a result of practical pressure measurement devices, we typically measure pressure with a gauge that displays the pressure relative to the local atmospheric pressure. As an example, a tire gauge reads zero pressure when not connected to the tire being measured.

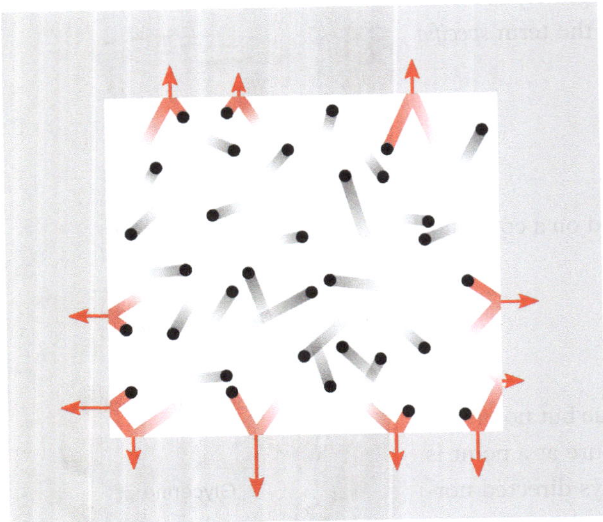

Figure 3-4 Pressure as exerted by atoms or molecules colliding with the inner walls of a container. Collisions and the resulting forces are shown in red. Pressure is the sum of the collisional forces.

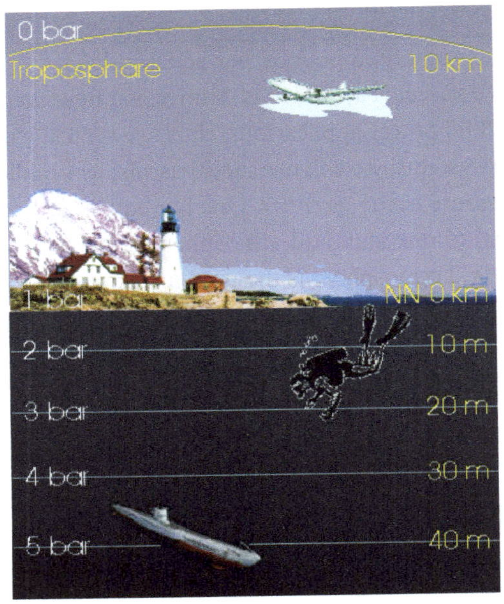

Figure 3-5 Pressure in bar above and below the surface of the ocean.

Absolute pressure is the sum of the gauge pressure being measured plus the local atmospheric pressure. In a perfectly evacuated space, the absolute pressure is zero. We can never have a negative absolute pressure. In this case, absolute pressure is similar to absolute temperature, in that zero is the lowest value achievable. Figure 3-7 graphically illustrates the relationship between gauge and absolute pressures.

The terms **vacuum and vacuum pressure** are employed in engineering and fluid power applications to represent the negative gauge pressure from the atmospheric pressure. In other words, if we were using 14.7 psi as the atmospheric pressure if we had a 4.7-psi vacuum, this could also be expressed as a gauge pressure of −4.7 psi. In this instance, we would have an absolute pressure of 10 psi. We can see that the term *vacuum* permits us to avoid using the negative when discussing gauge pressures that are less than atmospheric pressure. Students should be careful in reviewing the terms *gauge pressure*, *vacuum*, and *absolute pressure* to make sure they fully understand the relationships.

Figure 3-6 Tire pressure measurement.

As we use the units for pressure, we frequently append the unit with a small *g* to represent gauge pressure or a small *a* to represent absolute pressure. Since many systems record pressure-using gauges, if we see the term *pound per square inch* in the U.S. engineering systems, we assume that this is gauge; if we see the term *pascal* in the SI system, we assume the same. If the pressure being measured is absolute, we append or add on to the unit the letter *a* (*psia*) to express pound-per-square-inch absolute or the term *abs* (*Pa abs*) to express absolute pressure. You should be comfortable as an engineer in working with these various units of pressure, so please spend some time reviewing these terms to ensure clarity.

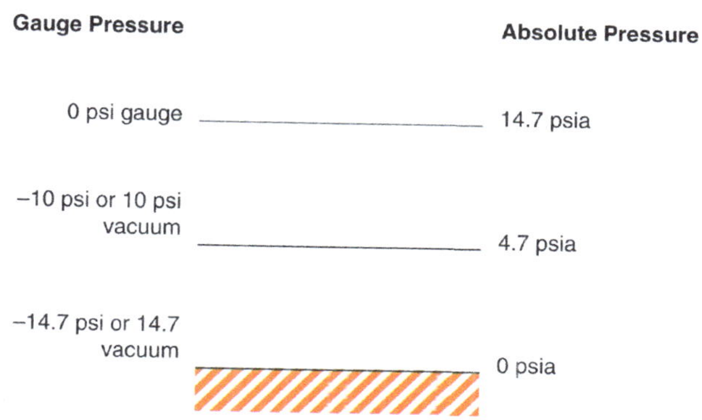

Figure 3-7 Absolute and gauge pressures.

Like mass and time, temperature is a fundamental dimension. However, it does not have a simple, concise definition. We all have experience with temperature differences as we express that one object is warmer or colder than another. From a microscopic or molecular point of view, we can think of temperature as being directly proportional to the square of the mean molecular speed of an ideal gas. This allows us to visualize higher temperature to mean faster moving particles in a molecule. In the U.S. customary system, we will use the temperature scale Fahrenheit (°F) and its absolute counterpart, the Rankine scale. It is important to note that the size of the temperature unit or the difference between any two temperature numbers in Fahrenheit and Rankine is identical. Thus, a 1°F change equals a 1° Rankine change. The difference between Fahrenheit and Rankine is a constant of 459.67, or 460°. Hence, 0°F equals 460° Rankine, as 0° Rankine equals −460°F; this point is defined as absolute zero. The SI units for temperature are Celsius, and its absolute counterpart is the Kelvin scale. Similar to the U.S. customary system, the size of the temperature unit or difference between any two temperature numbers in Celsius and Kelvin (K) is identical. Thus, a 1°C change equals a 1°K change. The difference between the Celsius and Kelvin scales is a constant of 273.15°. Consequently, the freezing point of water at 0°C equals 273°K, or 0°K equals −273°C: absolute zero.

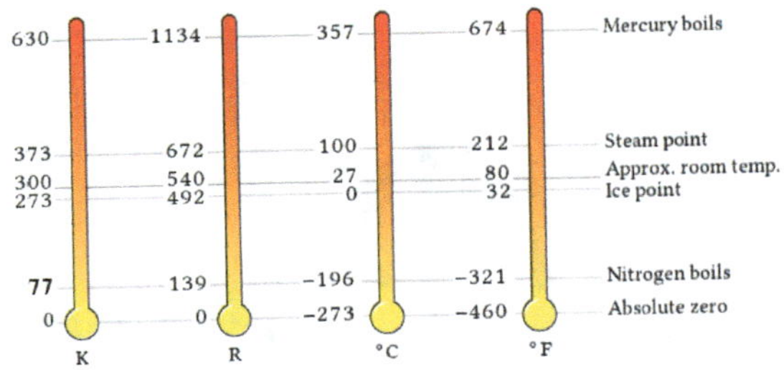

Figure 3-8 The Celsius and Kelvin scales, shown in a thermometer.

Viscosity and Bulk Modulus

Fluid viscosity is a measure of how the fluid can resist an applied shear stress (i.e., tangential stress). For liquids, many of us have seen viscosity as we look at a liquid when flowing or its resistance to flow. A common example would be our experience with ketchup and its seeming inability at times to flow out of the bottle onto our plate. We may also have seen viscosity in terms of our experience with maple syrup or honey and how they appear to be very thick and do not flow easily out of the container. As compared with ketchup, maple syrup, or honey, water has a low viscosity and flows easily.

Viscosity comes from the internal friction between neighboring pieces of the fluid that may be moving at different velocities and therefore restricting the flow of neighboring pieces. All fluids have some level of viscosity, or internal friction between neighboring particles. As seen in Figure 3-9, we can visualize the viscosity of a fluid by looking at moving and stationary plates with the fluid in between them.

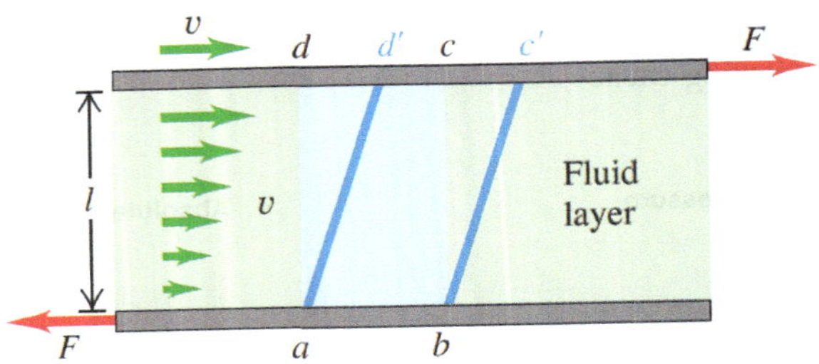

Figure 3-9 Apparatus for a two-moving-plate viscosity experiment.

We will assume that the plates are very large, so we need not concern ourselves with the fluid effects at the edges of the plate. As the upper plate moves, the fluid in contact with it will move with that plate. In contrast, fluid touching the lower stationary plate will remain stationary with it. A velocity gradient will exist in the fluid between the two plates. A shear stress is developed in the fluid, with a magnitude that depends on the fluid's viscosity.

We will denote this shear stress with the Greek letter *tau* (τ), defined as the force required to slide one layer of the fluid over another layer due to the velocity gradient that exists between the two plates. The shear stress is a force divided by an area with units of newton per meter squared or pascal (SI units) or pound per foot squared (U.S. customary units). The magnitude of the shear stress is directly proportional to the difference in velocity between different layers in the fluid based on the velocity gradient between the two plates.

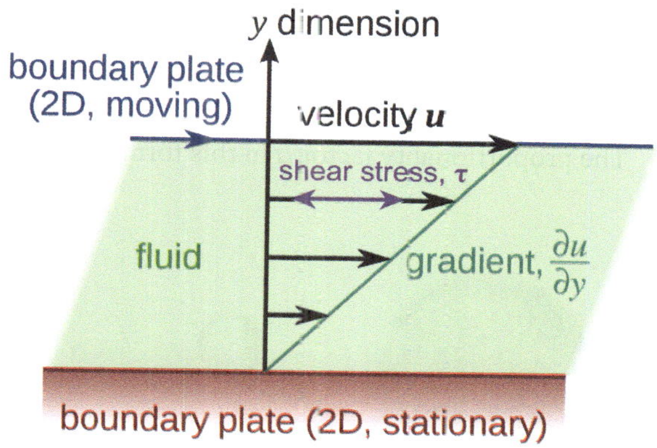

Figure 3-10 Laminar shear in a fluid.

Dynamic Viscosity

Figure 3-11 illustrates this concept of the velocity gradient in a fluid between a stationary plate and a moving plate. The velocity gradient is a measure of the velocity change as we move from the stationary plate to the moving plate. In most fluids, the shear stress in the fluid is directly proportional to this velocity gradient; consequently, we can state that the shear stress equals the velocity gradient multiplied by proportionality constant. This proportionality constant is called the fluid's *dynamic viscosity*, which is also known as *absolute viscosity*.

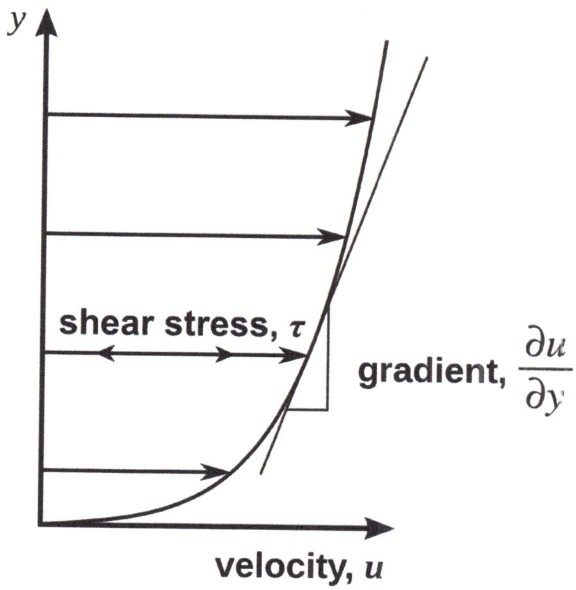

Figure 3-11 Laminar shear in a fluid.

F = magnitude of the force
µ = speed
A = area of each plate

y = separation between the plates

$$F = \mu A \frac{u}{y}$$

The proportionality factor μ in this formula is the dynamic viscosity.

Figure 3-12 Picture of the pitch-drop experiment started in 1930 at the University of Queensland (picture taken in 1990, two years into the life of the eighth drop).

We can think of this velocity gradient as also occurring when we stir a fluid with a rod. The velocity of the fluid by the rod will match that of the rod. The velocity of the fluid far away from the rod will approach zero as the fluid contacts a stationary boundary. We know from experience that a greater force is required to stir a very cold, thick, or high-viscosity liquid than to stir a low-viscosity liquid, such as water. Different unit systems are used to express viscosity. We can derive the units of dynamic viscosity by looking at the relationship between shear stress and the velocity gradient. The shear stress has a unit of newton per meter squared, and we divide this by the velocity gradient with units of meters per second per meter. This gives us dynamic viscosity units of newton seconds per meter squared. We know that newton per meter squared is also known as a *pascal*, and therefore dynamic viscosity can have units of pascal-seconds (Pa*s). Water has a dynamic viscosity of 0.001 Pa*s, whereas a typical hydraulic oil has a dynamic viscosity of 0.25 Pa*s, or approximately 250 times the viscosity of water. A tar pitch viscosity experiment is one of the longest-running experiments, with a single drop of pitch taking approximately 13 years to form and drop. Tar pitch has a viscosity 250 billion times that of water.

Table 3-1 lists the common dynamic viscosity units in the three most widely used unit systems.

Table 3-1. Common Dynamic Viscosity Units

Unit System	Dynamic Viscosity Units
International System (SI)	Pa*s or N*s/m^2
U.S. Customary System	lb*s/ft^2
Centimeter-Gram-Second (CGS) System (Obsolete)	Poise; dyne*s/cm^2 = 0.1 Pa*s Centipoise = poise/100 = 1 million Pa*s

The dimensions of force multiplied by time divided by length squared is common in each one of the systems. As a practicing engineer, you may see the units of centipoise or poise, as many published data are given in these units for many materials. These units are part of the obsolete CGS metric system derived from the base units of centimeter, dyne, gram, and second. One centipoise is equal to 0.001 Pa*s or 1 million Pa*s.

Kinematic Viscosity

In many fluid mechanics calculations, we will use the ratio of dynamic viscosity to the fluid's density. To allow us to do this easily, we will define a new term of kinematic viscosity: the Greek letter *mu* (μ). Kinematic viscosity is the dynamic viscosity divided by density. The common units for kinematic viscosity are listed in Table 3-2.

Table 3-2. Common Kinematic Viscosity Units

Unit System	Kinematic Viscosity Units
SI	m^2/s
U.S. Customary System	ft^2/s
CGS System (Obsolete)	Stoke; cm^2/s Centistoke = stoke/100 = 1 mm^2/s

The basic dimensions of kinematic viscosity are length squared divided by time or, in SI units, meter squared divided by seconds (m^2/s). Similar to dynamic viscosity, the obsolete units of Stoke and centistoke are listed because many common materials have published data with these units.

From your experience with viscosity, you are probably very familiar with how a fluid's viscosity of can vary with temperature. As examples, ketchup, maple syrup, and honey are generally difficult to pour when they are cold, a characteristic that indicates a high viscosity. As the temperature of these fluids is increased, the viscosity will decrease noticeably.

Gases behave differently from liquids, in that gases have increasing viscosity as temperature increases. Also, for gases, the change in viscosity is generally much smaller than that of a liquid in terms of viscosity change with temperature.

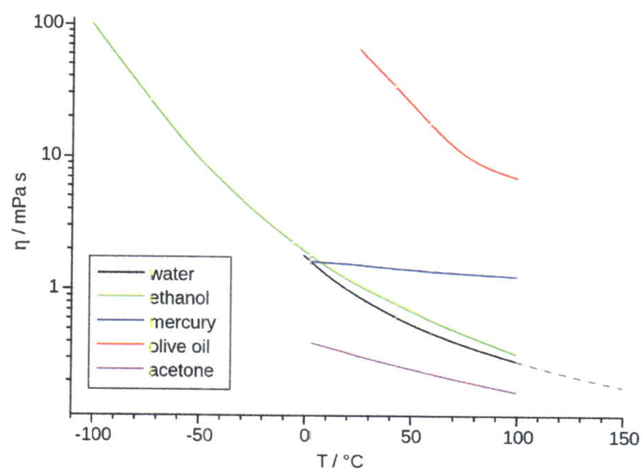

Figure 3-13 Viscosity of various liquids.

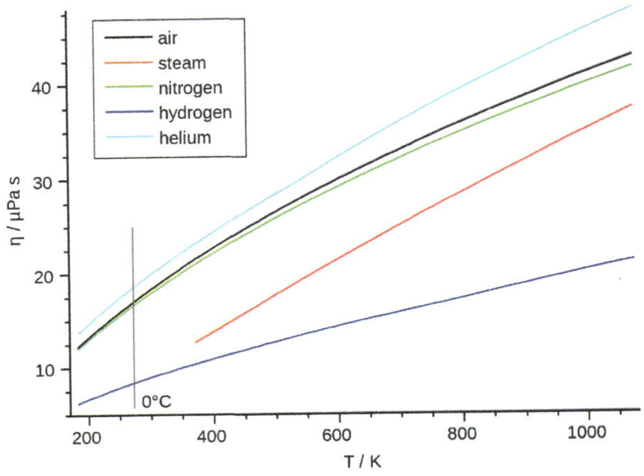

Figure 3-14 Viscosity of various gases.

In fluid power systems, viscosity is one of the most important fluid properties. Viscosity relates to the lubrication and the fluid's ability to be pumped and flow through the fluid power piping, tubing, cylinders, and valve control devices in the system. Special care is needed in designing a fluid power system when extremes in temperatures are anticipated. When the fluid power system needs to operate at very low temperatures, as an engineer, you must consider the fluid's ability to flow through the system. The performance of the fluid power pumps and valves can be remarkably different under the cold and warm operating range of a fluid power system due to the change in viscosity of the working fluid (especially liquids).

Bulk Modulus

The compressibility of a fluid is the change in volume of the fluid when it is subjected to a change in pressure. The quantity used to measure this change in volume when subjected to a change in pressure is the bulk modulus of elasticity, or simply the **bulk modulus**.

We will use the symbol β to represent the bulk modulus. The units for bulk modulus are the same units as the denominator, which is pressure. For our equation to be unitless, the units are therefore the same as pressure.

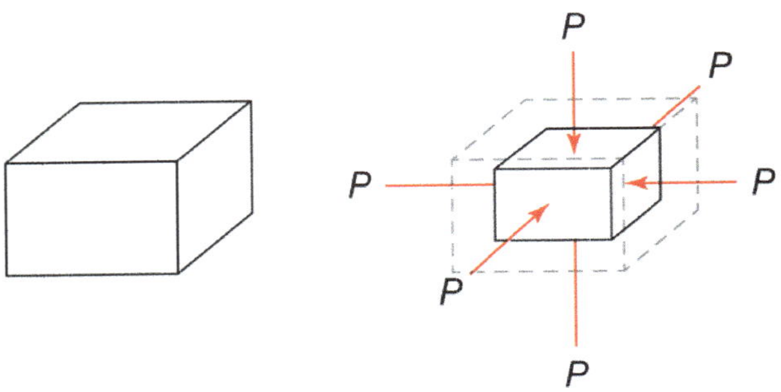

Figure 3-15 Bulk modulus, or pressure change required to change volume.

$$\frac{\beta}{\Delta P} = -\frac{V}{\Delta V}$$

For typical liquids, the bulk modulus is in the range of 100,000 psi to 4 million psi. From these very high values, we can conclude that liquid are very slightly compressible, or that it would take a very large change in pressure to produce only a small change in volume. For this reason, unless stated otherwise in this book, we will consider liquids to be incompressible, or to not exhibit any change in volume with pressure. It is important to note that the term *bulk modulus* is not usually applied to gas, as the principles of thermodynamics must be used to determine the change in volume of a gas with a change in pressure, as gases are readily compressible. As such, it is very important for practicing engineers to account for liquids under pressure that may have a small percentage of gas (i.e., air) entrapped in the liquid. A small percentage of air in a liquid can have an extremely significant effect on the bulk modulus for system pressures below 2,000 psi. A classic example of this effect is the need to bleed automobile brake lines of air to ensure that when the brake pedal is depressed, the system can develop full brake line pressures to stop the vehicle.

Key Equations

$$\vec{F} = m\vec{a}$$

$$\rho = m/V$$

$$\gamma = \frac{Weight}{Volume} = \frac{W}{V}$$

$$Specific\ Gravity = SG = \frac{\rho_{material}}{\rho_{reference}} = \frac{\gamma_{material}}{\gamma_{reference}}$$

$$Pressure_{Absolute} = Pressure_{gage} + Pressure_{Atmosphere}$$

$$T(Fahrenheit) = 1.8 * T(Celsius) + 32$$

$$F = \mu A \frac{u}{y}$$

$$KinematicViscosity(\text{v}) = \frac{AbsoluteViscosity}{Density} = \mu/\rho$$

$$\Delta P/\beta = -\Delta V/V_{initial}$$

$$\frac{Pressure_2 - Pressure_1}{\beta} = -(\frac{Volume_2 - Volume_1}{Volume_1})$$

Exercises

1. A fluid has a specific weight of $57lbf/ft^3$. What are the fluid's specific gravity and density in lbm/ft^3?
2. A tank truck has a capacity of 158,500 liters. It is filled with an oil additive with a specific gravity of 1.16. Determine the fluid's specific weight, density, and weight in SI units (N/m^3, kg/m^2, N; specific weight of water $\gamma w = 9810N/m^3$).
3. You measure the viscosity of olive oil at 40°C with two parallel plates separated by an oil film, as shown in Figure 3-16 with the following data: input force = 1.36 newton; plate area (A) = 2 m^2; velocity (u) = 2 cm/s; plate separation (y) = 1 millimeter (mm). Calculate the absolute viscosity (μ) of olive oil in units of pascal*seconds ({newton*second}/$meter^2$).

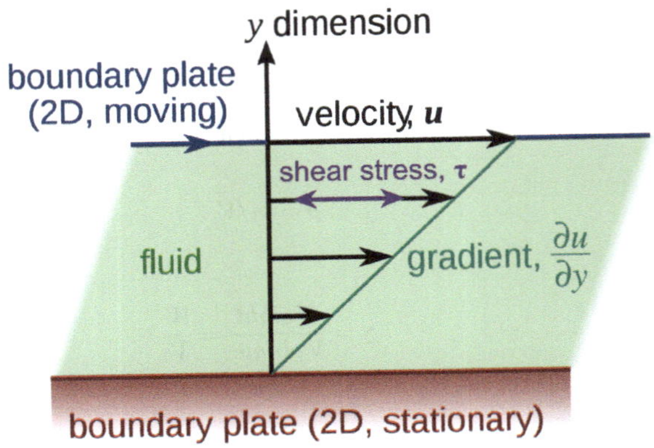

Figure 3-16 Laminar shear in Couette Flow

4. The pressure in a 4-inch-diameter hydraulic cylinder increases from 50 psi to 5,000 psi. Due to the oil's compressibility, the piston retracts by 0.010 inches (in). If the volume of the oil under compression is 15 cubic inches (in^3), what is the bulk modulus of the oil in units of pounds per square inch?
5. A 10-in^3 sample of oil is compressed in a cylinder until its pressure is increased from 10 to 2,000 psi. If the bulk modulus equals 250,000 psi, find the change in volume of the oil (hint: This is the same as the --delta volume, or the difference in volume from state 1 to state 2).
6. Engineering textbooks typically assume that hydraulic fluids are incompressible for ease in solving problems. Calculate the percentage change in the volume of oil to determine whether the bulk modulus can be neglected for a real fluid power hydraulic system. Comment on why or why not this is the case.
   a. A hydraulic cylinder has a 6-inch diameter and 20-inch stroke using hydraulic oil with a bulk modulus of 270,000 psi (0% air, at 80°F). Find the percentage change in the volume of oil if the system operates at 6,000 psi.
   b. As the fluid heats in the system and with air entrapment in the fluid, the oil bulk modulus changes to 96,000 psi (1% air, at 180°F). Find the percentage change in the volume of oil if the system operates at 6,000 psi.
   c. Comment on why or why not the bulk modulus can be ignored in a real fluid power hydraulic system.

Figure Credits

Fig. 3-11: Cutler, http://en.wikipedia.org/wiki/File:Laminar_shear_flow.svg, 2008.
Fig. 3-12: Copyright © by John Mainstone / The University of Queensland (CC BY-SA 3.0) at http://en.wikipedia.org/wiki/File:Pitch_drop_experiment_with_John_Mainstone.jpg.
Fig. 3-13: Copyright © by Prolineserver (CC BY-SA 2.5) at http://commons.wikimedia.org/wiki/File:Viscosity_of_liquids.svg.
Fig. 3-14: Copyright © by Prolineserver (CC BY-SA 2.5) at http://commons.wikimedia.org/wiki/File:Viscosity_of_gases.svg.
Fig. 3-15: Copyright © by Christophe Dang Ngoc Chan (CC BY-SA 3.0) at http://en.wikipedia.org/wiki/File:Isostatic_pressure_deformation.png.
Fig. 3-16: Copyright © by Farzad.Soleimani (CC BY-SA 3.0) at http://commons.wikimedia.org/wiki/File:Laminar_shear_in_Couette_Flow.png.

CHAPTER 4

Fluid Statics and Pascal's Principle

Overview and Learning Objectives

In this chapter, we will review how to compute the change in pressure due to a change in a liquid's elevation. We will apply this principle to common pressure measurement instruments.

The learning objectives of this chapter should allow students to do the following:

- Define the relationship between the change in height, or elevation, of a fluid and the change in its pressure.
- Describe how the shape of the container does not affect the fluid's pressure.
- Describe how a manometer works and how it is used to measure pressure.
- Describe how a U-tube manometer, a differential pressure manometer, a well-type manometer, and an inclined manometer operate.
- Describe how pressure can be expressed in terms of the height of a liquid.
- Describe a barometer and how it indicates the value of the local atmospheric pressure.
- Solve manometry problems and calculate forces created by fluid pressure on submerged surfaces.

Fluid Statics

If you have ever flown in a plane, you are probably familiar with how the change in the elevation of the plane affects the pressure on your eardrums. You may also have experienced this when swimming. As you dive deeper into the water, the pressure increases, which also affecting that on your eardrums. From these experiences, we can look at the general case of how pressure varies with the change in depth, height, or elevation within a fluid. In this book, we will use the term **elevation** to mean the fluid's vertical height from some reference point of interest. We will use the symbol H to represent the height change between two points of interest. Elevation will always be a positive number, as we look at the change in elevation or height in the upward direction. In other words, the highest point of interest in the fluid will have a larger elevation than the lower point. It is advisable to choose the lowest point as the reference point in a problem to eliminate the use of a negative change in height. We can define the change in pressure of a liquid at rest as the change in elevation multiplied by the specific weight of the liquid, as shown in the following equation:

$$\text{pressure} = \text{specific weight} * \text{height}$$

Some general items to note from this relationship are that this equation is valid for homogeneous liquids at rest (i.e., not moving). We can also see from this relationship that points on the same horizontal level will have the same pressure, as the change in height is zero (0) by definition for two points at the same horizontal level. We can also see that the change in pressure is directly proportional to the liquid's specific weight. As the specific weight increases, so does the pressure at a given height. In other words, a 33-foot change in the height of water will raise the pressure by approximately 1 atmosphere (atm), but a 33-foot change in the height of air will not have the same magnitude. As noted previously, the pressure varies linearly with the change in height or elevation of the liquid. This equation does not apply to gases because the specific weight of a gas changes with pressure. It is also important to note that a large change in elevation in a gas such as air is required to produce a significant change in pressure. Therefore, in this book, we will assume that the pressure in a gas is uniform in a container of varying height unless otherwise specified.

Pascal's Paradox

It is important to note that the change in pressure of a liquid depends only on the change in elevation or height of the liquid and the liquid's specific weight. Notice that the pressure does not depend on the size or shape of the liquid container. As a result, the pressure will be the same at the bottom of each container, even though they may have vastly different shapes. This phenomenon can contradict our visual assumption, so it is very important to reflect on this and understand that pressure is only a function of the height of the liquid. See the sketch of communicating vessels in Figure 4-1.

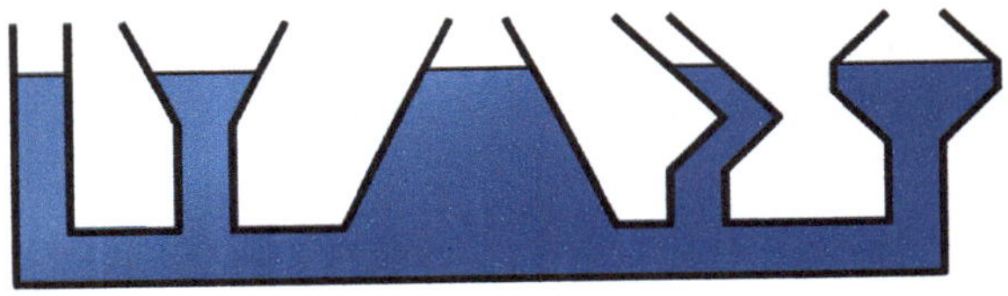

Figure 4-1 Communicating vessels; Pascal's paradox.

The level that the liquid reaches in each vessel is the same independently of the vessels' shape. The pressure at the bottom of each vessel and any point at the same elevation within each is the same. This phenomenon is called Pascal's paradox in honor of Blaise Pascal, a seventeenth-century scientist who contributed significantly to the knowledge of fluid mechanics. In practice, we can see this paradox in operation as we see water towers used to provide system water pressure in many towns and village locations. These towers provide a reserve supply of water, but their primary purpose is to maintain a sufficiently high pressure in the water system to satisfy delivery of water to the town's residents within the town.

In fluid power applications, a standing height containing a static liquid can be used to maintain a stable pressure at a pump or particular portion of a fluid power system. This standing height is positioned at a higher elevation relative to the other components in the system to ensure that adequate pressure is maintained in the system.

Figure 4-2 Darwin Airport Water Tower, owned by Power and Water Corporation.

Manometers

We can measure pressure using a manometer, which uses the relationship between a change in pressure and the change in elevation of a static fluid. An image of a common available manometer is shown in Figure 4-3.

The simplest kind of manometer is a U-tube manometer. One end of the U-tube is connected to the pressure to be measured (Po); the other end is left open to atmosphere (Pa). The tube contains a liquid called the gauge liquid, or gauge fluid. This gauge liquid does not mix with the fluid whose pressure is to be measured. Typical gauge liquids are water, mercury, or colored oils. With equal pressures on each side of the U-tube manometer, the fluid in it is at rest and at equal heights. As one end of the U-tube manometer is exposed to pressure, the fluid in it changes height. It is important to be able to calculate the pressure being measured from the manometer. The following step-by-step procedure will help you determine the pressure being measured:

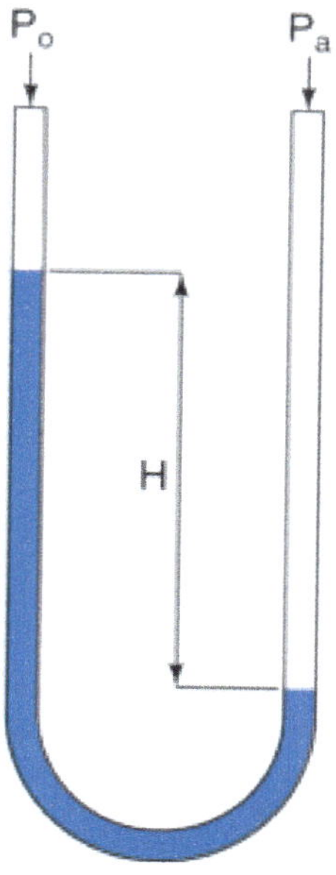

Figure 4-3 A schematic drawing of a U-tube differential pressure gauge (manometer).

PROCEDURE FOR DETERMINING MANOMETRIC PRESSURE

1. Begin from one end of the manometer and express the pressure with a subscripted 1. If this side of the manometer is exposed to atmospheric pressure, we will use 0 gauge pressure as the reference for this point.
2. Use our expression for the change in pressure equals the specific weight times the height to proceed from this starting point to include the column of each fluid separately. It is important to process each section separately.
3. When moving from one point to another in a downward direction (following gravity), the pressure increases, and the value of the change in pressure is added. Conversely, when we move from one point to the next in an upward direction (against gravity), the pressure decreases, and therefore the change in pressure is negative.
4. Continue to write an expression for the delta pressure equals the specific weight of the height for each fluid column separately until the opposite end of the manometer is reached. This will result in an expression for the pressure at the end point.
5. Solve this pressure equation for the desired pressure at the given point or the

difference in pressure between two points of interest by substituting known or given data into the equation.

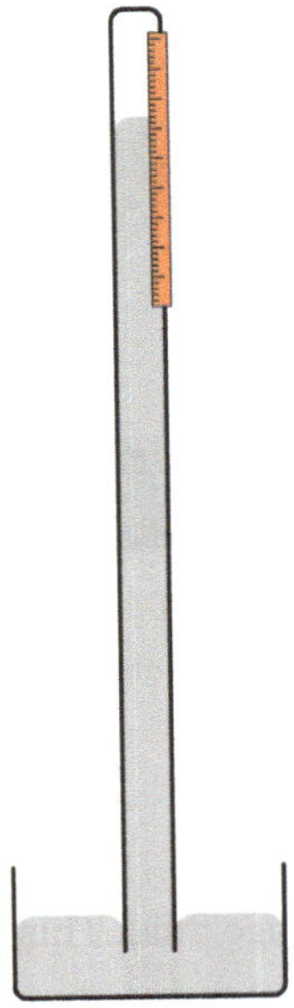

Figure 4-4 A diagram of a simple mercury manometer.

Barometers

Barometers are used for measuring atmospheric pressure. They originally did so by using a long tube closed at one end that was initially filled with mercury.

The open end of the tube was then placed under a surface of mercury and allowed to come to equilibrium, as shown in Figure 4-4. A change in the atmospheric pressure caused a change in the height of mercury in the column. This height in inches of mercury was often reported as the barometric pressure.

Since we know that the change in pressure equals the specific weight times the height, we can convert from inches of mercury to the true atmospheric pressure. Common units for a barometer are in inches of mercury or millimeters of mercury.

Figure 4-5 Old barometers, including Huygens' 1790 barometer, from the Musée des Arts et Métiers in Paris.

Nowadays, atmospheric pressure is typically measured using an aneroid barometer (Figure 4-6). This consists of a flexible-walled evacuated capsule, the wall of which deflects with changes in atmospheric pressure. The wall deflection is transmitted mechanically, through levers or gears, to an indicating needle. A mercury barometer is typically used to calibrate and check an aneroid barometer.

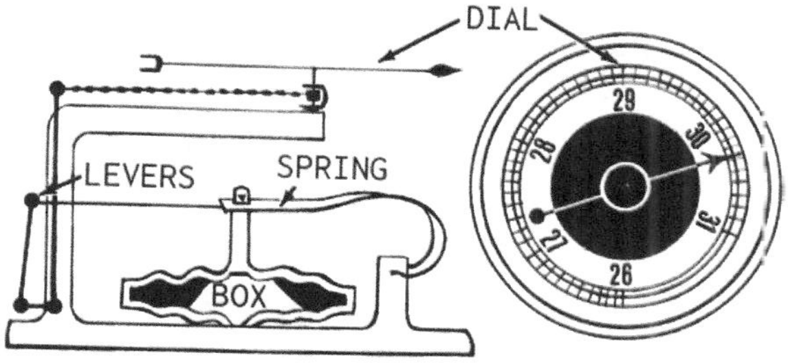

Figure 4-6 Schematic of an aneroid barometer.

Water Manometer

When measuring pressures in low-pressure applications, such as heating ventilation and air-conditioning ducts, the change in pressure reading is very small. Water manometers are sometimes used to measure these pressures. The pressure is expressed in units of inches of water or millimeters of water. Similar to a mercury barometer, we can convert between inches of water and pressure using our relationship of the change in pressure equals the specific weight times the height.

Forces on Surfaces

We will look at two different containers to understand the pressure exerted on the bottom of the containers. Recall that the pressure at the base of the containers is uniform across the entire bottom surface and acts perpendicular (i.e., 90 degrees) to the bottom container surface. The fluid in the container is not moving (at rest), so we are only concerned with the specific weight of the fluid and the height of the fluid inside the container. We will utilize our $P = F/A$ relationship to relate the pressure and force on the bottom of the container.

If an oil drum with a diameter of 0.5 meters (m) is filled with 0.25 m of water and 0.75 m of oil, calculate the force on the bottom of the drum. Note that the top of the drum is exposed to atmospheric pressure; the drum is not pressurized.

Figure 4-7 Barrel of oil.

PROBLEM-SOLVING PROCESS

1. State what is known or given in the problem with both the symbol for the item, the value given, and, *most importantly, the unit of the value.*
    a. Drum diameter: 0.5 m
    b. Water height: 0.25 m
    c. Oil height: 0.75 m
2. State the quantities that you are asked to determine. It is very important at this step to understand what units are being asked for in the solution (U.S. customary or International System of Units [SI]).
    a. Force at bottom of drum – (SI units – newton)
3. Draw and label a sketch of the system where possible. The sketch should contain, if required, the pressures, forces, and components to fully describe it.
    a. Oil depth: 0.75 m; specific gravity (SG): 0.9
    b. Water depth: 0.25 m; SG: 1.0
4. List any initial assumptions that allow you to solve the problem.
    a. Water-specific weight: 9.81 kilonewtons (kN)/meters squared (m^3)
    b. Atmospheric pressure: 0 kilopascals (kPa; gauge)
5. Analyze the problem and identify the key definitions and principles that you would like to apply toward the solution.
    a. Pressure = specific weight * height =
    b. Area = pi * (diameter^2/4) =
    c. Force = pressure * area =
6. Write the symbolic or algebraic solution to your problem. It is important to *delay* substitution of

numerical values as late as possible in the process to allow you to make sure you are not mixing unit systems.

7. Substitute the numerical given values as indicated for all of the physical data given. Please note: As you write numerical values, you must *always include the units* associated with them. Although this will appear to add extra time to obtaining your solution, you will find that this helps you not mix or handle units inappropriately.
8. Check the units associated with each calculation in your final answer.
9. Examine your answer critically. Does this answer appear to be reasonable and consistent with your expectations? If you calculate a number that is extremely small or large, this may indicate that you have made an error.

$$Pressure = Patm + \gamma_{oil} * H_{oil} + \gamma_{water} * H_{water}$$

$$Pressure = Patm + SG_{oil} * \gamma_{water} * H_{oil} + \gamma_{water} * H_{uaer}$$

$$Pressure = 0kPa(gauge) + 0.90 * 9.81kN/m^3 * 0.75m + 9.81kN/m^3 * 0.25m$$

$$\text{Pressure} = 9.074\text{kN/m}^2(\text{gauge})\text{orkPa}(\text{gauge})$$

$$\text{Area} = \prod * D^2/4$$

$$\text{Area} = 3.14 * ((0.5meters)^2/4)$$

$$\text{Area} = 0.196\text{m}^2$$

$$\text{Force} = \text{P} * \text{A}$$

$$\text{Force} = (9.074\ \text{kN/m}^2) * (0.196\text{m}^2)$$

$$\text{Force} = 1.78\text{kN}$$

Keep in mind that as discussed in Pascal's paradox, there will be no difference between the forces on the bottom of any drum, as long as the fluid, height, and diameter on the bottom are the same.

Forces on Submerged Surfaces

A retaining wall, as shown in Figure 4-8, is a common application of a vertical or an inclined wall exposed to varying pressures due to the fluid (liquid) being retained. The pressure varies from 0 gauge pressure at the surface of the fluid to a maximum pressure at the base of the retaining wall.

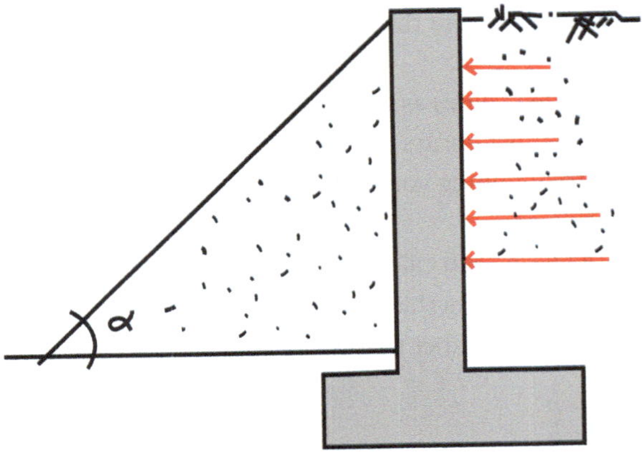

Figure 4-8 Vertical retaining wall.

The force resulting from the fluid (liquid) pressure will act to push the retaining wall over. The force on the wall is distributed from zero force (at the top, liquid surface) to maximum force (at the bottom) as you move down the wall. The shape of this pressure distribution on the vertical retaining wall is a right triangle, as the pressure varies linearly from the top to the bottom.

We can use the pressure, specific weight, and height relationship to find the total resultant force on the wall. This total resultant force can be calculated using the average pressure times the wall area.

Figure 4-9 Pressure distribution, right triangle.

$$Force - resultant = Pressure_{Average} * area = p_{avg} * A$$

The average pressure due to the linear change in pressure from the top of the wall to the bottom is at the middle of the wall and can be calculated by using half of the height.

$$Pressure_{average} = specificweight * \left(\frac{height}{2}\right) = \gamma * \frac{h}{2}$$

Note that the height is the depth of the fluid (liquid) being analyzed. Substituting or average pressure equation into or resultant force equation yields:

$$Force_R = \gamma * \frac{h}{2} * A$$

As we look at the pressure distribution (triangular shape), we notice that the largest pressures act at the base of the wall rather than at the upper part. From our knowledge of statics, we know that the location where the resulting force acts (i.e., the center of pressure) is at the centroid of the pressure distribution triangle. In other words, we could replace the triangular-shaped pressure distribution with a single resulting force at the centroid and have the same effect on the wall in terms of summation of forces and moments. The centroid for a triangle is one-third of the distance from the base, as shown in Figure 4-10.

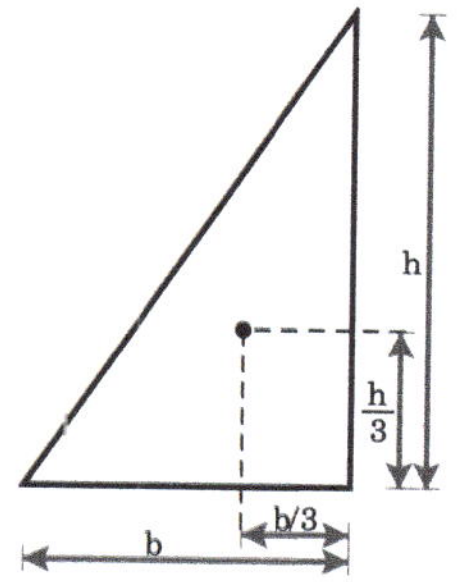

Figure 4-10
Centroid for triangle.

Following is the general process for determining the resultant force and location on a vertical rectangular wall:

1. Calculate the magnitude of the resulting force on the wall.
2. Locate the center of pressure at the centroid of the triangular pressure distribution (h/3 from the bottom of the wall).
3. Show the magnitude of the resulting force on the wall at the centroid acting perpendicular to the wall.

Buoyancy

Buoyancy is the force opposing gravity that extends upward to support an immersed object. Buoyancy forces act vertically through the centroid of the object, whether floating or fully submerged. The magnitude of the buoyancy force is equal to the weight of the fluid displaced.

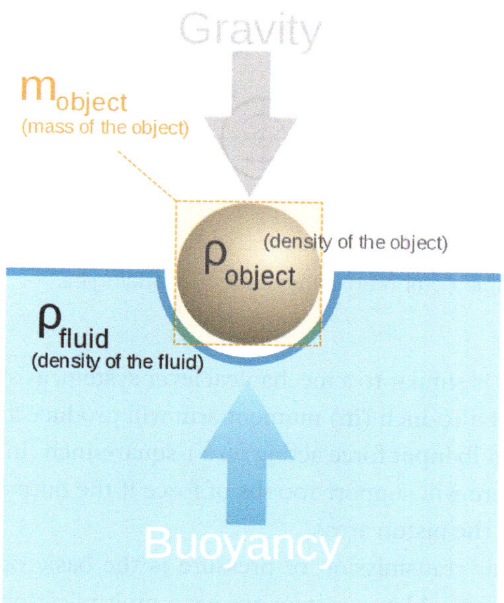

Figure 4-11 Buoyancy graphic.

$$Force_{bouyant} = SpecificWeightofFluid * DisplacedVolumeofFluid = F_b = \gamma * V_d$$

Applications of Force Multiplication

A key outcome of fluid power systems is the multiplication of forces. A simple hydraulic jack is a good example of how fluid power produces a proportionately larger force from a relatively small input.

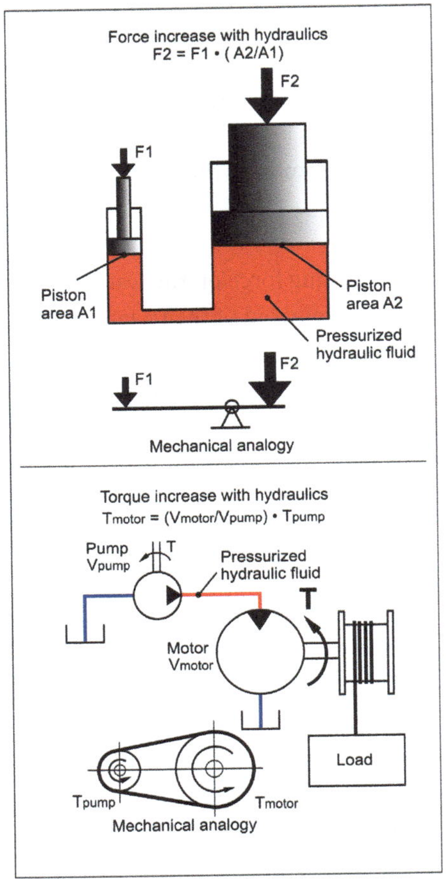

Figure 4-12 Hydraulic force principle compared with mechanical analogies.

This multiplication of forces is similar to a mechanical lever system, as shown in Figure 4-12. For example, a 10-pound (lb) input force acting at a 10-inch (in) moment arm will produce a 100-lb output force acting at a 1-in moment arm. This is similar to a 10-lb input force acting on a 1-square-inch (in^2) input area, developing a pressure of 10 lbs/in^2 in the oil. This pressure will support 100 lbs of force if the output piston has an area of 10 in^2. Note that the forces are proportional to the piston areas.

As shown in Figure 4-13, this transmission of pressure is the basic operating principle of a fluid-power hydraulic jack. Similar to the mechanical lever system, we get a multiplication of forces, but we should note that displacement on the input side must be 10 times greater than the output displacement. From our work equation, we know that work is force times distance.

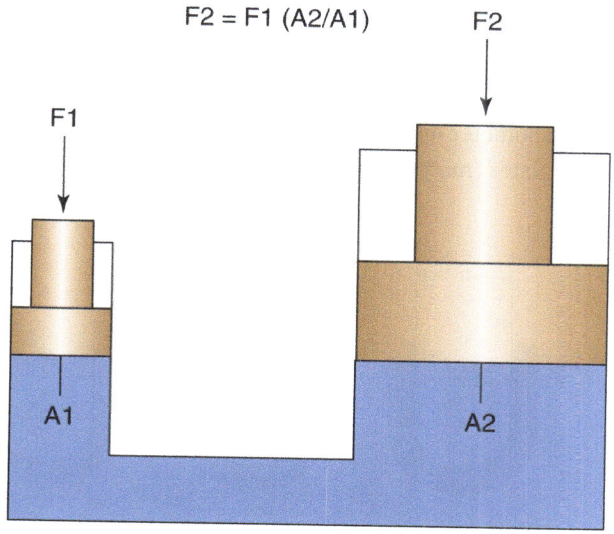

Figure 4-13 Hydraulic force ratio principle.

The work or energy we input to achieve the force multiplication would result on equal work or energy output. A fluid power system cannot create energy, as energy must be conserved; therefore, as we achieve greater output force, we also achieve less displacement. The basic law that describes pressure transmission is Pascal's law. Pascal's law demonstrates the basic principle of how a pressure is transmitted *undiminished in all directions throughout and acts perpendicularly to the surfaces in contact with the fluid.*

Key Equations

$$Force_R = \gamma * \frac{h}{2} * A$$

(retention wall)

$$Force_{bouyant} = SpecificWeightofFluid*DisplacedVolumeofFluid = F_b = \gamma*V_d$$

Exercises

1. What is the pressure at the bottom of a 32-m-high water tower in gauge pressure and absolute pressure (answers in kPa and bar)?
2. Find the gauge and absolute pressure in pounds per square inch (psi) and psi absolute (psia) on a drilling blowout preventer at a depth of 8,000 feet (ft) of ocean seawater (answer in psi and psia). The SG of seawater is approximately 1.03 (specify the weight of water).
3. A news broadcaster reports that the barometric pressure is at 30.04 in of mercury. What is the pressure in psia?
4. The pressure in an unknown fluid at a depth of 8.0 ft is measured to be 3.7 psi in gauge (psig). Compute

the fluid's SG.

5. For the simple U-tube manometer in Figure 4-14, determine the pressure in the second figure if the units of measure are inches of mercury.
6. For the simple U-tube manometer in Figure 4-14, determine the vacuum in the third figure if the units of measure are inches of water (material is water; answer in psi).

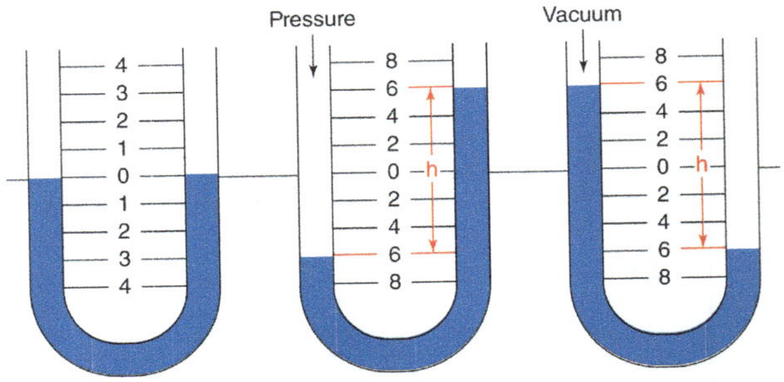

Figure 4-14

7. Write the equation and find P1. (Start with atmospheric pressure and follow the process.)

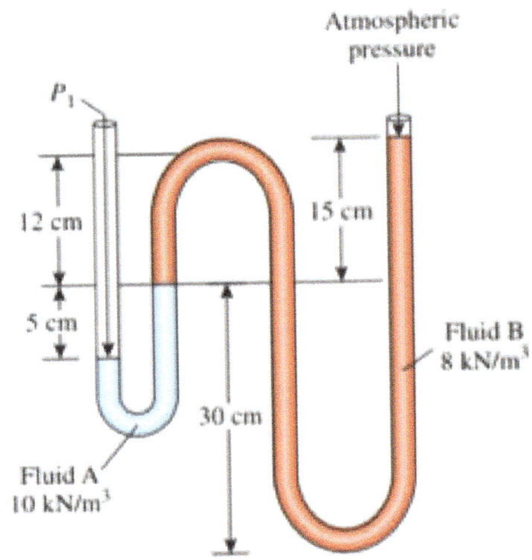

FIGURE P1–78

Figure 4-15

8. A 110-ft-long pipe is inclined at a 30-degree angle with the horizontal. The pipe is filled with hydraulic oil with an SG of 0.92. What is the pressure at the base of the pipe if the top of the pipe is vented to the atmosphere in psi and psia?
9. What is the pressure at the base of the pipe if the top of the pipe is sealed and pressurized to 12 psi in problem 7 above? Answer in psi and psia. (Key: Draw a sketch of the system.)
10. Depicted in Figure 4-16 is a common arrangement designed to provide lubrication to turbo machinery

in the case of pump failure. It also portrays the operation of a typical gas-loaded accumulator in hydraulic systems. The accumulator is precharged with 150 psig of nitrogen (N_2), and the oil "free" (or level) surface inside the accumulator is located 45 ft (height) above the main supply line to the system. The oil has an SG of 0.88. What is the pressure reading indicated by the gauge (in psig)?

Note: The pump is not operating, and the system is not taking flow.

(Specific weight of freshwater, γ_w = 62.4 pound-force [lbf]/cubic feet [ft^3])

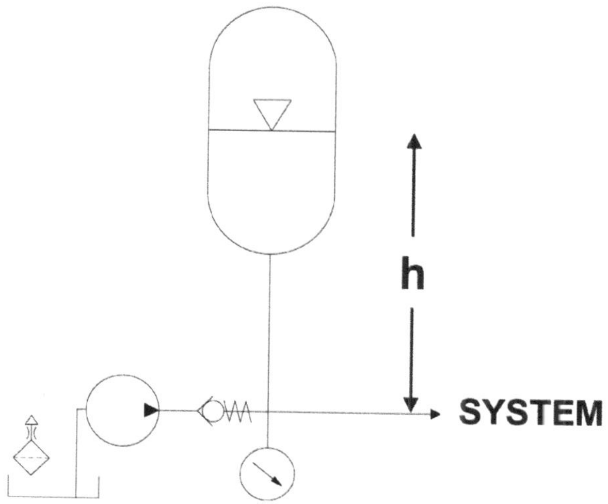

Figure 4-16

11. A conical-shaped container has 2.4 m of oil (SG = 0.9) and 1.5 m of water (SG = 1). It has a 1.2-m-diameter top and a 3.0-m bottom, with a constant slope between these diameters over the total height of 3.9 m. It is open to the atmosphere at the top. Calculate the force on the bottom of the container (answer in kN).
12. A standard barrel (sample below) has 4 ft of water (SG = 1.00). It has a 2-ft diameter (top to bottom) and a height of 4 ft. It is open to the atmosphere at the top (not pressurized). Calculate the force on the bottom of the container (answer in lbf). The specific weight of water on Earth at 5°C is 62.43 lbf/ft^3 or 9.807 kN/m^3.

Figure 4-17

13. A tank container has the dimensions listed in Figure 4-18. The air pressure **inside** the top of the tank is 52 kPa (gauge). Calculate **the force** (differential force) on the bottom of the container (answer in kN). Note: **The tank top is sealed, not open to atmosphere, and is pressurized to 52 kPa.** The specific weight of water on Earth at 5°C is 62.43 lbf/ft^3 or 9.807 kN/m^3.

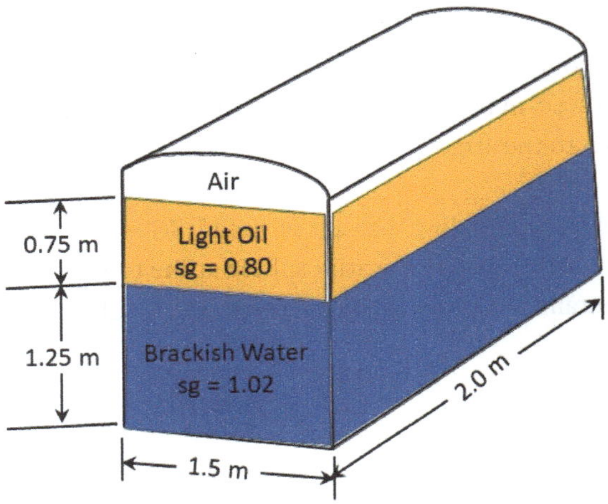

Figure 4-18

14. A car plunges into a lake during an accident and lands on its wheels with the top edge of the door 10 m from the water's surface. The door is 1.2 m tall and 1 m wide. Assume that the door is a vertical rectangular plate, with atmospheric air pressure inside the car (no water leakage). The specific weight

of the water is 9.810 kN/m^3. Can the driver open the door?

15. Using the diagram in Figure 4-19, calculate the force needed in pounds at the pump side of the system (point A) if the load to be lifted is 65,000 lbs (lbf; point B), the ram diameter is 8 in (point B) in diameter, and the pump piston diameter is 2 in (point A).

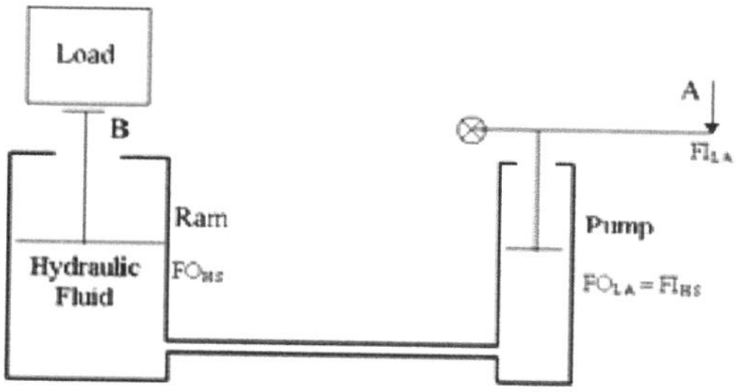

Figure 4-19

Figure Credits

Fig. 4-1: Zeichner: Bd, https://commons.wikimedia.org/wiki/File:Communicating_vessels.svg, 2004.
Fig. 4-2: Copyright © by Bidgee (CC BY 3.0) at http://commons.wikimedia.org/wiki/File:Darwin_Airport_Water_Tower.jpg.
Fig. 4-3: Copyright © by Ruben Castelnuovo (CC BY-SA 3.0) at http://en.wikipedia.org/wiki/File:Utube.PNG.
Fig. 4-4: Danomagnum, http://en.wikipedia.org/wiki/File:MercuryBarometer.svg, 2009.
Fig. 4-5: Copyright © by Anton Lefterov (CC BY-SA 3.0) at http://en.wikipedia.org/wiki/File:Old-barometers.jpg.
Fig. 4-6: Pearson Scott Foresman, https://commons.wikimedia.org/wiki/File:Aneroid_Barometer_(PSF).png.
Fig. 4-7: Lexicon, https://commons.wikimedia.org/wiki/File:Renault_F1_team_ELF_oil_drum.JPG, 2006.
Fig. 4-8: Mariana Ruiz Villarreal, http://commons.wikimedia.org/wiki/File:01453_1106_19_%28jpg%29.svg, 2007.
Fig. 4-9: Dante, http://commons.wikimedia.org/wiki/File:Isosceles-right-triangle.svg, 2007.
Fig. 4-10: K. Aainsqatsi, http://commons.wikimedia.org/wiki/File:Triangle_centroid_2.svg, 2007.
Fig. 4-11: Copyright © by Yupi666 (CC BY-SA 3.0) at https://en.wikipedia.org/wiki/File:Buoyancy.svg.
Fig. 4-12: Copyright © by Lidingo (CC BY-SA 3.0) at http://en.wikipedia.org/wiki/File:Hydraulic_Force_Torque_275px.png.
Fig. 4-13: Darbyshmr, http://en.wikipedia.org/wiki/File:Hydraulic_Force,_language_neutral.png, 2008.
Fig. 4-15: Yunus Cengel and Michael Boles, "Write the equation and Find P1," https://studysoup.com/tsg/49920/thermodynamics-an-engineering-approach-8-edition-chapter-1-problem-78p. Copyright © 2015 by StudySoup, Inc.
Fig. 4-17: Copyright © by Meggar (CC BY-SA 3.0) at https://commons.wikimedia.org/wiki/File:Drum_(container).jpg.
Fig. 4-19: Copyright © by The Technology Interface.

CHAPTER 5

Fluids in Motion: Fundamental Conservation Principles

Overview and Learning Objectives

This chapter reviews general and specific items in conservation of mass, conservation of energy, and conservation of linear momentum. We will *not* consider situations involving nuclear transformations or relativistic effects. We will consider fluid power systems and the concept of flow rate at its relationship to the average velocity of a flowing fluid. We will also review the fundamental principles of conservation of energy and conservation of momentum in fluid power systems. We will apply these principles to fluid volume at rest and in motion. We will look at pressure forces—in this case, shear forces and gravitational bodies.

After reviewing this chapter, students should be able to do the following:

- Calculate volume and mass flow rates for fluid power circuits.
- Define in their own words the principle of conservation of mass in fluid power systems.
- Have an understanding of the conservation of energy principle and how it applies to fluid power systems.
- Explain the physical significance of conservation of energy in a fluid power circuit.
- Describe the forces that act on the fluid element when at rest and when in motion.
- Define in their own words and with symbols the conservation of momentum principles.

Conservation of Mass

The law of **conservation of mass states** that the mass of any system will remain constant and not change over time, as long as mass is not added or removed from the system. In other words, the quantity of mass in a system does not change and is conserved over time. In terms of flowing fluids, Leonardo da Vinci expressed for incompressible fluid flow that "a river of uniform depth will have a more rapid flow at the narrower section than at the wider, to the extent that the greater surpasses the lesser." He called this the *law of continuity*.

Figure 5-1 River flow.

We will use the conservation of mass as we look at fluid systems by defining the **mass flow rate** as the mass of the fluid flowing per unit of time. The mass flow rate will have a unit of kilogram per second (kg/s) in the International System of Units (SI) or pound per second (lb/s) in the U.S. customary system. The common symbol for mass flow is $\dot{m}$. This can be measured in a fluid system by collecting the fluid flow, weighing it, and measuring the time it takes for the fluid collection to occur. In large fluid systems, it would be very difficult to collect a flowing fluid in large buckets to know the flow rate through a pipe at a given time. The equation for mass flow is shown below:

$$\dot{m} = \rho * v * A = \rho * Q$$

where:

- ρ = mass density of the fluid,
- v = velocity field of the mass elements flowing,
- A = cross-sectional area, and
- Q = volumetric flow rate.

The above equation is valid when the velocity direction is perpendicular to the area of interest. In other words, the velocity vector is normal to the cross-sectional area. We will use this with our idea of **conservation of mass/continuity of mass flow** to state that the mass flow in any system in which mass is not being added or subtracted is constant. Phrased differently, in any steady-state system, the mass flow entering the system is equal to that which is leaving the system. If more than one flow stream enters or exits the system, the total mass flow entering the system equals the total mass flow exiting the system. This equation is called the *continuity equation for mass*.

$$\dot{m} = \rho_1 * v_1 * A_1 = \rho_2 * v_2 * A_2$$

When dealing with liquids, if the density is constant, the mass flow rate simplifies to the volume continuity equation. If the density does not change (is constant), we call this *incompressible flow* because liquids are treated as incompressible fluids. We will utilize the symbol Q as the volumetric flow rate, or the volume of fluid passing a given point per unit. The units for volumetric flow rate in the SI system will be cubic meters per second (m^3/s); in the U.S. customary system, we will use cubic feet per second (ft^3/s), cubic inches per second (in^3/s), or gallons per minute (gpm).

$$Q = v_1 * A_1 = v_2 * A_2$$

where:

- v = velocity field of the mass elements flowing,
- A = cross-sectional area, and
- Q = volumetric flow rate.

We can rewrite the volume continuity equation for liquids as:

$$\frac{v_1}{v_2} = A_2/A_1$$

Or:

$$\frac{v_1}{v_2} = (D_2/D_1)^{\wedge}2$$

This equation demonstrates that the smaller the inside diameter of the pipe is, the greater the velocity in the pipe is. Note that the inside diameter relationship is squared and thus has a significant effect on the pipe's velocity.

Conservation of Energy

The **conservation of energy law** was developed in the 1800s. The law states that the amount of energy in a system remains constant over time. This means that the energy is being conserved and does not change. For fluid systems, this means that the energy can change form within the system, but the total energy remains constant. Specifically, the kinetic energy can transform into potential energy, but the total energy can neither be created nor destroyed.

In fluid systems, the potential energy of a fluid is due to its elevation and/or pressure.

Potential Energy due to Elevation

If we have a fluid of weight *W* at a height or elevation with respect to a reference point, the weight will have a potential energy of weight times elevation. The units of potential energy in the SI system are newton-meters (N-m); in the U.S. customary system, they are pounds-foot (lb-ft).

$$Potentialenergyduetoelevation = Weight * Elevation$$

$$PE(Elevation) = W * z$$

where:

- *W* = weight of the fluid at a point in a streamline, and
- *z* = elevation.

Pressure

The pressure of a fluid of weight *W* is the weight times the pressure over the specific weight. The units of this energy in the SI system are N-m; in the U.S. customary system, they are lb-ft.

$Pressure = Weight * \frac{Pressure}{Specific\ Weight}$

$PE\,(Pressure) = W * \frac{P}{\gamma}$

where:

- *W* = weight of the fluid at a point in a streamline,

- P = pressure at the chosen point, and
- γ = the material's specific weight.

Kinetic Energy

We know from the study of physics that kinetic energy can be expressed as shown in the following equation:

$$KineticEnergy=\frac{1}{2}*mass*(velocity)^2$$

The kinetic energy of a fluid that is moving with a velocity can be expressed as one-half the weight over gravity times velocity squared.

$$FluidKineticEnergy=\frac{1}{2}*mass*(velocity)^2=\frac{1}{2}*\frac{Weight}{g}*(velocity)^2$$

$$FluidKineticEnergy=\frac{1}{2}*\frac{W}{g}*(v)^2$$

If we examine the units in the equation, we find that the units of kinetic energy in the SI system are N-m or lb-ft in the U.S. customary system.

Using the principle of the conservation of energy, we can state that for a mass piece of fluid, the total energy contained in the fluid remains constant. This principle assumes that no energy is added to the fluid via pumps or removed from the fluid due to friction or other means as the mass piece of fluid flows through the system. Mathematically, we have:

$PE\,(Elevation)+PE\,(Pressure)+Fluid\ Kinetic\ Energy\,(Motion)=Constant$

$$W*z+W*\frac{P}{\gamma}+\frac{1}{2}*\frac{W}{g}*(v)^2=Constant$$

We can simplify this equation by dividing out the common weight term, as a constant divided by a common weight is still a constant to obtain:

$z+\frac{P}{\gamma}+\frac{(v)^2}{2g}=Constant$

where:

- g = acceleration due to gravity,
- Z = elevation of the point above a reference plane, with the positive z-direction pointing upward; so, in the direction opposite tohe gravitational acceleration,
- P = pressure at the chosen point,
- γ = specific weight of the fluid at all points in the fluid, and
- v = fluid flow speed at a point on a streamline.

This equation is a common form of the total energy equation for incompressible flow. We use the term **incompressible flow** to denote that even though the pressure can vary in the fluid, the density must remain constant along the streamline. A second key assumption for this equation is that the friction forces due to internal fluid viscosity must be negligible.

We will use the term **fluid parcel** (fluid element or small chunk) as a very small volume of fluid within the larger fluid flow (see Figure 5-2). The mass and weight of the fluid parcel remain constant as it flows downstream. The shape of the fluid parcel and its volume may change within the flow, but its mass always remains constant.

It is important to note that the total energy possessed in a fluid parcel remains constant; however, the energy can change form from the potential energy due to elevation, pressure, or kinetic energy. For example, the fluid parcel may decrease in elevation as it flows from a storage tank through the system and thus have less potential energy due to elevation. This change in energy, however, would result in an equal amount of energy added to the fluid in terms of potential energy due to pressure or kinetic energy. We will expand this equation in the next section to account for the energy added to a fluid from a pump or the energy removed from a fluid due to motors or friction in the pipeline as fluid flows through real-world systems.

Bernoulli's Equation

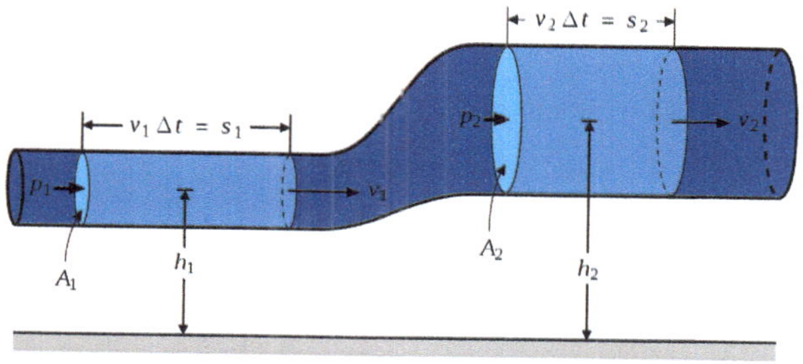

Figure 5-2 Two "parcels' of fluid in a pipe.

As we stated previously, the total energy between any two points in our fluid system is constant. Energy can change forms from potential energy to kinetic energy or vice versa. As shown in *Figure* 5-3, the flow goes from the larger section of tubing on the right to the smaller tubing section on the left; the reduction in area produces an increase in velocity, and the increase in velocity leads to a reduction in pressure as indicated by the differential pressure gage. It could also be said that because the flow is restricted on the left side of the tubing, a high velocity on the left side requires a high pressure on the right side of the tubing. We must emphasize that the fluid is moving!

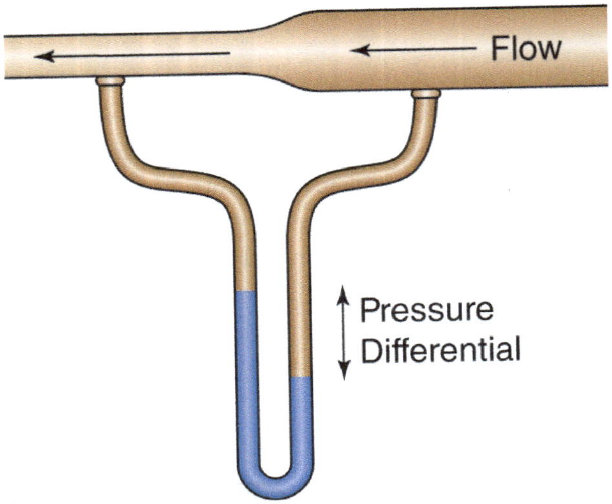

Figure 5-3 Photo of a Venturi tube with labels.

This principle is named after Dutch scientist Daniel Bernoulli, who published this principle in 1738. This principle and subsequent equation are one of the fundamental relationships for analysis of fluid circuits. The application of this principle assists us in designing components in fluid power circuits such as pumps, piping, valves, and motors for acceptable system performance.

We can look again at the fluid parcel at two points in a pipe, as shown in *Figure 5-2*. At our first point, the fluid parcel has a certain elevation, pressure, and kinetic energy with respect to the common reference plane. At our second point in the pipe, the fluid parcel again has certain but different elevation, pressure, and kinetic energy with respect to the same common reference plane. These energies are shown in *Table 5-1*.

Table 5-1. Types of Energy

Types of Energy	Point 1	Point 2
Elevation – potential energy	$W * Z_1$	$W * Z_2$
Pressure – potential energy	$W * (P_1/\text{specific weight})$	$W * (P_2/\text{specific weight})$
Kinetic energy	$W * (V_1\text{^}2/2g)$	$W * (V_2\text{^}2/2g)$

We know that the total energy contained in the fluid parcel between these two points is equal, assuming that there are no losses and no energy is added or removed from the fluid as it travels between these two points. If we cancel the weight term from both sides of the equation, we have the original Bernoulli equation for an ideal frictionless fluid system that does not contain pumps or motors between the two points:

$$z_1 + \frac{P_1}{\gamma} + \frac{(v_1)^2}{2g} = z_2 + \frac{P_2}{\gamma} + \frac{(v_2)^2}{2g}$$

As we examine the units for each term in the equation, we find that each term has a unit of length. This unit of length or height we call the *head*. Thus, we will use the terms *elevation head*, *pressure head*, or *velocity head*.

This equation does not take into account frictional losses that can take place between points 1 and 2 or additions of energy from pumps or removal of energy from motors that may exist between our two points. Therefore, we can modify the original Bernoulli equation to account for these conditions. The equation as shown below uses the conservation of energy law to state that the total energy in a parcel of fluid at one point plus the additional energy added by or removed by a machine (pump or turbine; h_m) minus the energy lost due to friction in the fluid (h_f) equals the total energy possessed by a parcel of fluid when it arrives at point 2. This equation is the modified Bernoulli equation, also called the *general energy equation.* It is important to remember that each term in the general energy equation has units of length. In fluid power, we will use the term *head* to represent these lengths.

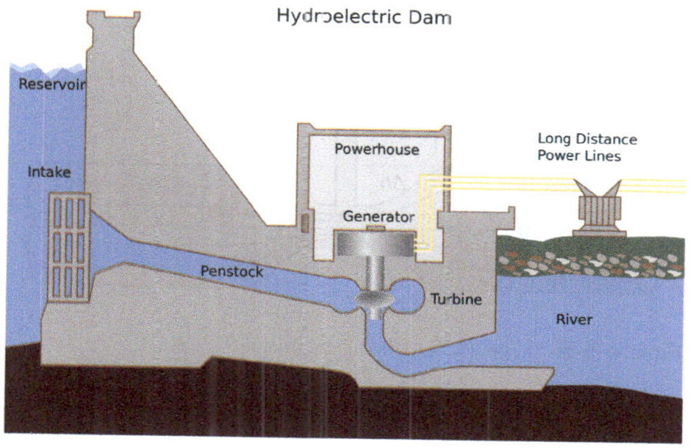

Figure 5-4 Hydraulic head.

$$z_1 + \frac{P_1}{\gamma} + \frac{(v_1)^2}{2g} - h_f + h_m = z_2 + \frac{P_2}{\gamma} + \frac{(v_2)^2}{2g}$$

where:

1. *hf* = friction head loss (energy lost due to the friction of the fluid against the pipe wall); this can also take into account the entrance losses or losses due to diameter changes within the pipe system, and
2. *hm* = losses or gains from other sources such as pumps/turbines (machines). If the machine is a pump (adding fluid energy), *hm* is positive. If the machine is a turbine or motor (removes fluid energy), *hm* is negative.

Bernoulli's Equation Application: Venturi Tube

As an example, we will look at a special device whose diameter decreases and then increases, also known as a Venturi tube. As shown in *Figure 5-6*, the pressure coming into the funnel section is larger than the pressure in the reduced diameter section. According to Bernoulli's equation, we know that the total energy is constant between these two points. Since the elevation between these two points does not change, we can see that the fluids velocity must increase as it passes through the smaller diameter section while its pressure decreases. This also follows the continuity equation, which states that the mass flow is constant. The mass flow is a function of density area and velocity: If the density of our fluid does not change between the incoming point and the decreased diameter point, then the velocity must increase to account for the decreased area resulting from the decreased diameter.

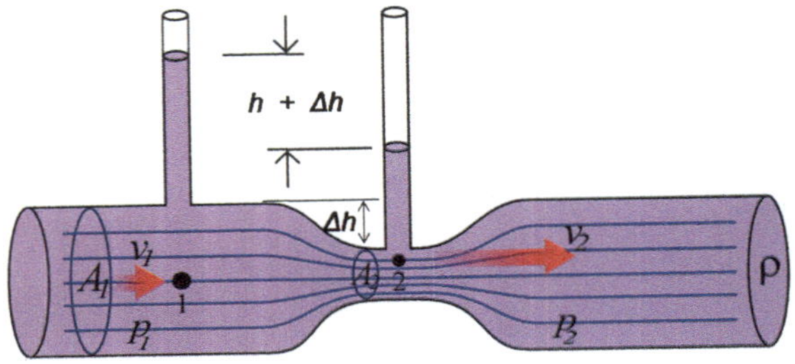

Figure 5-5 Venturi effect graphic.

Conservation of Linear Momentum

The conservation of linear momentum is a basic concept in physics. From physics, we recall that momentum is the mass of an object or particle multiplied by the velocity of the same object or particle:

$$Momentum = mass * velocity = m * v$$

Within a domain space, the amount of momentum is conserved (does not change, remains constant). Like matter and energy, momentum cannot be created or destroyed but only changed through the interaction of forces. Different from matter and energy, momentum is a vector quantity with both magnitude and direction. Therefore, we must consider that momentum is conserved in all three physical directions at once.

As we look at the fluid parcel traveling through a pipe, we can see that the fluid parcel has a density and velocity that are a result of its time and position. With this density and velocity, the momentum of this parcel of fluid per unit volume is:

$$momentumperunitvolume = density * velocity = \rho * v$$

In our closed fluid system, this total linear momentum is a constant quantity. This constant quantity does not change; it is conserved.

Key Equations

$$m = \rho_1 * v_1 * A_1 = \rho_2 * v_2 * A_2$$

$$Q = v_1 * A_1 = v_2 * A_2$$

- where:
  - v = velocity field of the mass elements flowing,
  - A = cross-sectional area, and
  - Q = volumetric flow rate.

$$z_1 + \frac{P_1}{\gamma} + \frac{(v_1)^2}{2g} = z_2 + \frac{P_2}{\gamma} + \frac{(v_2)^2}{2g}$$

Exercises

Continuity: Conservation of Mass

1. How large of a pipe (inside diameter; give your answer in inches) is required to maintain a velocity below 20 feet per second if the flow rate is 30 gallons per minute (gpm)?
2. How long would it take (time in seconds) for a 30-gpm pump to extend a 6-inch (in)-diameter cylinder through 10 in?
3. How long would it take (time in seconds) for a 50-gpm pump to retract an 8-in diameter cylinder with a stroke of 24 in if the rod diameter is 4 in?

Conservation of Energy: Bernoulli's Equation

4. Fluid is flowing horizontally at 100 gpm from a 3-in-diameter pipe (point 1) to a 2-in-diameter pipe (point 2). If the pressure at point 1 is 15 pounds per square inch (psi), find the pressure at point 2 (answer in psi). The specific gravity of the hydraulic oil is 0.9.

Figure Credits

Fig. 5-1: Copyright © by Burim (CC BY-SA 3.0) at http://commons.wikimedia.org/wiki/File:River_flow.JPG.
Fig. 5-2: Copyright © by MannyMax (CC BY-SA 3.0) at http://en.wikipedia.org/wiki/File:BernoullisLawDerivationDiagram.svg.
Fig. 5-3: Copyright © by ComputerGeezer (CC BY-SA 3.0) at http://commons.wikimedia.org/wiki/File:VenturiFlow.png.
Fig. 5-4: Tennessee Valley Authority, http://en.wikipedia.org/wiki/File:Hydraulic_head.PNG, 2004.
Fig. 5-5: HappyApple, http://en.wikipedia.org/wiki/File:Venturifixed2.PNG, 2006.

CHAPTER 6

Fluid Power and Systems: Basic Components and Symbology

Overview and Learning Objectives

Transmitting power with the use of pressurized fluids is more common than people realize and has a very wide range of applications. Equipment such as excavators, dump trucks, lifts, aircraft wing surface controls, and power brakes and the resurgence of hydraulic hybrid vehicles are some of the examples in heavy construction and mobile applications in which the massive power density of hydraulics is unsurpassable. This power density, or power-to-weight ratio, is also an advantage in other fields, such as biomechanics, in which current scientists are designing and developing novel fluid power–assisted artificial/mechanical braces, to improve the quality of life of patients in remission and disabled individuals.

In this chapter, we discuss what hydraulic and pneumatic power is. Restating the concept of pressure energy and invoking Pascal's principle and the equation of continuity, we derive the relationship between fluid power and the pressure of the fluid and the volumetric flow rate. We also discuss the differences between the U.S. customary system and the International System of Units (SI) in terms of power magnitude as well as the concept of efficiency in fluid power systems.

In addition, we extend the discussion on the basic components of a fluid power system and introduce the use of symbols by the International Organization for Standardization (ISO) to represent the components of a fluid power system in an engineering diagram or a drawing. In the fluid power industry, engineers design, create, test, use, and modify integrated systems. From reservoirs and pumps to valves, hoses, piping, actuators, and motors, fluid power is used everywhere. As part of a global industry, fluid power engineers communicate using standardized drawings and language. Therefore, the knowledge of fluid power ISO symbols is fundamental to designing, analyzing, operating, and troubleshooting hydraulic and pneumatic circuits.

After completing this chapter, students should be able to do the follows:

1. Relate Pascal's principle to pressure energy and flow work.
2. Define fluid power and calculate the power transmitted by a pressurized fluid.
3. Differentiate among U.S. customary and SI units when describing and performing calculations on fluid power.
4. Use the general concept of efficiency in the context of fluid power.
5. Describe various designs for the components of a fluid power system.
6. Identify the basic components of a basic fluid power system using ISO symbols and diagrams.

Hydraulic Power: Pressure and Work, Flow and Power, and Efficiency

Pressure and Work

As stated in Chapter 5, the energy associated with the pressurized fluid is given by:

$$PE_{PRESSURE} = W \cdot \frac{P}{\gamma}$$

Written in this traditional way, it is difficult to grasp the concept of pressure energy in a meaningful way. In contrast, the kinetic energy associated with the velocity of an object—in this case, the flowing fluid parcel or the potential energy associated with their changes in elevation—is easier to visualize. Nevertheless, a simple manipulation of the pressure energy relationship reveals the direct connection of pressure, work, and fluid power.

In Chapter 3, we learned that one of the most important properties of any fluid is the **specific weight**. The **specific weight** is simply the gravitational force exerted on a fluid parcel of a given density. Using the definition of density for a fluid parcel of volume V, we can write the specific weight as:

$$\rho = \frac{M}{V} \rightarrow \rho \cdot g = \frac{M}{V} \cdot g \rightarrow \gamma = \frac{M \cdot g}{V}$$

$$\gamma = \frac{W}{V}$$

where:

- γ = specific weight of the fluid parcel,
- ρ = density of the fluid parcel,
- M = mass of the fluid parcel,
- V = volume of the fluid parcel, and
- g = acceleration of gravity.

The specific weight definition and pressure energy relationship can be combined to remove the weight of the fluid parcel from the pressure energy relationship. When the volume of a given fluid parcel is known, its weight also can be identified, and the pressure energy relationship can be rewritten as:

$$\gamma = \frac{W}{V} \rightarrow W = \gamma \cdot V$$

$$PE_{PRESSURE} = W \cdot \frac{P}{\gamma} \rightarrow PE_{PRESSURE} = (\gamma \cdot V) \cdot \frac{P}{\gamma} = V \cdot P$$

Or:

$$PE_{PRESSURE} = P \cdot V$$

In terms of SI units, pressure energy is measured in:

$$P \cdot V = Pa \cdot m^3 = \frac{N}{m^2} \cdot m^3 = N \cdot m = Joule$$

In terms of U.S. customary units, pressure energy is measured in:

$$P \cdot V = psf \cdot ft^3 = \frac{lb_f}{ft^2} \cdot ft^3 = lb_f \cdot ft$$

Note: The typical unit of pressure in the U.S. customary system is pounds per square inch (psi; pound-force [lbf]/square inches (in^2) or 144 lbf/square feet [ft^2]).

This new relationship shows a direct correlation of the energy in a fluid parcel of a certain volume V and its surrounding pressure, P. Now, we proceed to demonstrate how this pressurized parcel of fluid is able to perform mechanical work. Pressure energy and mechanical work can be related by analyzing a hydraulic cylinder. Recalling Pascal's principle that pressure is transmitted undiminished in a confined body of fluid, supplying pressurized fluid to a hydraulic cylinder generates a force that acts perpendicular to the surface of the piston due to the pressure in the fluid. Figure 6-1 presents a schematic of a hydraulic cylinder with cross-sectional area A exerting a force F under the action of a fluid under pressure P.

In Chapter 2, we recalled that mechanical work is defined as the product of the force F required to move an object and the distance Δs traveled by the object under the action of that force:

$$Work = F \cdot \Delta s$$

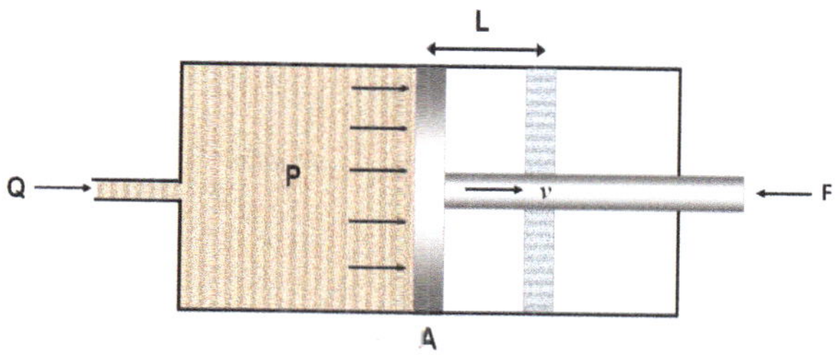

Figure 6-1 Schematic of a hydraulic cylinder.

Neglecting the friction at the cylinder walls and assuming the pressurized fluid is supplied at a rate that ensures the piston is pushed slowly at a constant velocity, then the force exerted by the pressurized fluid on the piston is numerically equivalent to that exerted by the rod. From the definition of pressure, force-per-unit area, the force on the surface of the piston exerted by the pressurized fluid is given by:

$$P = \frac{F}{A} \rightarrow F = P \times A$$

where P is the pressure in the fluid and A is the area of the piston. If the piston travels a distance L under the action of the pressurized fluid, the mechanical work is given by:

$$Work = F \cdot \Delta s = P \times A \cdot L$$

The product A·L is the amount of fluid supplied to displace the piston. In other words, A·L is the volume of fluid V with pressure P fed to the hydraulic cylinder to perform the mechanical work F·Δs. The expression that follows is the same obtained when combining the definitions of pressure energy and specific weight:

$$Work = F \cdot \Delta s = P \cdot V$$

Flow and Power

Pressure energy (P·V) is also referred to as flow work, or the work required to move the fluid against the pressure of the surroundings. Fluid power is simply the rate at which this pressure energy or flow work is transmitted. In terms of the size and velocity of the piston and invoking the continuity equation for an incompressible fluid:

$$Q = v \times A$$

$$Power = \frac{Work}{time} = \frac{F \cdot \Delta s}{\Delta t} \leftarrow F = (P \times A) \leftarrow v = \frac{\Delta s}{\Delta t}$$

$$Power = \frac{Work}{time} = (P \times A) \cdot v = P \cdot (v \times A) \leftarrow Q = (v \times A)$$

$$Power = P \cdot Q$$

A similar expression can be derived if we directly consider the rate at which the fluid volume is provided to the cylinder (i.e., the **volumetric flow rate** Q):

$$Q = \frac{\Delta V}{\Delta t}$$

$$Power = \frac{Work}{time} = \frac{P \cdot \Delta V}{\Delta t} = P \cdot \frac{\Delta V}{\Delta t} = P \cdot Q$$

$$Power = P \cdot Q$$

The product P·Q is normally referred to as **hydraulic power**. This is used strictly to calculate the power transmitted or converted by incompressible fluids.

In terms of SI units, hydraulic power is measured in:

$$P \cdot Q = Pa \cdot \frac{m^3}{s} = \frac{N}{m^2} \cdot \frac{m^3}{s} = \frac{N \cdot m}{s} = \frac{Joule}{s} = watt$$

In terms of U.S. customary units, hydraulic power is measured in:

$$P \cdot Q = psf \cdot \frac{ft^3}{s} = \frac{lb_f}{ft^2} \cdot \frac{ft^3}{s} = \frac{lb_f \cdot ft}{s}$$

Note: The typical unit of hydraulic power in the U.S. customary system is horsepower (hp; 550 lbf·foot [ft]/second[s]).

Efficiency

Figure 6-2 presents a sketch on how the energy or power is converted and transmitted in a typical fluid power system. As discussed in previous chapters, in a fluid power system, the actual input energy and power are provided by an external motor commonly an electric motor in stationary applications and an internal combustion engine in mobile applications. This external motor converts either electricity or fuel into mechanical energy or power to drive the pump or compressor which turns the mechanical energy and power into fluid energy and power. This fluid energy and power is sent and directed to the required locations via conduits, hoses and control valves. Once at the hydraulic or pneumatic motor or actuator, the fluid energy and power is converted back to mechanical energy and power, either linear or rotational. The final conversion process is defined by the external load, lifting, breaking, and propelling. In all energy conversion processes, there are friction and heat losses.

As also discussed, the term **efficiency** is used in engineering to evaluate how effective an energy conversion process is. It is defined as the ratio of the output from the system and the input to it. In Figure 6-2, that which has been put into the system is the energy and power provided by the external motor, either V·I (electrical power) or (heat rate of combustion). What the system puts out is mechanical energy, either F·v (linear) or T·ω (rotational). The overall efficiency of the process is given by:

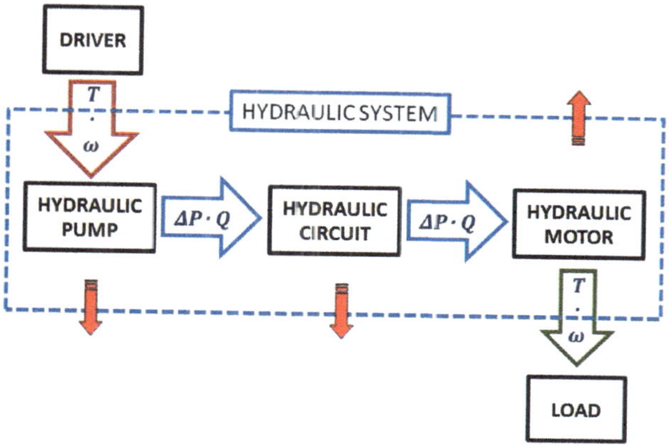

Figure 6-2 Conversion of energy or power in a typical fluid power system.

$$\eta_O = \frac{Power_{Output}}{Power_{Input}}$$

Linear Power	**Rotational Power**
$\eta_O = \frac{F \cdot v}{V \cdot I} or \frac{F \cdot v}{\dot{m} \cdot \Delta H_C}$	$\eta_O = \frac{T \cdot \omega}{V \cdot I} or \frac{T \cdot \omega}{\dot{m} \cdot \Delta H_C}$

We could also look at the efficiency of each individual component in the entire system. Although it is not part of the hydraulic system per se, the efficiency of the external motor providing the energy and power to the pump is given by:

$$\eta_{EM} = \frac{T_{EM} \cdot \omega_{EM}}{V \cdot I} or \frac{T_{EM} \cdot \omega_{EM}}{\dot{m} \cdot \Delta H_C}$$

The hydraulic pump efficiency on converting mechanical energy/power into fluid energy/power is:

$$\eta_P = \frac{\Delta P_P \cdot Q_P}{T_{EM} \cdot \omega_{EM}} or \frac{\Delta P_P \cdot Q_P}{T_P \cdot \omega_P}$$

Neglecting the losses in the conduits, hoses, and control valves, the efficiency of the hydraulic motor on converting fluid energy/power into mechanical energy/power is:

$$\eta_M = \frac{T_M \cdot \omega_M}{\Delta P_M \cdot Q_M}$$

It should be noted that all these components are connected in series; the energy and power get converted on one component flowing or being transmitted to the next. It is easy to demonstrate that the overall efficiency of the entire system is given by the product of the individual efficiency of all components. For an electric device as the external motor:

$\eta_O = \eta_{EM} \cdot \eta_P \cdot \eta_M$

$$\eta_O = \frac{T_{EM} \cdot \omega_{EM}}{V \cdot I} \cdot \frac{\Delta P_P \cdot Q_P}{T_p \cdot \omega_P} \cdot \frac{T_M \cdot \omega_M}{\Delta P_M \cdot Q_M}$$

with:

- TEM = TP for direct coupling with no losses, ωEM = ωP,
- ΔPP = ΔPM for no hydraulic losses in the circuit (e.g., conduits, hose, control valves, etc.), and
- QP = QM for no leakage in the circuit (e.g., conduits, hose, control valves, etc.).

$$\eta_O = \frac{T_M \cdot \omega_M}{V \cdot I}$$

This ratio is identical to the one outlined at the beginning of the discussion early in this section.

Pneumatic Power: Pressure and Work, Gas Laws, and Flow and Power

Pressure and Work

As we have discussed extensively, the main difference between hydraulic and pneumatic systems is the compressibility of the fluid, in the case of the latter type of system. This characteristic has a significant impact on the performance of any component in the fluid power system. This effect is associated with the relationship between pressure and work. The amount of work performed by a pressurized fluid depends on the nature of the expansion process of the fluid within the actuator or motor.

If a compressible fluid performs work at constant pressure like in an in-ground automotive lift, where the pneumatic cylinder extends under constant load to lift a car, the amount of work can be approximated by the same equation derived for an incompressible fluid. The amount of work is given by the product of the pressure P and the volume V of gas required to move the load:

$$Work = P \cdot V$$

This formula can be rearranged to prove that the work performed by the pressurized gas is, in this situation, equivalent to the definition of **mechanical work**, the product of the force required to move the object, and the distance traveled. If we use the in-ground lift as an example:

$$P = \frac{W}{A}$$

where W is the weight of the automobile and A is the cross-sectional area of the lift pneumatic cylinder. Upon rearrangement:

$$Work = P \cdot V = \frac{W}{A} \cdot V = \frac{W}{A} \cdot (A \cdot H) \leftarrow V = A \cdot H$$

$$Work = W \cdot H \rightarrow F \cdot \Delta s$$

where the product A·H represents the volume of air (at pressure P) needed to lift the car a height H. The product of the weight of the car and that height lifted W·H is exactly the product of the force F required to move an object the distance Δs.

On the other hand, estimating the work performed by a pressurized compressible fluid when the fluid expands during the process requires knowledge of thermodynamics, which is beyond the scope of this textbook. In Chapter 9, we include explicitly the theoretical equations used to estimate both work extracted from an expanding gas and the work required compressing a gas.

Gas Laws

Late in the sixteenth and seventeenth centuries, Robert Boyle and Jacques Charles performed experiments with gases at low pressure and found that the pressure, volume, and temperature of a gas are interrelated. Joseph Gay-Lussac early in the eighteenth century confirmed and published Charles' work. Boyle found that the pressure of gases is inversely proportional to their volume. Charles determined that the volume of a gas is proportional to its temperature. The combination of these relationships is now commonly known as the **general gas law** or the ideal or perfect gas law when the nature of the gas is considered.

BOYLE'S LAW

If the gas in a pneumatic cylinder expands or is compressed very slowly, the temperature of the gas remains constant, as shown in Figure 6-3. In that case, the relationship between the pressure and the volume of the gas is given by:

$$P_2 \cdot V_2 = P_1 \cdot V_1$$

Or:

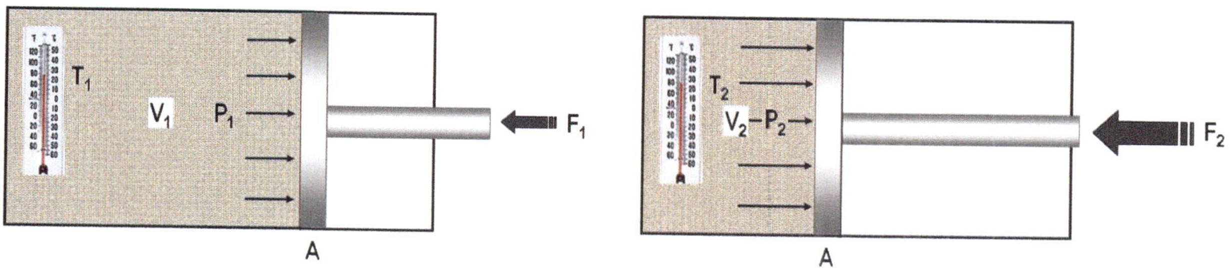

Figure 6-3 Slow compression of a gas in a pneumatic cylinder (T1 = T2).

$$\frac{P_2}{P_1} = \frac{V_1}{V_2}$$

where P_1 and V_1 represent the pressure and volume of the gas at the initial state and P_2 and V_2 represent the pressure and volume of the gas at the final state.

In a real process, the gas expands or is compressed rapidly, causing a change in its temperature. Nevertheless, if enough time is provided, the gas heats or cools down, and the gas temperature goes back to the initial value defined by the temperature of the surroundings. In this case, Boyle's law is still applicable.

CHARLES' LAW

If the gas in the same pneumatic cylinder expands under the action of an external heat source, the volume of the gas is directly proportional to its temperature provided that the load or pressure remains constant, as seen in Figure 6-4. In that case, the relationship between the temperature and the volume of the gas is given by:

$$\frac{T_2}{T_1} = \frac{V_2}{V_1}$$

where T_1 and V_1 represent the temperature and volume of the gas at the initial state and T_2 and V_2 represent the temperature and volume of the gas at the final state.

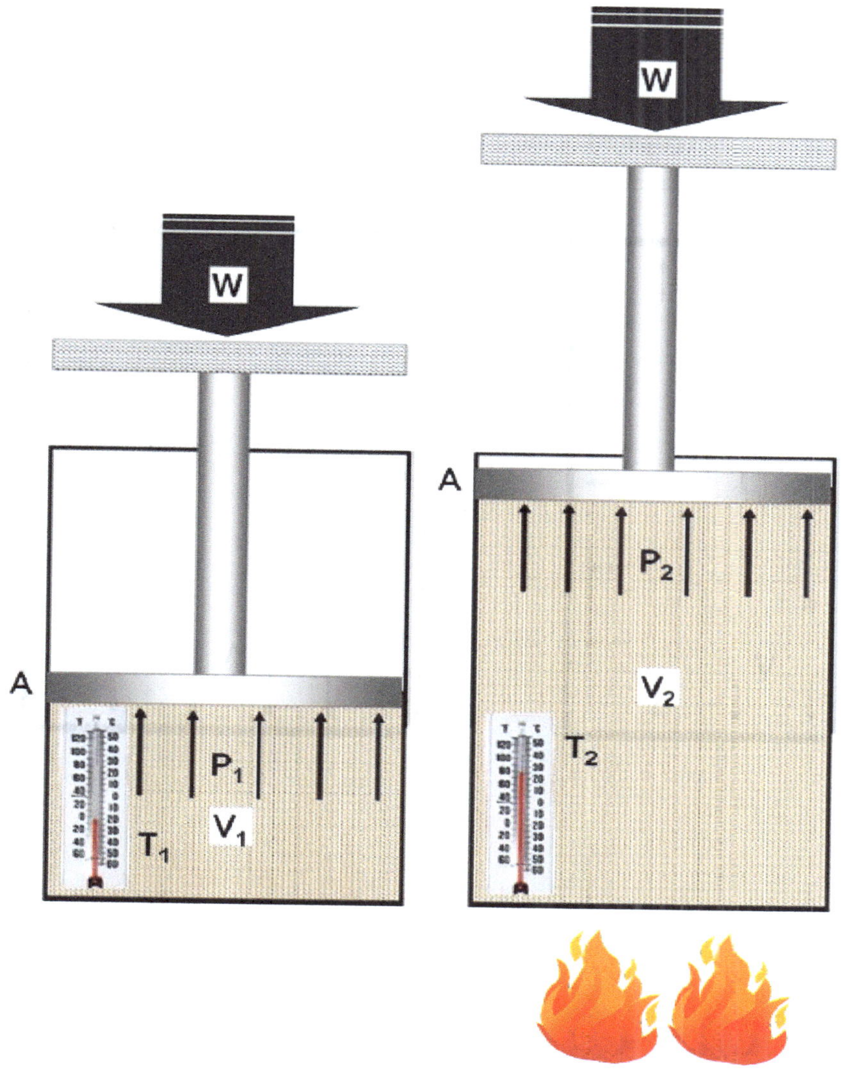

Figure 6-4 Heating of a gas in a pneumatic cylinder (P1 = P2).

It should be pointed out that Charles' law is valid for a relatively fast heating of the gas, where the process can be considered adiabatic. An **adiabatic** process is one in which there is no heat transfer to the surroundings. This implies that the cylinder must be well insulated. In a real process, there is always some heat dissipation, so using Charles' law carries some error.

GAY-LUSSAC'S LAW

If a gas trapped in a stationary cylinder is heated by an external heat source, the pressure of the gas is directly proportional to the temperature, provided that the volume remains constant (Figure 6-5). In that case, the relationship between the temperature and the pressure of the gas is given by:

$$\frac{T_2}{T_1} = \frac{P_2}{P_1}$$

where T_1 and P_1 represent the temperature and pressure of the gas at the initial state and T_2 and P_2 represent the temperature and pressure of the gas at the final state.

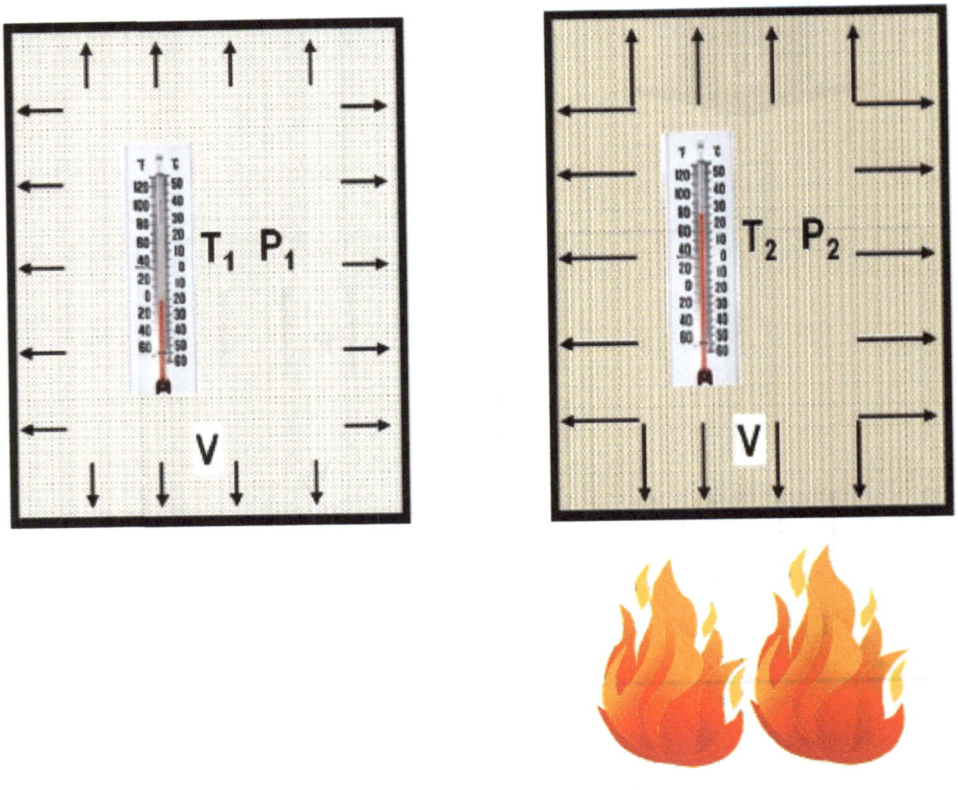

Figure 6-5 Heating of a gas in a stationary pneumatic cylinder (V1 = V2).

Like Charles' law, Gay-Lussac's law is valid for an adiabatic process. Again, this implies that the cylinder is well insulated and there is no heat transfer to the surroundings.

GENERAL GAS LAW

The laws of Boyle, Charles, and Gay-Lussac can be combined in a single equation relating all three properties of the gas, pressure, temperature, and volume. This relationship is commonly known as the **general law of gases.** Combining and rearranging the three laws:

Starting with Boyle's law:

$$\frac{P_2}{P_1} = \frac{V_1}{V_2}$$

Rewriting Charles' law:

$$\frac{T_2}{T_1} = \frac{V_2}{V_1} \rightarrow \frac{V_1}{V_2} = \frac{T_1}{T_2}$$

Combining Boyle's law while stating Gay-Lussac's law:

$$\frac{P_2}{P_1} = \frac{V_1}{V_2} = \frac{T_1}{T_2} \leftarrow \frac{T_1}{T_2} = \frac{P_1}{P_2} \leftarrow \frac{T_2}{T_1} = \frac{P_2}{P_1}$$

Therefore:

$$P_2 \cdot V_2 \cdot T_1 = P_1 \cdot V_1 \cdot T_2$$

And finally:

$$\frac{P_2 \cdot V_2}{T_2} = \frac{P_1 \cdot V_1}{T_1}$$

This equation is valid to relate the final and initial state of the gas under expansion or compression. It does not consider the nature of the process that changed the state of the gas. The applicability of the general law is restricted to the calculation of the gas properties between the two states.

Warning: In all gas laws, **absolutes values for both pressure and temperature are required.** Recall that pressure and temperature in a gas are related to the collisions of the molecules with the container and themselves. The way in which both pressure and temperature are measured has no influence in the molecular behavior.

Flow and Power

Invoking the general continuity equation, the volumetric flow of a compressible fluid can be estimated if the gas conditions, pressure, and temperature at the particular location in the fluid power system are known. Restating the general continuity equation:

$$\dot{m} = \rho_1 \cdot v_1 \cdot A_1 = \rho_2 \cdot v_2 \cdot A_2$$

Recalling also the volumetric flow definition:

$$\dot{m} = \rho_1 \cdot Q_1 = \rho_2 \cdot Q_2 \leftarrow Q = v \times A$$

And:

$$\frac{Q_2}{Q_1} = \frac{\rho_1}{\rho_2}$$

Although determining the density of the gas on each state is beyond the scope of this book, the density of any fluid is directly proportional to the volume:

$$\rho = \frac{M}{V} \rightarrow \rho \cdot V = M$$

Or for a fixed mass of gas:

$$\rho_1 \cdot V_1 = \rho_2 \cdot V_2$$

$$\frac{\rho_1}{\rho_2} = \frac{V_2}{V_1}$$

Therefore, the volumetric flow rate of a compressible gas can be determined using the general gas laws discussed in the previous sections. The density of the gas is inversely proportional to the volume that the gas occupies. The change in volume of any gas—and hence the **volumetric flow**—can be calculated if the pressure and temperature of both states are known.

In terms of power, we already hinted that the work performed by the expansion of a compressible fluid or the work required to compress it depends on the nature of the process. The incompressibility nature of the fluid in hydraulics allows a straightforward calculation of the flow work as P·V. In a compressible fluid, as we just discussed in gas laws, pressure and volume are related. In other words, the pressure is variable and might be a function of the volume of the fluid. In such cases, estimating the work and power requires knowing a priori the nature of the gas and the equation of state that better approximates the actual expansion or compression process.. Further discussion on the volumetric flow rate, the work, and the power associated with the handling of compressible fluids under pressure is covered in a generalized manner in Chapter 9.

Fluid Power Symbology

To design, analyze, and troubleshot fluid power systems and circuits, engineers generate drawings and diagrams using specific symbols conceived to represent each component. Most of the fluid power industry in the United States adheres to the standard **ISO 1219**, fluid power systems and components—graphical symbols and circuit diagrams. Part 1 of this standard establishes basic elements for symbols and specifies rules for devising fluid power symbols for use on components and in circuit diagrams.

Components of a Fluid Power System

Hydraulics

As we already discussed, the first component of a hydraulic system is a **tank reservoir** that holds all the hydraulic fluid used in the system, including the amount required to compensate **or** the changes in volume in the system as cylinders extend or retract, changes in volume due to temperature and system leakage. In small systems, the reservoir can be as simple as a metallic box or an oil-resistant plastic container with a cap. In large systems, reservoirs can be made of treated steel, aluminized steel, or aluminum to prevent corrosion. They can include individual separate pumping systems (different from the pump or pumps feeding the main system) with filters, where the fluids can be pumped in and out or kidney-looped to clean the oil without cross-contamination.

Figure 6-6 presents the most common ISO symbols used to represent tanks or reservoirs in hydraulic power systems. Figure 6-6a displays the simplest symbol and represents an atmospheric reservoir. Figure 6-6b shows the symbol used to denote the most common hydraulic tank, a vented cap reservoir. Figure 6-6c represents a pressurized or sealed reservoir typically used to ensure positive pressure at the suction of the pump or minimize fluid contamination by foreign material or moisture.

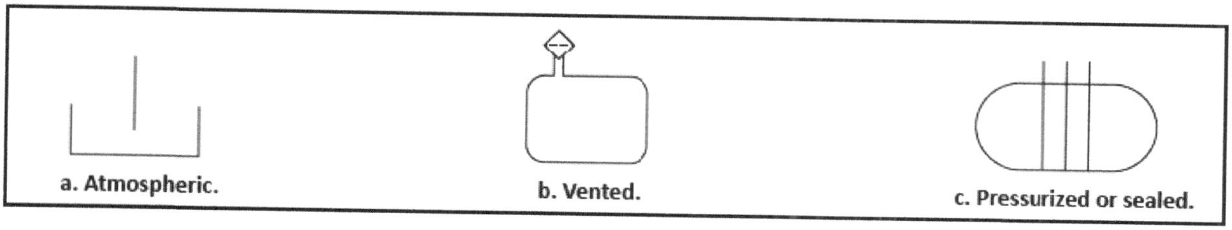

Figure 6-6 ISO symbols for hydraulic fluid reservoirs.

The second component of the hydraulic power system is the **pump**. The pump supplies the fluid at the flow rate required for the system having enough power to overcome the pressure induced by the load or loads in the system. Hydraulic pumps have an enormous power density capability; they can be of fixed or variable volumetric displacement, unidirectional or bidirectional. The design and operation of hydraulic pumps are discussed in detail in Chapter 9.

Figure 6-7 presents the most common ISO symbols used to represent pumps in hydraulic power systems. Figure 6-7a displays the simplest symbol, representing a unidirectional fixed-displacement hydraulic pump. Figure 6-7b shows the symbol used to denote a unidirectional variable-displacement hydraulic pump. Figure 6-7c represents a bidirectional variable-displacement hydraulic pump. It should be noted that the arrows on the symbols point to the direction on the flow and are completely filled (black arrowheads). The completely filled arrowheads indicate to the fluid power engineer that the system is a hydraulic power system.

As previously discussed, hydraulic pumps are driven by an external power source, or prime mover. This external power source can be an electric motor (Figure 6-8a) or a gas or diesel engine (Figure 6-8b).

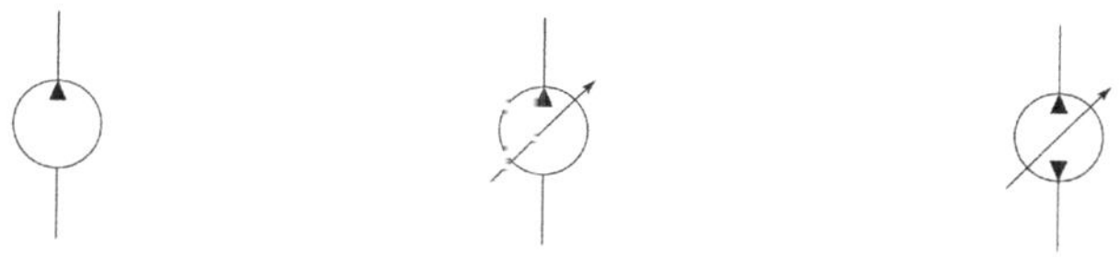

Figure 6-7 ISO symbols for hydraulic pumps.

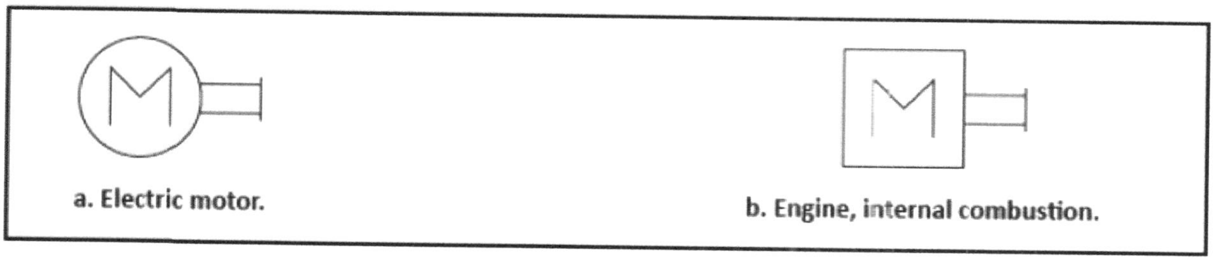

Figure 6-8 ISO symbols for external power sources or prime movers.

The pressurized hydraulic fluid provided by the pump is directed toward the actuators and motors in the fluid power system using **directional control valves**. The design characteristics of the directional valves depend on the nature of the circuit or actuator to be activated. In addition to directional valves, there are other hydraulic valves that can have multiple design features, activation mechanisms, and interconnections. The ISO symbols used to represent hydraulic valves are rich in information. Every line and feature in an ISO symbol of a hydraulic valve has a meaning and defines a specific characteristic of the valve.

Most hydraulic valves utilize a spool that slides inside a housing between two or multiple positions. The positions of a fluid power valve are represented by squares or boxes in an ISO diagram. Figure 6-9a presents three boxes representing one position of a directional valve. The first box is empty, so it is simply a place holder for the normal position of the valve. The second box represents a position of the valve that has two ports normally closed. The third box represents another possible position of the valve, now with the two ports connected to indicate the valve is normally open; the double-headed arrows in the box indicate that the hydraulic fluid can flow in both directions. Figure 6-9b combines two boxes to represent the simplest valve, a two-position, two-ports-normally-closed, spring-return, push button–activated hydraulic directional valve. Similarly, Figure 6-9c has a valve with two boxes representing the two positions of three ports normally open spring return push-button activated hydraulic directional valve. Single boxes are also used to represent fully functioning hydraulic valves. Valves that typically control or protect the hydraulic circuit on their own are represented by single boxes. Figure 6-9d represents an adjustable spring-loaded pressure relief valve.

Symbols in Figure 6-9 represent a very small sample of hydraulic valve designs. Different hydraulic applications have different requirements in terms of flow and pressure operational characteristic. Valve designs must accommodate the flexibility demanded by each hydraulic system. The combinations of the number of positions, the number of ports, the methods of activation and control of hydraulic valves are almost limitless.

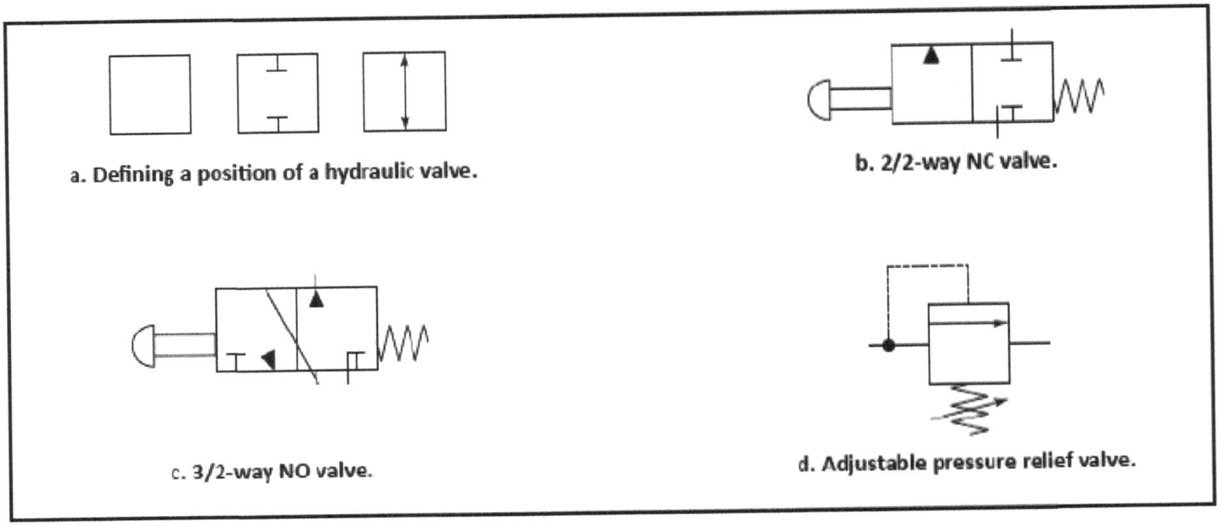

Figure 6-9 ISO symbols for hydraulic valves.

As the end purpose of a fluid power system is to transmit and use power, the final major component of the system is a **motor or an actuator**. Actuators normally generate linear motion as in hydraulic cylinders. There are essentially two types of hydraulic cylinders: single-acting (Figure 6-10a) and double-acting (Figure 6-10b). Single-acting cylinders exert the power stroke in only one direction, needing a nonhydraulic mechanism to return to the normal position. Double-acting cylinders exert power strokes in both directions, using only the pressurized fluid to extend or retract.

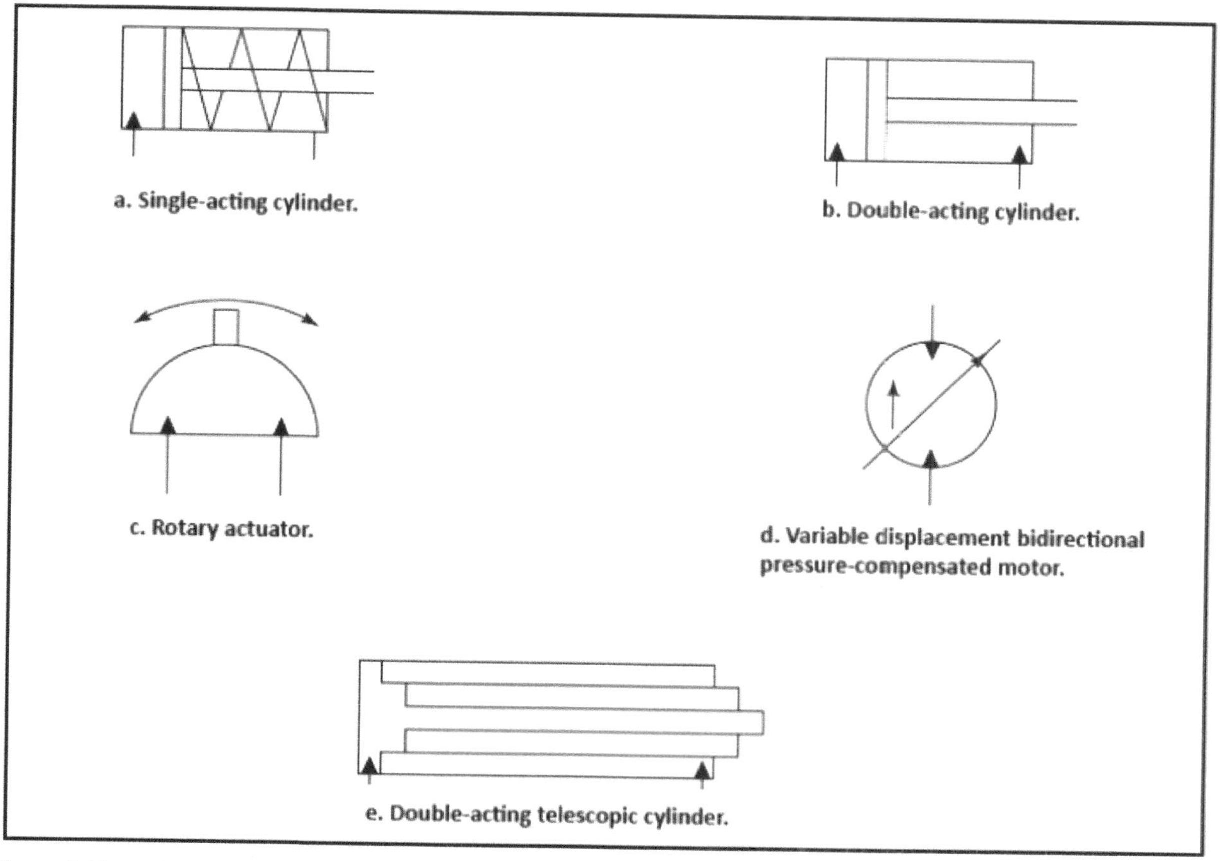

Figure 6-10 ISO symbols for hydraulic actuators and motors.

Although there are single-revolution rotary actuators (Figure 6-10c), most hydraulic rotary actuators are called motors. Like hydraulic pumps, hydraulic motors can be fixed- or variable-displacement, unidirectional or bidirectional. Others include features such as pressure compensation to reduce loads on shaft and bearings. Figure 6-10d presents the symbol of a variable-displacement, bidirectional, pressure-compensated hydraulic motor. Like valves, actuators and motors can have multiple design features. For instance, the symbol in Figure 6-10e represents a double-acting telescopic cylinder.

Finally, all components of the fluid power hydraulic system are connected by **tubing, pipes, or hoses**. They are needed to transport the fluid to and from the components. Generally, the fluid power application, either stationary or mobile, defines the types of conduits required to transport the pressurized fluid. Tubing and pipes mostly used in stationary applications are represented by continuous solid lines (Figure 6-11a). Flexible conduits and hoses mostly used in mobile applications are represented by curved lines as seen in Figure 6-11b. Other connectors such as three-way fittings (see Figure 6-11c) or double couplers with check valves (Figure 6-11d) are also necessary to build fluid power distribution lines.

Pneumatics

Figure 6-11 ISO symbols for hydraulic connectors.

Regarding the arrangement of the fluid power system, pneumatic installations are similar to those of hydraulic applications. The first major difference between a pneumatic system and a hydraulic system is that in a pneumatic system, the fluid is compressible; in most cases, the fluid is air directly drawn from the atmosphere. The second difference is a consequence of the first one. There is not a supply tank or reservoir. Instead, there is a receiver that stores the pressurized air or gas supplied by the compressor. The rest of the components of a pneumatic system such as the external power source, control valves, actuators, motors, tubing, piping, and hoses are similar to those of hydraulic systems.

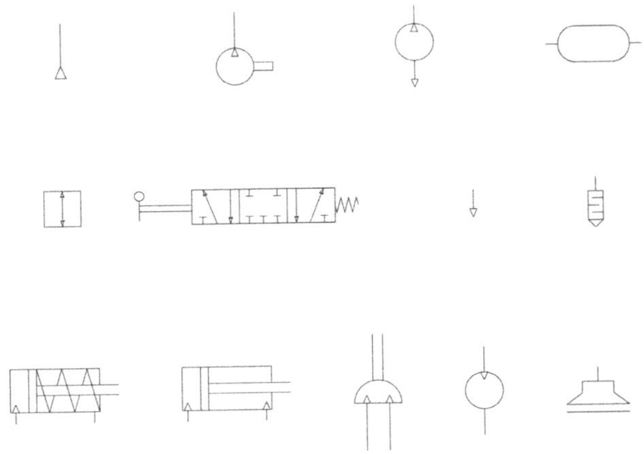

Figure 6-12 ISO symbols for pneumatic components.

As the fluid in most pneumatics systems is atmospheric air, the most pressing design concern is the moisture present in the air. The changes in pressure and temperature cause the moisture in atmospheric air to condense inside all components of the fluid power system. Receivers, long pipes, and most components should include automatic water traps, automatic drains, or manual valves to remove the condensed water from the system. In terms of the symbols used to represent and draw pneumatic components, there are no substantial differences between hydraulic and pneumatic systems. Indeed, most of the symbols are interchangeable between both fluid power applications. The way fluid power engineers identify a pneumatic system is by the open or unfilled arrowheads of the directional arrows in the symbols of most pneumatic components.

Figure 6-12 presents a sequence of ISO symbols used to represent the most commonly used pneumatic components. As a way to validate the similarities with the symbols used to represent hydraulic components, most symbols included in the figure are for components equivalent to those presented in the previous hydraulic section. Symbols that are exclusive of pneumatic systems are an air compressor without inlet port (Figure 6-12c), an exhaust (Figure 6-12g), a silencer (Figure 6-12h), and a vacuum cup (Figure 6-12m).

Reading and Creating a Fluid Power ISO Diagram

Fluid power diagrams are a graphic representation of a fluid power system. Simple geometric symbols drawn to American National Standards Institute—but preferably to **ISO**—standards represent the system components and their controls and connections. Graphics and diagrams of circuits are used for design purposes and are extremely valuable in troubleshooting situations.

In Figure 6-2, we presented a block diagram showing how fluid power is converted and transmitted in a typical fluid power system. From the external motor that provides the input power to the pump, to the motor or actuator that converts the fluid power back to mechanical power. Figure 6-13 presents an ISO symbol diagram with the components of a basic hydraulic circuit and highlighting the power conversion processes, as was done in Figure 6-2.

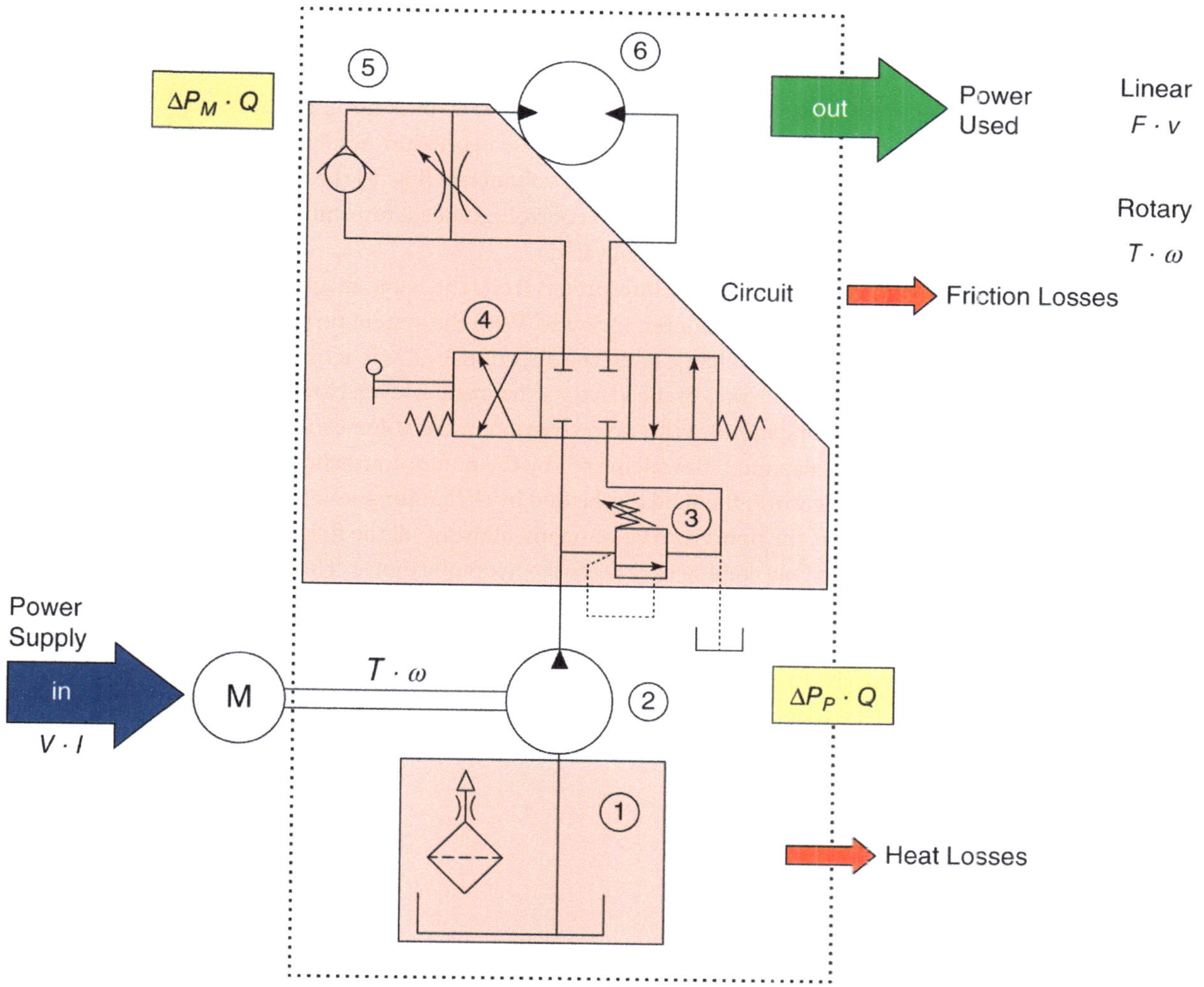

Figure 6-13 ISO symbol diagram of a basic hydraulic power system.

The components labeled in the hydraulic circuit of Figure 6-13 are the following:

1. Vented reservoir
2. Pump: hydraulic, unidirectional, and electric motor–driven
3. Adjustable pressure relief valve; spring-loaded

4. Directional valve: 4/3 (ports/positions) normally closed, spring-centered, lever-activated
5. One-way restrictor valve: a throttle valve paired with a nonreturn valve
6. Motor: hydraulic and bidirectional or reversible

The sequence of operation interpreted from the diagram, Figure 6-13, is as follows. The system is turned on by powering the electric motor. The pump draws fluid from the reservoir and pressurizes the discharge lines. Because the directional valve is closed, the pressure increases rapidly and activates the pressure relief valve, which redirects the totality of the flow back to the reservoir. Once the left position (box or envelope with the crossed arrows) of the valve is actuated via the manual lever, one port of the valve directs the fluid to the one-way restrictor valve. Once the fluid reaches this valve, the position of the nonreturn valve forces the fluid to go through the throttle valve. The throttle valve allows regulating the volumetric flow by returning excess fluid to the reservoir through the pressure relief valve once it is activated due to the rise in pressure associated with the throttle restriction. The one-way restrictor valve supplies then a regulated volumetric flow to the hydraulic motor, which turns one way—let's suppose clockwise—performing work at a controlled speed, returning the nonpressurized fluid to the reservoir through the second port in the directional valve.

An alternate sequence of operation can be interpreted from the same diagram to operate the hydraulic motor in the opposite direction, in this case, counterclockwise. With the system turned on and the pump drawing fluid from the reservoir to pressurize the discharge lines, the directional valve is actuated to activate the right position (box or envelope with the straight arrows) of the valve via the manual lever. Now, the second port in the valve sends directly the fluid to the hydraulic motor. With no restriction in the volumetric flow supplied, the hydraulic motor would be able to receive the maximum flow delivered by the pump, converting power while rotating in the opposite direction. After performing work, the fluid discharged by the motor goes to the one-way restrictor valve. Once in the one-way restrictor valve, the nonreturn valve opens, allowing all the fluid to pass through; the throttle valve has no effect in the volumetric flow coming back from the hydraulic motor. The fluid returns to the reservoir this time through the first port in the directional valve.

In summary, Figure 6-13 represents a basic hydraulic circuit in which the speed of the hydraulic motor can be controlled in one direction only. In the opposite direction, the speed of the motor is defined by the flow supplied by the hydraulic pump. The design of the directional valve, normally closed, will cause high pressure on the discharge lines once the pump is turned on. Therefore, the circuit includes a pressure relief valve to protect the system from excessive pressures.

Key Equations

Power = P * Q

In terms of SI units, hydraulic power is measured in:

$$P \cdot Q = Pa \cdot \frac{m^3}{s} = \frac{N}{m^2} \cdot \frac{m^3}{s} = \frac{N \cdot m}{s} = \frac{Joule}{s} = watt$$

In terms of U.S. customary units, hydraulic power is measured in:

$$P \cdot Q = Psf \cdot \frac{ft^3}{s} = \frac{lb_f}{ft^2} \cdot \frac{ft^3}{s} = \frac{lb_f \cdot ft}{s}$$

Note: The typical unit of hydraulic power in the U.S. customary system is hp (550 lbf-ft/s).

$$\eta_O = \frac{Power_{Output}}{Power_{Input}}$$

$$\frac{P_2 \cdot V_2}{T_2} = \frac{P_1 \cdot V_1}{T_1}$$

Exercises

Hydraulic Power: Pressure and Work, Flow and Power, and Efficiency

1. What is the work performed by a 4.5-inch (in) bore diameter cylinder when extending the full stroke of 48 in under a supply pressure of 1,500 psi? Provide answer in lbf-ft.
2. A volumetric flow of 50 gpm is supplied to a 2.5-in bore diameter cylinder with a 36-in stroke. If the supply pressure is 1,500 psi, what is the maximum hydraulic power the cylinder can develop? Provide answer in hp.
3. The boom cylinders (2) of a 350D LC John Deere excavator have a bore diameter of 145 millimeters (mm) and a stroke of 1,520 mm. The hydraulic pressure of the system can be boosted up to 36.3 megapascals (MPa) when breaking hard ground or lifting wet sand or gravel. What is the maximum extending speed of both cylinders (extending simultaneously)? Assume that all the engine power is converted into hydraulic power. (Answer = 0.17 meters per second.)
4. What is the hydraulic power delivered by an 85% efficient pump if it is driven by a 15-hp electric motor?
5. The John Deere 350D LC excavator has two main pumps with a capacity of 76.1 gpm each. If the maximum pressure in the hydraulic circuit is 5,260 psi, find the following:
    a. The hydraulic power provided by both pumps when in operation, in hp
    b. The efficiency (in percentages) of each pump if the rated power of the engine driving the pumps is 271 hp. Note: The engine is designed to drive one pump at a time.

Pneumatic Power: Gas Laws

6. A 30-gallon (gal) central pneumatic compressor has an auto shut-off pressure of 180 psi gauge (psig). What would be the volume occupied by the pressurized air in the tank if it was released to the atmosphere, 14.5 psi absolute (psia)? Provide answer in cubic feet (ft^3).
7. What is the volume of atmospheric air, 14.7 psia, required to fill a 17-liter (L) scuba tank at 3,442 psig? Provide answer in ft^3.
8. In a particular outdoor facility, the temperature varies from 100 degrees Fahrenheit (°F) to 55°F at night. A 250-gal air tank was pressurized up to 90 psig at noon. What is the pressure variation due to the change in temperature? Assume the air in the tank is in equilibrium with the environment in terms of temperature. Provide the answer in psi.

9. Atmospheric air needs to be stored in a 1,000-L tank. The atmospheric conditions are 100 kilopascals and 5 degrees Celsius (°C). What is the volume of free air required to fill the tank if the air is compressed to 1 MPa at 55°C? Provide the answer in cubic meters.

Fluid Power Symbology

10. Figure 6-14 shows an ISO drawing of a hydraulic conveyor application (courtesy of Famic Technologies). Fully describe each component identified with a numerical label (a total of 10 components).

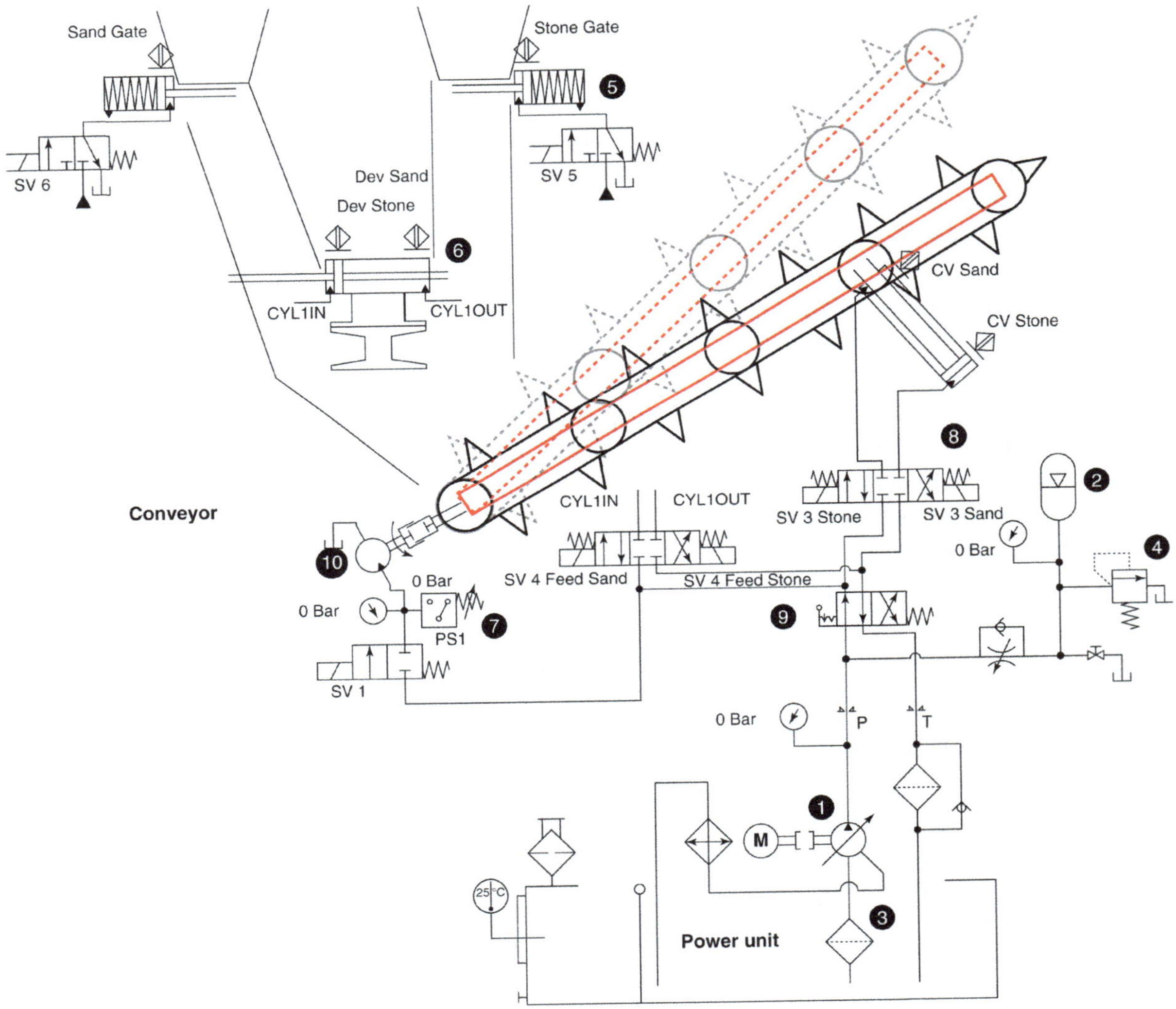

Figure 6-14 ISO drawing of a hydraulic conveyor application.

Fluid Power ISO Symbols

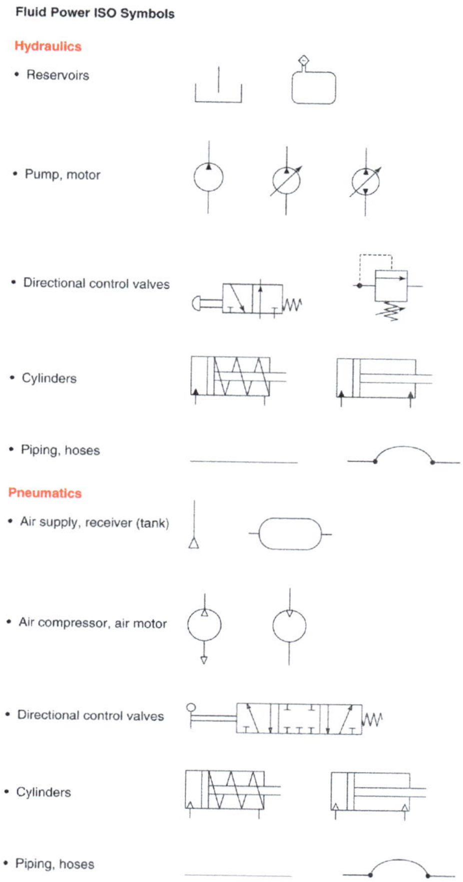

Figure 6-15 Fluid power ISO symbols.

Figure Credits

Fig. 6-4a: Clker-Free-Vector-Images, https://pixabay.com/en/fire-flame-hot-burning-burn-heat-312258/, 2014.
Fig. 6-5a: Clker-Free-Vector-Images, https://pixabay.com/en/fire-flame-hot-burning-burn-heat-312258/, 2014.
Fig. 6-14: Copyright © by Famic Technologies. Reprinted with permission.

CHAPTER 7

Basic Control Theory and Components for Fluid Power Systems

Overview and Learning Objectives

A key component of any fluid power system is the control system embedded in the overall fluid power system. The control system switches the fluid flow between devices and is the "software" in the fluid power system. The control system can be simple or very complex, but in either case, the fluid power control system is an integral and important part of any fluid power system.

Upon completion of this chapter, students should be able to do the following:

- Discuss the two common classes of control systems.
- Explain control logic functions and establish truth tables for the following functions: OR/NOR, AND/NAND, NOT, and FLIP-FLOP (LATCH) devices.
- Perform Boolean algebra operations to simulate control of fluid power circuits.
- Understand the operation of key electrical logic control components.
- Read and understand the operation of electrical ladder diagrams.
- Relate ladder diagrams and fluid power International Organization for Standardization (ISO) diagrams to describe circuit operation.

Basic Control Systems

The fluid power circuits reviewed in previous chapters comprise a variety of components, including compressors/pumps, receivers, pressure regulators, hoses/tubing, cylinders, motors, and control valves. These components, along with the control system, are designed to perform a function. The term **control system** includes:

1. Control algorithms/logic devices that translate the user interfaces into predefined actions

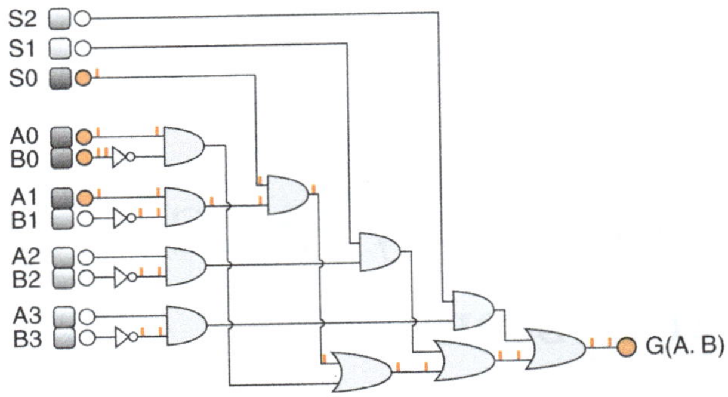

Figure 7-1 Logic diagram with AND and OR logic.

Figure 7-2 Siemens Simatic programmable logic controller.

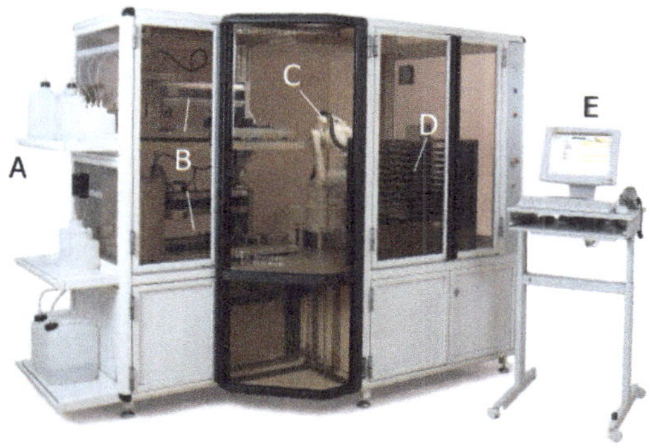

Figure 7-3 Robotic system: a) chemical reservoirs; b) robots for the excision of spots and the pipetting; c) industrial robot for aid work; d) gel tray; e) storage controller.

2. Manual components (e.g., levers, buttons, switches, and touchscreens) that can act as the user interface to the fluid power device

Figure 7-4 Push-button Start/Stop power switch.

Figure 7-5 Touchscreen user.

3. Sensors that monitor position, temperature, pressure, or other needed input

Figure 7-6 Proximity-type limit switch.

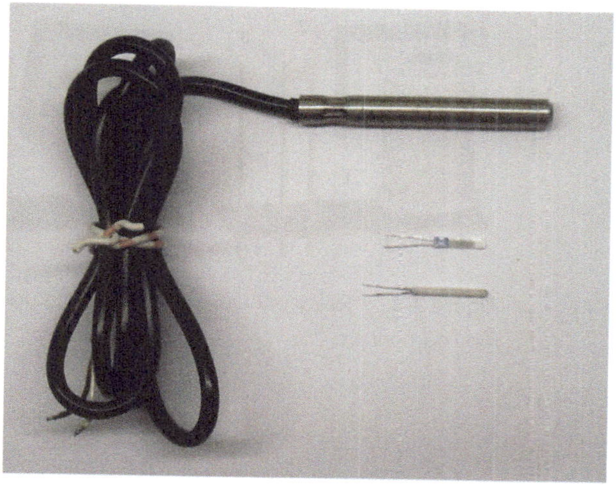

Figure 7-7 Three types of temperature sensors.

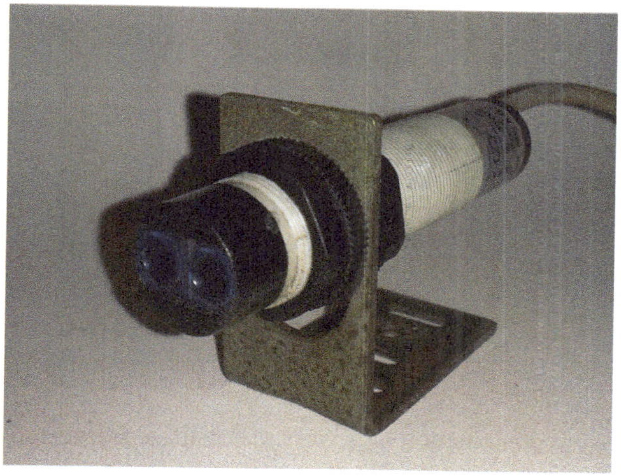

Figure 7-8 Photoelectric sensor, photoelectric switch.

The last component group of control systems components includes solenoids and relays.

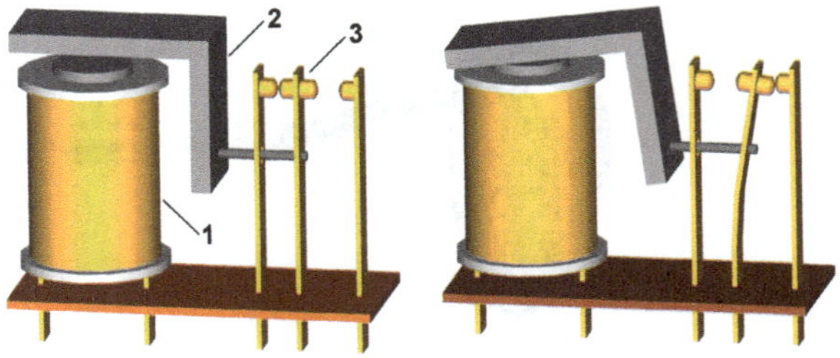

Figure 7-9 Relay principle: 1) coil; 2) armature; 3) moving contact.

Figure 7-10 Electronic component; relays.

The control system is a combination of devices that manage the operation of the fluid power system. This management is accomplished by directing, regulating, and monitoring the components' operation in the fluid power system. For example, upon receiving a signal from the operator commanding cylinders to advance in a defined sequence, the control system monitors the position of the cylinders and regulates the pressure within the fluid power system.

Control systems have two common classes with many different combinations of these basic classes. The first common class of control systems is *logic or sequential controls*. The second common control system is *feedback control*. As mentioned, most control systems are combinations of the two common classes.

Logic or Sequential Controls

Logic or sequential controls perform a logical operation based on Boolean functions to produce a single logic-based output. In other words, the output of a logic control device is either on or off and can be rapidly switched from one state to the other. Logical operations can be implemented using transistors, diodes that act as electronic switches, electromagnetic relays, or fluid power logic devices. The selection of the specific logical operation device depends on the operating environment, specifically temperature extremes, vibration, explosive potential, and mechanical shock. If these extreme temperatures, vibration, or explosive conditions are present, a fluid-based logic system may be ideally suited, as this type of system is not affected by heat and does not generate electrical sparks or electrical/magnetic fields. Logical operations or logic gates can be in series (i.e., cascaded) or parallel to allow the construction of a complex logic model. These complex logic models would include such examples of computer microprocessors that may contain hundreds of millions of individual logic operations/gates packaged in a single application-specific integrated circuit.

As mentioned, logic controls are based on Boolean functions and Boolean algebra. **Boolean functions** are a subarea of algebra in which the values of the functions can have two values/states (e.g., 1 or 0, true or false, on or off). These functions reduce logic circuits to their simplest form and consist of standard functions.

AND Function

The first standard Boolean algebra function is the **AND function**. This function allows two or more input signals to be evaluated. All inputs must be at the "1" or "true" state to allow a 1 or true output. As shown in Figure 7-11 and in the truth table in Table 7-1, both inputs—A and B—must be at the true state to allow a true output. The function symbols shown in Figure 7-11 are American National Standards Institute (ANSI) symbols. In Boolean algebra, the AND function can be expressed as multiplication of the A and B inputs. Please note that the inputs can be more than two, but two inputs are shown for simplicity.

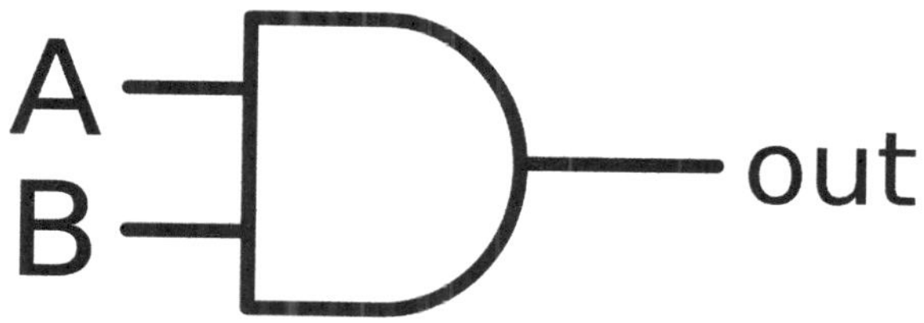

Figure 7-11 Schematic symbol for an AND-gate according to ANSI/Institute of Electrical and Electronics Engineers (IEEE) standards.

Table 7-1. AND Function Truth Table

INPUT		**OUTPUT**
A	B	A AND B
0	0	0
0	1	0
1	0	0
1	1	1
Boolean AND Function	*Output (Y) = A * B*	

OR Function

The OR Boolean algebra function is similar to the AND function. The **OR function** allows evaluation of two or more input signals. Different from the AND function, only one of the inputs must be at the 1 or true state to allow a 1 or true output. As shown in Figure 7-12 and truth table (Table 7-2), either input A or B must be at the true state to allow a true output. In Boolean algebra, the OR function can be expressed as addition of the A and B inputs. Again, note that the inputs can be more than two; two inputs are shown merely for simplicity.

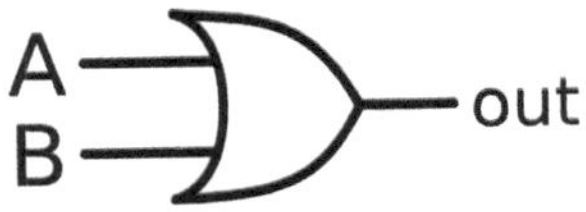

Figure 7-12 Schematic symbol for an OR-gate according to ANSI/IEEE standards.

Table 7-2. AND Function Truth Table

INPUT		**OUTPUT**
A	B	A OR B
0	0	0
0	1	1
1	0	1
1	1	1
Boolean OR Function	*Output (Y) = A + B*	

NOT Function

The NOT Boolean algebra function delivers the opposite state from the input state. In electronic circuits, the **NOT function** is commonly called an inverter because it inverts or flips the input signal to the opposite state. As shown in Figure 7-13 and the truth table (Table 7-3), the output is the opposite of the input state. The circle on the NOT symbol is called a bubble. This bubble can be added to the AND or OR function to deliver the opposite or inverted output.

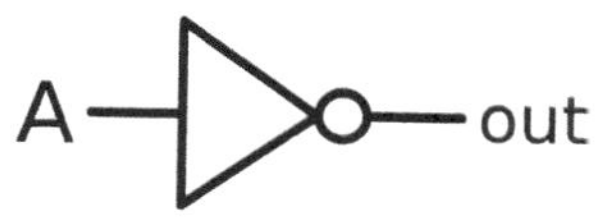

Figure 7-13 Schematic symbol for a NOT-gate according to ANSI/IEEE standards.

Table 7-3. NOT Function Truth Table

INPUT	OUTPUT
A	NOT A
0	0
1	1
Boolean NOT Function	

Not AND (NAND) Function

The **NAND function** is the NOT of the AND function. The NAND is the inverted, or opposite, AND function. In other words, either input signals can be at the 0 or false state to allow a 1 or true output, as shown in *Figure 7-14* and the truth table (*Table 7-4*).

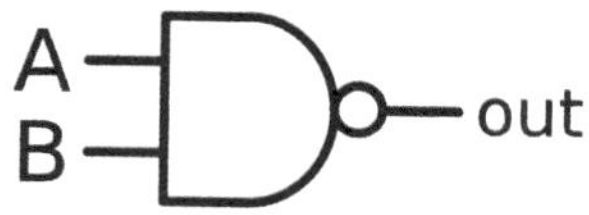

Figure 7-14 Schematic symbol for a NAND-gate according to ANSI/IEEE standards.

Table 7-4. NAND Function Truth Table

INPUT		OUTPUT
A	B	A NAND B
0	0	1
0	1	1
1	0	1

Table 7-4. NAND Function Truth Table

1	1	0
Boolean NAND Function	$Output\ (Y) = \overline{A * B}$	

Not OR (NOR) Function

The **NOR function** is the NOT of the OR function. The NOR is the inverted OR function. As shown in *Figure 7-15* and the truth table (*Table 7-5*), both inputs of A and B must be at the false state to allow a true output.

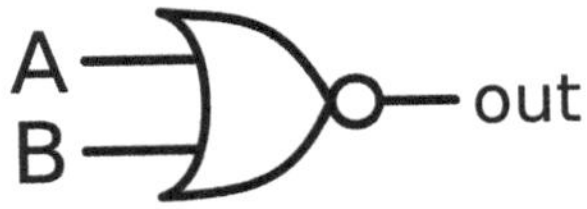

Figure 7-15. Schematic symbol for a NOR-gate according to ANSI/IEEE standards.

Table 7-5. NOR Function Truth Table

INPUT		OUTPUT
A	B	A NOR B
0	0	1
0	1	0
1	0	0
1	1	0
Boolean NOR Function	$Output\ (Y) = \overline{A + B}$	

In addition to truth tables, we can also use Venn diagrams to represent Boolean algebra operations. These diagrams are helpful in visualizing the Boolean algebra functions. The red shaded-in areas in the figure represent the 1 or true output of the Boolean algebra operation. The three Venn diagrams in Figure 7-16 represent the AND function (X*Y), OR function (X+Y), and NOT function.

The logical functions can be implemented on simple chips. As an example, Texas Instruments' 7400 quad two-input NAND chip and 7408 quad two-input AND are shown in Figures 7-17 and 7-18.

Logic circuits can be built by combining the logic functions. For instance, if we have three inputs, A, B, and C, and desire a 1 or true output only when C is true and A or B is true, we would have the following logic circuit with one OR function and one AND function. Figure 7-19 shows the two-step process in building the logic function. In the top figure, we have the Boolean algebra function we desire [C*(A + B)]. In the middle of the figure, we have the inputs and functions from within our brackets (A + B). We then combine the inputs and function external to our bracket function.

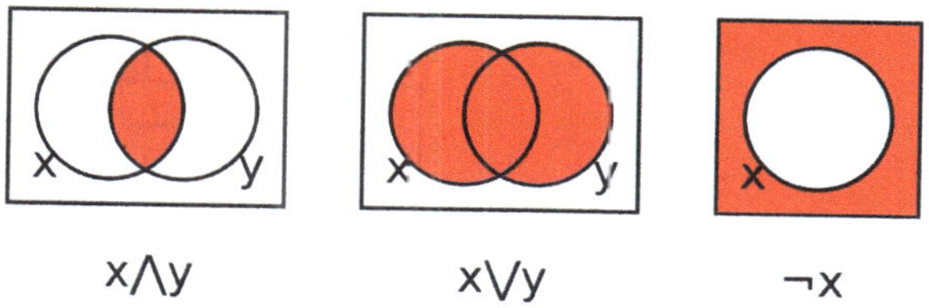

Figure 7-16 Schematic symbol for a NOR-gate according to ANSI/IEEE standards.

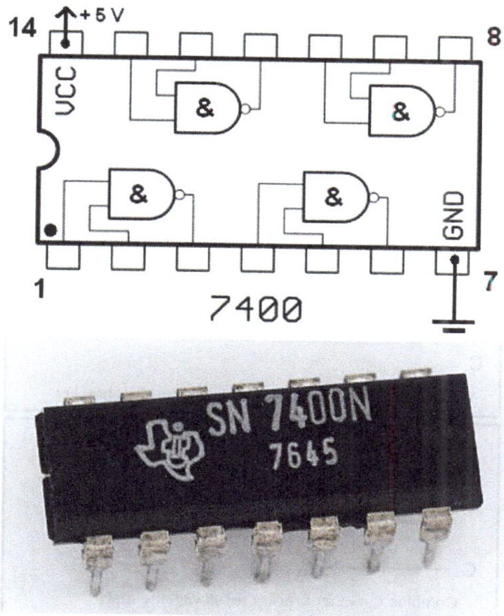

Figure 7-17 Chip contains four logical elements NOT-AND (NAND).

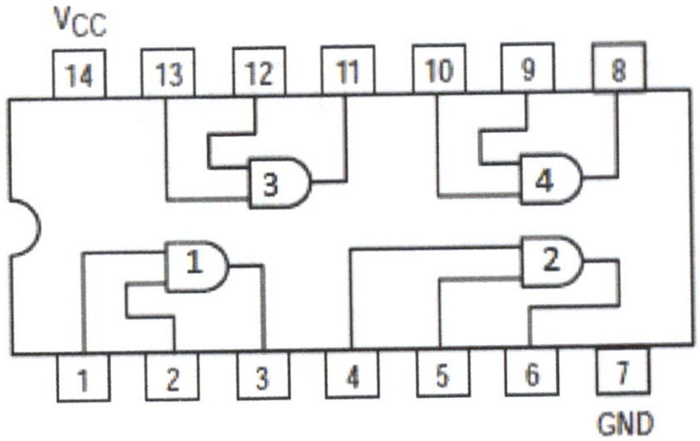

Figure 7-18 Chip contains four logical elements AND.

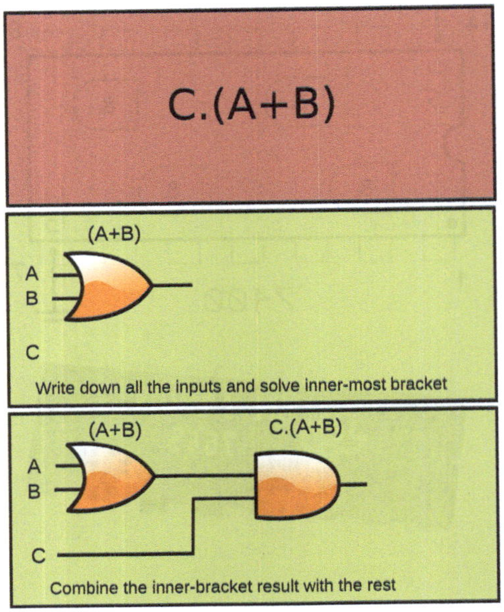

Figure 7-19 How to convert Boolean algebra into a logic gate diagram.

Flip-Flop/Latch or Memory Function

The **flip-flop/latch or memory function** has two stable states and can be used to remember or store state information of a component or the circuit itself. The flip-flop circuit changes output states with input signals applied to one or more control inputs. The flip-flop will have one or two outputs. This simple circuit is a basic memory or storage element in logic circuits. The flip-flop can be thought of as a fundamental building block of digital integrated circuits used in computers, telecommunications, and other electronic systems. The fundamental flip-flop is a simple SR latch, with the S representing set and the R standing for reset. This flip-flop can be constructed from a pair of cross-coupled NOR logic gates, as shown in Figure 7-20.

The SR flip-flop (latch) function symbol as used on a Texas Instruments 4043 chip is shown in Figure 7-21.

With the set S and reset R inputs at the 0 or unenergized state, the output remains at a constant state. If the set input is pulsed to an energized value (1) while the reset input remains unenergized (0), the Q output is set or energized (1). When the set output returns to an unenergized state, the output remains energized or is remembered/latched to an energized state. If the reset input is energized while the set input is unenergized, the output is unenergized or reset and stays reset. The input of both an energized set and energized reset combination is a restricted or forbidden combination, as this combination breaks the logical sequence. This restricted or forbidden combination would be inappropriate in a circuit because the output would either switch in an energized or unenergized state, depending on the timing relationship between the signals. The operation described above is shown in the truth table (Table 7-6).

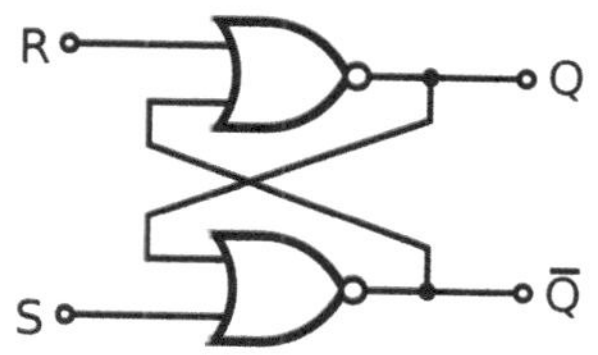

Figure 7-20 Set-reset NOR logic flip-flop (latch).

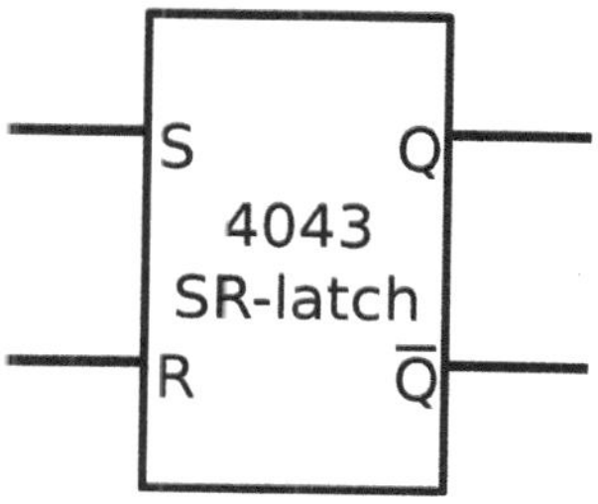

Figure 7-21 Texas Instruments 4043 latch chip.

Table 7-6. SR Flip-Flop (Latch) Function Truth Table

S	R	Action	Q
0	0	Hold state	0
0	1	Reset	0
1	0	Set	1
0	0	Hold state	1
1	1	Not allowed	1

As discussed in the beginning of this chapter, the logic is only one component of the typical fluid power control system. The other areas of the control system include the input devices, sensing devices, and relays/solenoids that switch power. These components are typically electrical control devices that interface with industrial computers, known as programmable logic controllers (PLCs).

Basic Electrical Devices

Six basic electrical devices are typically used to control fluid power circuits. These devices and their symbols are reviewed in the following sections.

Push-Button Switch

A **push button** is a simple switch for controlling an aspect of a fluid power circuit. It is employed as a user interface to the fluid power circuit or machine. Push buttons come in a normally open (no electricity flowing) or normally closed (electricity flowing) configuration and are only momentarily activated by the user (the user pushes the button but does not have to hold it). We will typically use the abbreviations NO for normally open and NC for normally closed.

Figure 7-22 NO and NC push buttons (with the equivalent electrical symbols).

Limit Switch

A **limit switch** performs a similar function to that of push-button switches; however, **limit switches** are typically actuated by mechanical devices rather than users. A limit switch can consist of a contact-type sensor in which an actuator arm is mechanically linked to a set of contacts. Limit switches can also consist of noncontact-type sensors in which a proximity sensor detects the local proximity of an item. Like push buttons, limit switches come in an NO (no electricity flowing) or NC (electricity flowing) configuration.

Figure 7-23 NO or NC limit switch (with the equivalent electrical symbols).

Pressure Switch

A pressure switch is a type of switch that closes or opens electrical contacts when a defined set pressure has been attained. In the operation of a fluid power cylinder, the pressure peaks or spikes when the cylinder has reached its end of stroke. Therefore, a pressure switch can be used in place of limit switches to determine when a fluid power cylinder has reached its end of stroke in cases where attaching a physical limit or proximity switch would not be possible. Like previous switches, pressure switches come in an NO or NC configuration.

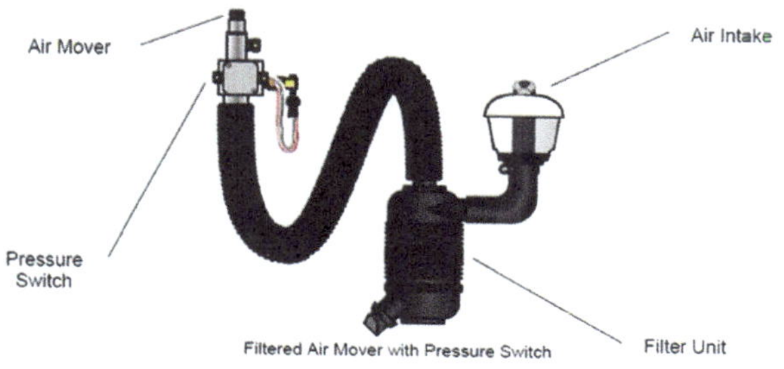

Figure 7-24 NO or NC pressure switch (with the equivalent electrical symbols).

Temperature Switch

A **temperature switch** senses a change in temperature and opens or closes an electrical set of contacts when a preset temperature is reached. Temperature switches can be used to protect a fluid power circuit from serious damage when operating components malfunction or excessive heat is generated in the system. The temperature switch typically would shut off the entire system to allow inspection of malfunctioning components or for system cooldown. Temperature switches also come in an NO or NC configuration.

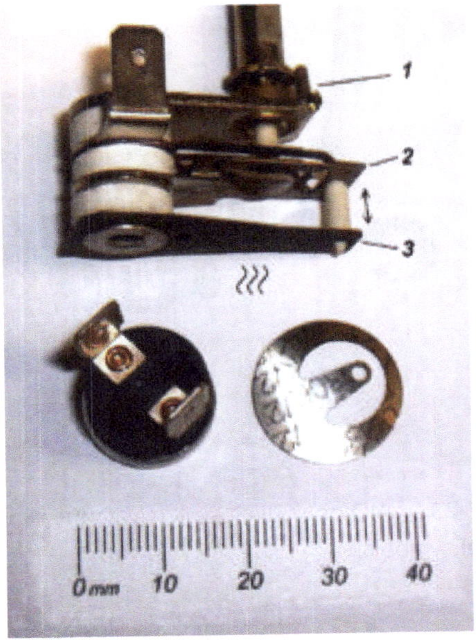

Figure 7-25 NO and NC temperature switches (with the equivalent electrical symbols).

Solenoid

A **solenoid** is a wound coil of wire wrapped around a metallic core that produces a uniform magnetic field when an electric current is passed through the coil. A **solenoid** can be thought of as a device that converts electrical energy to linear motion to actuate a pneumatic or hydraulic fluid power valve. A solenoid could simply open and close a fluid power valve, as shown in Figure 7-26, or interface with the spool within the fluid power directional control valve to change its position.

Relay

A **relay** is an electrically operated switch whose contacts open or close when an electromagnetic coil is energized. Relays are typically used to control a higher power circuit with a lower power signal. They also allow all electrical isolation between controlled high-power circuits and the low-power control signal. Relays can control a single controlled circuit or multiple controlled circuits with a single control signal. A simple single contact relay is shown in Figure 7-27. When the coil is energized, the armature is magnetically attracted to the electromagnetic coil and the contacts switch position. When the coil is de-energized, the spring returns the armature to its initial position.

A solid-state relay (SSR), or solid-state contactor, is an electronic component that does not have any moving components and therefore has an increased level of reliability. These solid-state components are activated by alternating current (AC) or direct current (DC) control signals for logic control, or microprocessor sources. As labeled in Figure 7-28, a 3-18-volt DC signal (low voltage) controls a 240-volt AC signal (medium voltage).

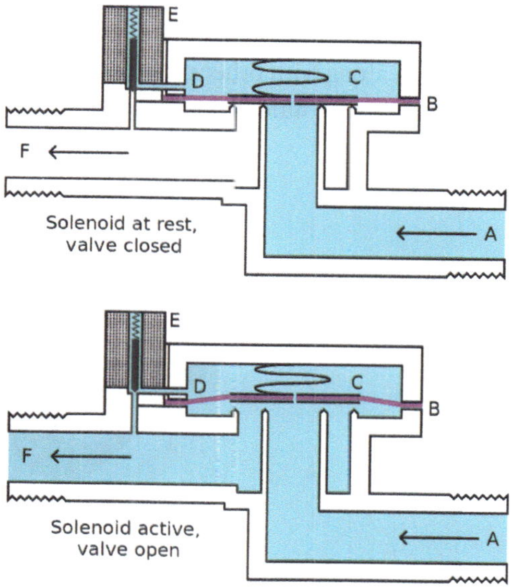

Figure 7-26 Solenoid-activated fluid power valve: a) input side; b) diaphragm; c) pressure chamber; d) pressure relief conduit; e) solenoid; f) output side.

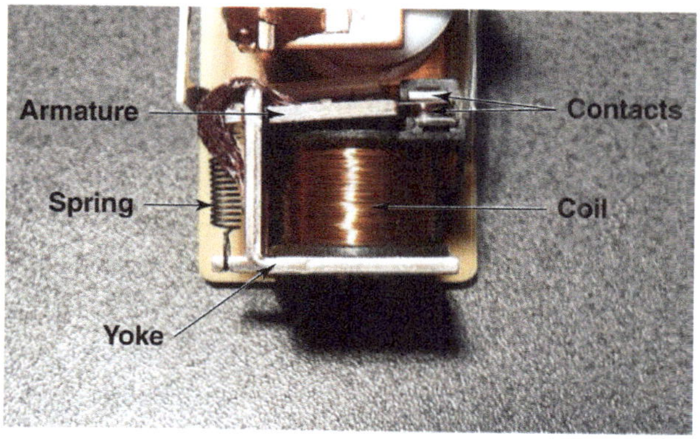

Figure 7-27 Picture of a simple electromechanical relay with the main parts named.

Relays, along with timers, were historically used for the control and sequencing of fluid power systems. For complex systems, hundreds to thousands of relays were required to accomplish the necessary control system logic. In the 1970s, industrial process computers were developed that would tolerate industrial environments and easily manage Boolean algebra input and output interfaces with fluid power and automation systems. These industrial control computers were PLCs. The abbreviation PLC and the term programmable logic controller are registered trademarks of the Allen Bradley brand of Rockwell Automation. These PLCs were programmed in ladder logic, which was developed from the early relay logic schematics. Current PLCs can be programmed with various languages, such as ladder logic, BASIC (Beginner's All-Purpose Symbolic Instruction Code), C, and functional block diagrams, along with other programming languages. We will use ladder logic to illustrate how electrical logic systems interface with fluid power systems.

Figure 7-28 NO and NC relay coil and relay contacts (with the equivalent electrical symbols).

Ladder Logic

As mentioned previously, **ladder logic** derived from the construction and design of relay logic systems for automation, manufacturing, and process controls. Each electrical device in the relay logic system (e.g., switches, solenoids, relays) was represented by an individual symbol on the ladder logic diagram. Connections between individual devices represented the wires in the relay logic that connected the individual devices. Ladder logic programming is a graphical programming language based on switches, solenoids, and relay logic. The graphical presentation of the components resembles a ladder, with two vertical rails on each side and a series of horizontal rungs between the rails. Each horizontal rung of the ladder can be thought of as representing a single rule that executes simultaneously and immediately with all of the other rungs. **Ladder logic** can be viewed as a programming language that connects input and feedback logical switches that are arranged in series and parallel (Boolean algebra AND OR functions) with output solenoids/coils that interface with fluid power components (e.g., control valves, motors, etc.).

Key Equations

Boolean AND function $Output\ (Y) = A*B$
Boolean OR function $Output\ (Y) = A + B$
Boolean NOT function $Output\ (Y) = \bar{A}$
Boolean NAND function $Output\ (Y) = \overline{A*B}$

Boolean NOR function $Output\ (Y) = \overline{A+B}$

Exercises

Boolean Algebra

1. Prove that A + (A *B) = A using a truth table.
2. Prove that (not(A + B)) = (notA * notB). Be careful with the "not" order.
3. Prove that A * (B + C) = (A * B) + (A * C) (distributive law).
4. Generate the truth table for the function A*notB + notA*B and draw the logic circuit diagram using the ANSI/IEE functions of OR, NOT, and AND.
5. Look up DeMorgan's laws on Wikipedia. Why does it matter if AND and OR functions can be interchanged by using the NOT function? Hint: If you could design systems with only one function (i.e., OR-shuttle valves), would it help in terms of inventory, supply, service, etc.?

Figure Credits

Fig. 7-1: Copyright © by Kanavis2 (CC BY-SA 3.0) at http://commons.wikimedia.org/wiki/File:A_mayor_que_B_grande.png.
Fig. 7-2: Mixabest, http://en.wikipedia.org/wiki/File:Siemens_Simatic_S7-416-3.jpg, 2010.
Fig. 7-3: Copyright © by Jörg Bernhardt (CC BY-SA 3.0) at http://commons.wikimedia.org/wiki/File:Spot_cutting_and_pipetting_robot_multi-language.jpg.
Fig. 7-4: Michael Holley, http://commons.wikimedia.org/wiki/File:Start_Stop_Power_Switch.jpg, 2012.
Fig. 7-5: Copyright © by Maurizio Napolitano (CC BY 2.0) at http://commons.wikimedia.org/wiki/File:Touch_screen.jpg.
Fig. 7-6: Copyright © by Joongbae Kim (CC BY-SA 3.0) at http://commons.wikimedia.org/wiki/File:Proximity_sensor.jpg.
Fig. 7-7: dirkhb, http://commons.wikimedia.org/wiki/File:Pt100_Sensors.png, 2006.
Fig. 7-8: Copyright © by Honmingjun (CC BY-SA 3.0) at http://commons.wikimedia.org/wiki/File:Photoelectric_sensor.jpg.
Fig. 7-9: Copyright © by Bisgaard (CC BY-SA 3.0) at http://commons.wikimedia.org/wiki/File:Relay_principle_horizontal.jpg.
Fig. 7-10: Copyright © by Fdominec (CC BY-SA 3.0) at http://commons.wikimedia.org/wiki/File:Electronic_component_relays.jpg.
Fig. 7-11: Towopedia, http://commons.wikimedia.org/wiki/File:Logic-gate-and-us.svg, 2012.
Fig. 7-12: Copyright © by Booyabazooka (CC BY-SA 3.0) at http://commons.wikimedia.org/wiki/File:Or-gate-en.svg.
Fig. 7-13: Copyright © by Booyabazooka (CC BY-SA 3.0) at http://commons.wikimedia.org/wiki/File:Not-gate-en.svg.
Fig. 7-14: Copyright © by Booyabazooka (CC BY-SA 3.0) at http://commons.wikimedia.org/wiki/File:Nand-gate-en.svg.
Fig. 7-15: Copyright © by Booyabazooka (CC BY-SA 3.0) at http://commons.wikimedia.org/wiki/File:Nor-gate-en.svg.
Fig. 7-16: Tilman Piesk, http://en.wikipedia.org/wiki/File:Vennandornot.svg, 2010.
Fig. 7-17: Copyright © by Audrius Meskauskas (CC BY-SA 3.0) at http://en.wikipedia.org/wiki/File:7400.jpg.
Fig. 7-18: Losaaamedios, http://commons.wikimedia.org/wiki/File:7408.jpg, 2011.
Fig. 7-19: Copyright © by Arturo Urquizo (CC BY-SA 3.0) at http://commons.wikimedia.org/wiki/File:CPT-logic-gate_building_circuit.svg.
Fig. 7-20: Heron, http://commons.wikimedia.org/wiki/File:SR-NOR-latch.png, 2006.
Fig. 7-21: Copyright © by Gr416 (CC BY-SA 3.0) at http://commons.wikimedia.org/wiki/File:SR_latch.svg.
Fig. 7-22a: Copyright © by oomlout / Flickr (CC BY-SA 2.0) at http://commons.wikimedia.org/wiki/File:5_x_Pushbuttons.jpg.
Fig. 7-23a: Copyright © by S.J. de Waard (CC BY-SA 3.0) at http://en.wikipedia.org/wiki/File:Eindschakelaar_op_de_Mallegat-sluis_in_Gouda.JPG.
Fig. 7-24a: Copyright © by Lindsey Sh (CC BY-SA 3.0) at http://commons.wikimedia.org/wiki/File:Filtered_Air_Mover_with_Pressure_Switch.jpg.
Fig. 7-25: Copyright © by Ulf Seifert (CC BY-SA 3.0) at http://commons.wikimedia.org/wiki/File:Bimetswitch.jpg.
Fig. 7-26: SimonOrJ, http://en.wikipedia.org/wiki/File:Solenoid_Valve.svg, 2012.
Fig. 7-27: David Boettcher, http://en.wikipedia.org/wiki/File:Relay_Parts.jpg, 2008.
Fig. 7-28: Mike1024, http://en.wikipedia.org/wiki/File:Solid_state_relay.jpg, 2005.

CHAPTER 8

Fluid Power Circuits and Applications

Overview and Learning Objectives

In this chapter, we discuss the design and operation of both hydraulic and pneumatic circuits. A fluid power circuit is the arrangement of several fluid power components, such as machinery, actuators, pumps, control valves, and ancillary equipment, to perform a specific operation. The chapter focuses primarily on analyzing the performance of the circuit's desired and/or required function. Operational safety is discussed briefly in the pneumatic circuits section. The analysis on circuit performance is done both the traditional way, using graphic symbols to describe the circuit flow paths and deriving expressions for some of the operating conditions, and the modern way, using state-of-the-art simulation software to facilitate the analysis of multiple circuits and comparison with real data.

After completing this chapter, students should be able to do the following:

- Differentiate single- and double-acting cylinders.
- Describe the operation of a complete fluid power circuit using International Organization for Standardization (ISO) diagrams.
- Troubleshoot fluid power circuits.
- Determine the operation conditions, flow, and pressure of flow restrictors.
- Differentiate meter-in from meter-out flow control strategies.
- Analyze fluid power circuits that require sequencing and synchronization.
- Analyze fluid power circuits that require remote and safe operation.
- Explain the differences between hydraulic and pneumatic circuits.
- Use state-of-the-art simulation software to create, describe, analyze, and troubleshoot fluid power circuits.

Hydraulic Circuits

Single-Actuator Operation and Control

MANUAL OPERATION: HYDRAULIC JACK

Figure 8-1 Garage jack. Single-actuator applications; bottle-type hydraulic jack.

Figure 8-1 shows one of the most common and simplest applications of a single-acting hydraulic cylinder. Figure 8-1a presents a bottle-type hydraulic car jack, typically available in most auto repair shops and the home garages of auto enthusiasts.

Figure 8-2 presents the ISO drawing and the operation schematics of a bottle-type hydraulic jack like the image shown in Figure 8-1.

The bottle jack is a hydraulic circuit with several components, including a reservoir to store the hydraulic fluid, a manual piston pump to provide the pressurized fluid and power, a single-acting cylinder to lift or push converting hydraulic power into mechanical work (force and displacement), and a flow-control valve (typically a needle valve) to slowly release the pressurized fluid back to the reservoir. The pressure in the hydraulic circuit—in this case, 103 pounds per square inch (psi; 710 kilopascals [kPa]) is determined by force required to lift the object or push down the press beam and the internal diameter of the piston (technically, the cylinder diameter). Following the fluid path, the oil is taken by the pump from the reservoir (blue arrow lines, indicating low pressure) and delivered to the cylinder at the required pressure (red arrow lines, indicating high pressure). In both cases, lifting and lowering, the high pressure in the circuit is contained by the release valve, either closed or slightly opened. The small difference in the pressure values displayed in figures 8-2b and 8-2c is associated with the characteristics of the fluid paths simulated. The software used to virtually run the circuits includes pressure losses due to viscous friction in the lines and pressure drops across control valves.

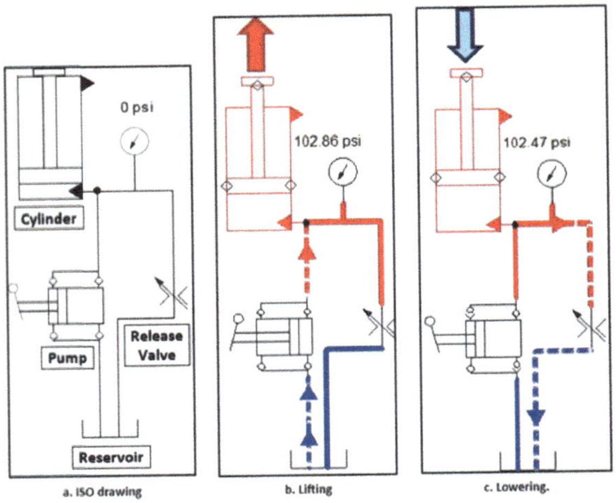

Figure 8-2 Hydraulic bottle jack circuits, ISO drawing and operation (Automation Studio (AS) v.5.7E).

Single- Versus Double-Acting Cylinders

Figure 8-3 compares the hydraulic circuits of a single-acting, spring-return cylinder and a double-acting cylinder (DAC). In a single-acting cylinder, pressurized fluid is supplied to extend the piston-rod assembly but retracts under the action of an internal spring (Figure 8-3a). In a DAC, pressurized fluid is supplied to extend and retract the piston-rod assembly, as shown in Figure 8-3b. The nature of the fluid supply requirements defines the characteristics of the directional valves. To activate the single-acting cylinder, a three-way, two-position directional valve is used. To activate the DAC, a four-way, three-position directional valve is used. A three-position valve is used to maintain the DAC extended and loaded without operator intervention. All other components in both hydraulic circuits are the same. It should be noted that for simplicity, in the schematics the pressure relief valve was left out, although it is mandatory in any fluid power circuit.

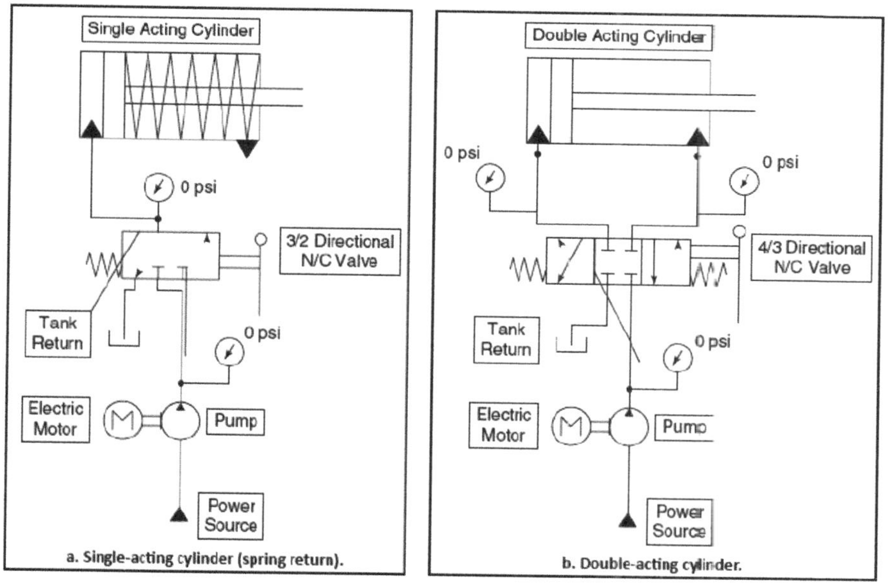

Figure 8-3 Single-cylinder hydraulic circuits: Single-acting versus double-acting cylinders.

Figure 8-4 compares the operation of the hydraulic circuits in Figure 8-3, a single-acting, spring-return cylinder versus a DAC. In Figure 8-4a, both cylinders have the same diameter and extend under the same loading conditions. The operating pressures are shown. Notice there is a roughly 10-psi (65-kPa) difference in net pressure. The piston in the single-acting cylinder must compress the spring to get the cylinder fully extended.

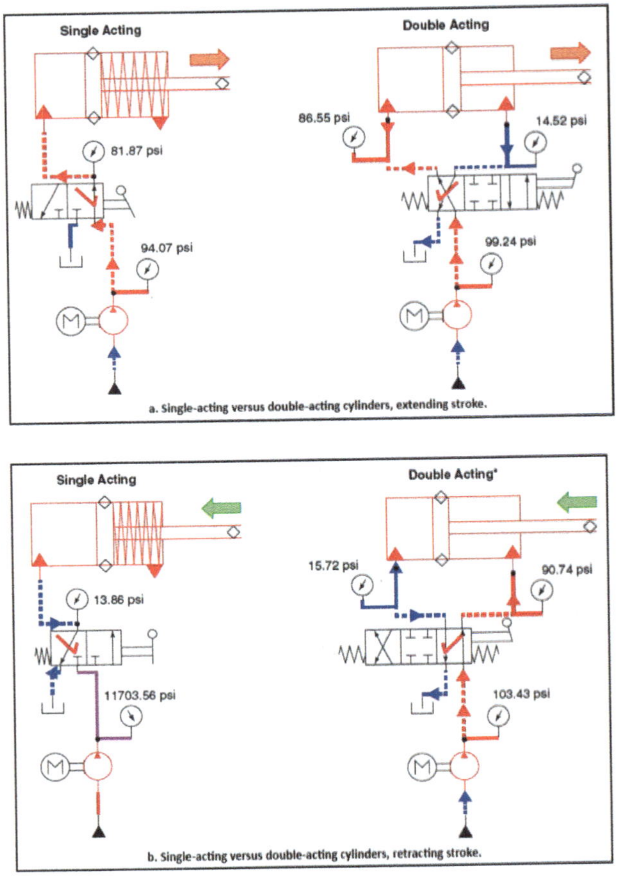

Figure 8-4 Hydraulic circuit operation: Single-acting versus double-acting cylinders (AS v.5.7E).

In Figure 8-4b, both cylinders are retracting; the DAC does it under the same loading conditions as in Figure 8-4a. The back pressure shown in the case of the single-acting cylinder corresponds with the return line friction losses and the pressure drop across the directional valve. The pressure difference between the rod end and back end of the DAC is slightly higher, 75 psi (520 kPa), due to the difference in cross-sectional areas. It should be pointed out the extremely high pressure indicated by the pump discharge gauge is unrealistic. As indicated above, the mandatory pressure relief valve is not included for simplicity.

For completeness, we include in Figure 8-5 the operation of both single-actuator circuits, when neither the single-acting cylinder nor the DAC is actuated but the hydraulic pump is on. With both directional valves closed, the pressure builds up in the circuit, activating the pressure relief valve and redirecting the pump flow back to the reservoir. The single-acting pressure relief valve will maintain the circuit pressurized at the full-flow pressure setting.

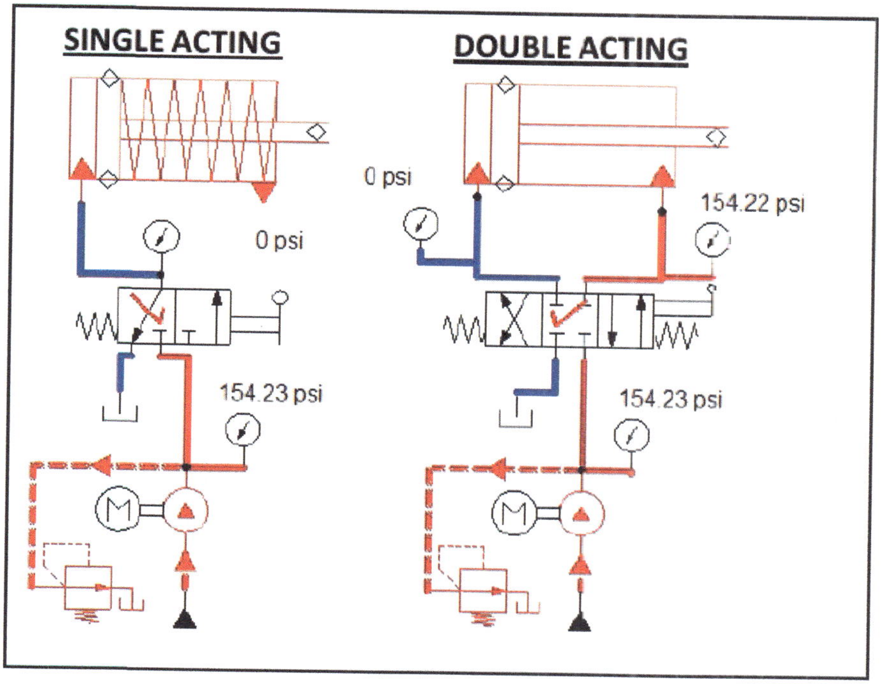

Figure 8-5 Hydraulic circuit operation with the pump idling, no fluid delivered (Automation Studio v.5.7E).

One interesting application of a single DAC hydraulic circuit is the upgrading of the lock doors' opening and closing mechanisms in the Erie Canal in New York. Figure 8-6a shows the Erie Canal Lock 33 in Henrietta, New York. Figure 8-6b is a close-up of the original lock door and driving rack beam. In the original design, which is still in use in some of the locks, an electric motor drives a large speed reduction gear assembly, which drives the rack beam linked to the lock door (see Figure 8-6c). In many locks, these mechanisms have been replaced with small hydraulic systems that drive DACs. The DACs were installed, replacing the original large rack beam on each lock door (see Figure 8-6d). The operating system of a lock door is an example of an application that requires a large amount of force but one that is applied continuously at slow speed. In this case, a complex mechanical system with several gears and moving parts was replaced by a more compact and simpler hydraulic system. The pictures depicted in the figures were taken by Dr. Galen R. Frysinger, a retired chemistry and physics scientist, who now spends most of his time traveling to interesting places in the world.

a. Erie Canal door lock.

b. Opening/closing rack (red beam).

c. Original opening/closing mechanism, multiple pinion, and reduction gears in series, ending with the door pinion and rack.

d. Updated opening/closing mechanism, synchronized hydraulic rams.

Figure 8-6 Double-acting cylinder application. The Erie Canal lock system; the travels of Tug 44.

Speed Control

USING A FLOW RESTRICTOR

The operational speed of a hydraulic actuator with fixed volumetric displacement, either a cylinder or a rotary motor, can be regulated only by adjusting the amount of flow that is fed to it. For practical purposes, hydraulic fluids are incompressible, and as such, all the volumetric flow sent by the directional valve to the actuator will be used to activate it. In a typical hydraulic circuit, the main pump will feed multiple actuators. Although a pump can be designed and operated to supply variable flow rate, it is not practical to adjust the flow supplied to the actuators by continuously adjusting the pump's operating conditions. Instead, a variable opening valve or flow restrictor can be installed in series with the actuator to limit the amount of flow that reaches each actuator.

As already mentioned, hydraulic fluids are nearly incompressible; hence, reducing the opening area of a valve feeding the actuator will not reduce the amount of flow going through it. This valve becomes a flow restrictor when the circuit is designed to provide an alternate route for the excess flow. As we will discuss in chapters 9 and 10, the great majority of hydraulic pumps are of the positive-displacement type. Consequently, every circuit must have a pressure relief valve installed. Activating a flow restrictor will immediately increase the pressure upstream from the restrictor. The flow passing through the restrictor will remain the same until the increase in pressure is large enough to activate the pressure relief valve. Once the pressure reaches the setting of this valve, the excess flow will be diverted back to the reservoir. It is then that the flow restrictor actually regulates the flow, feeding the actuator.

Figure 8-7 shows two similar circuits in which the speed of a double-acting control cylinder is regulated by one flow restrictor. In both circuits, the flow restrictor is one-way and is regulating the extension speed of the cylinder. The only difference among the circuits is the fact that in the circuit in Figure 8-7a, the flow restrictor is located upstream from the DAC, whereas in the circuit in Figure 8-7b, the restrictor is located downstream from the cylinder.

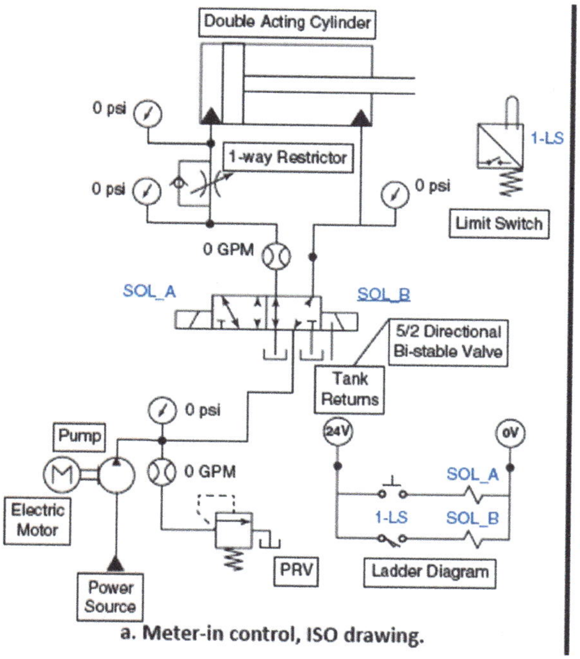

Figure 8-7 Speed control of a double-acting cylinder using a one-way flow restrictor.

Figure 8-8 compares the expected operation (via simulation) of the flow restrictor–driven hydraulic circuits of Figure 8-7. The cylinder in both arrangements extends under the same loading conditions. The main difference among the circuits is the operating pressure of the cylinders. In the meter-in arrangement in Figure 8-8a, the back-end pressure in the cylinder is determined by the loading, roughly 72 psi (500 kPa), because the rod end is connected back to the atmospheric reservoir. In contrast, in the meter-out arrangement, the back-end pressure in the cylinder would be roughly the setting of the pressure relief valve, 250 psi (3,625 kPa), as in Figure 8-8b. In this case, what the loading in the cylinder determines is the pressure difference across the piston, which remains close to 72 psi (500 kPa). Metering out restricts the flow, leaving the rod end of the cylinder increasing the pressure on that end, and consequently, extending the cylinder requires pushing not only the load but also the fluid itself out of the cylinder to overcome the pressure increase created by the restrictor. In both cases, metering in and out, the cylinder extends at the same speed, as implied by similar flow rates, approximately 15 gallons per minute (gpm; 60 liters per minute [lpm]). The excess flow—approximately 17 gpm (65 lpm) in both cases—goes back to the reservoir via the pressure relief valve.

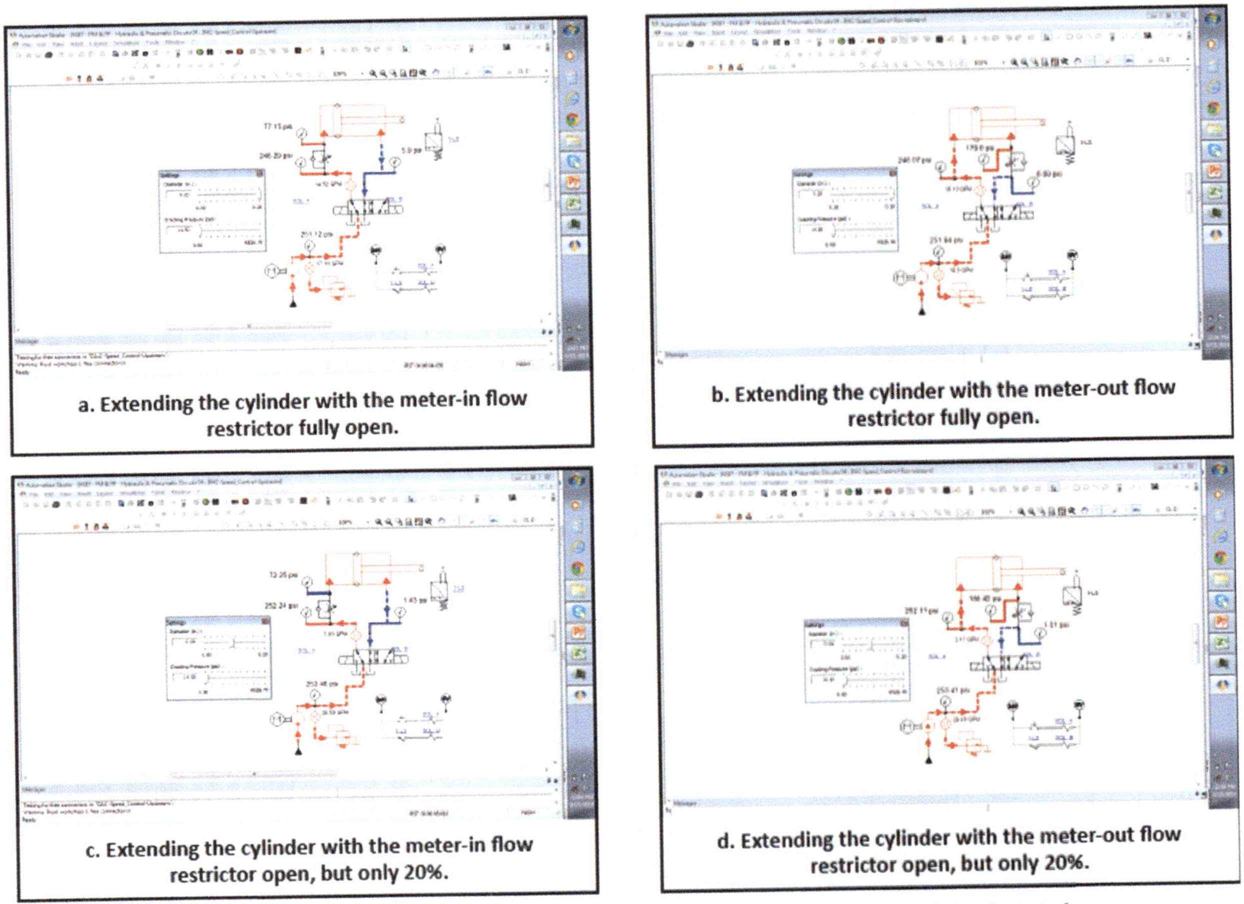

a. Extending the cylinder with the meter-in flow restrictor fully open.

b. Extending the cylinder with the meter-out flow restrictor fully open.

c. Extending the cylinder with the meter-in flow restrictor open, but only 20%.

d. Extending the cylinder with the meter-out flow restrictor open, but only 20%.

Meter-In Speed Control

Meter-Out Speed Control

Figure 8-8 Comparing the operating conditions of meter-in and meter-out speed control methods (AS v.5.7E).

Also in Figure 8-8, both metering approaches are compared when the restrictor is open by only 20%. Under the same loading conditions, the operating pressures remain approximately the same. In the meter-in arrangement (Figure 8-8c), the back-end pressure is predicted to be roughly 77 psi (530 kPa), whereas in the meter-out arrangement, the back-end pressure in the cylinder should be roughly 250 psi (3,625 kPa; Figure 8-8d). With a higher restriction, the flow sent to extend the cylinders is significantly reduced to approximately 3 gpm (11 lpm) in both arrangements, indicating that most of the pump flow is diverted as excess flow and goes back to the reservoir via the pressure relief valve, approximately 30 gpm (110 lpm).

The pressure drop across a flow restrictor valve is inversely proportional to the valve coefficient, as expressed below (derivation included in Chapter 10):

$$\Delta P = SG \times \left(\frac{Q}{C_v}\right)^2$$

1. ΔP = pressure drop across the flow restrictor valve,
2. SG =specific gravity of the fluid,
3. Q = volumetric flow rate passing through the flow restrictor valve, and
4. C_V = valve (flow restrictor or control) coefficient.

The valve coefficient of any valve of a flow restrictor is directly proportional to the size and throttled opening of the valve, as expressed below:

$$C_v \propto A_o$$

with A_O being the actual open area of the valve (restrictor orifice fully open or throttled).

As the flow restrictor is opened—creating a larger flow area—the actual valve coefficient also increases, generating a reduced pressure drop across the valve. Conversely, as the flow restrictor is closed—reducing the flow area—the actual valve coefficient also drops, increasing the pressure drop across the valve. This relationship can be expressed visually as follows:

$$A_o \uparrow \rightarrow C_v \uparrow \rightarrow \Delta P \downarrow$$

$$A_o \downarrow \rightarrow C_v \downarrow \rightarrow \Delta P \uparrow$$

With the basic operating parameters of the flow restrictor established, we proceed to examine the effect on the circuit and actuators in the hydraulic system of using a flow restrictor. The DAC of any of the circuits in Figure 8-8 pushes and/or pulls a specific load in regular operation. During extension, the piston in the cylinder provides the necessary force to overcome the load with an increase in the pressure of the cylinder back end as the piston opposes the flow provided by the pump of the hydraulic system. In addition, the piston-cylinder arrangement needs to overcome any pressure required to displace the fluid on the rod-end side of the cylinder. Hence, to evaluate the effect of the flow restrictor in the system, we start with the balance of forces in the cylinder (see Figure 8-9).

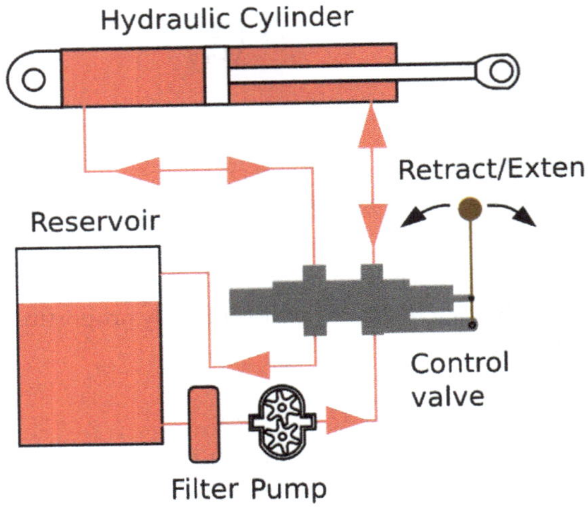

Figure 8-9 Force analysis on a double-acting cylinder installed in a simple hydraulic circuit.

The forces in the cylinder are related as:

$$F_B - F_R = F_L$$

1. F_B = force exerted by the pressure on the piston in the back end of the cylinder,
2. F_R = force exerted by the pressure on the piston in the rod end of the cylinder, and
3. F_L = load driven by the cylinder.

Recalling the definition of pressure—force divided by area—the pressure in the back of the cylinder can be estimated:

$$F_B = P_B \cdot A_B$$

and

$$F_R = P_R \cdot A_R$$

$$P_B \cdot A_B - P_R \cdot A_R = F_L$$

$$P_B \cdot A_B = F_L + P_R \cdot A_R$$

$$P_B = \frac{F_L + P_R \cdot A_R}{A_B}$$

1. P_B = pressure on the piston at the back end of the cylinder,
2. A_B = cylinder effective area at the back end of the cylinder,
3. P_R = pressure on the piston at the rod end of the cylinder, and
4. A_R = cylinder effective area at the rod end of the cylinder.

The effective areas can be calculated once the dimensions of the cylinder are known, while the pressure at the rod end of the cylinder depends on the design of the hydraulic system. In most hydraulic circuits, the returns from all actuators go back to the reservoir via the directional valves. Thus, when the flow restrictor is located upstream of the cylinder, it can be safely assumed that the return pressure in the rod end of the cylinder is minimal. With these premises, the pressure in the back of the cylinder in the meter-in arrangement turns out to simply be the load-to-piston-area ratio, as shown below:

$$P_R \approx 0 \rightarrow F_R \cong 0 \rightarrow P_B \cong \frac{F_L}{A_B}$$

$$A_B{=}A_P = \pi \cdot \frac{{D_P}^2}{4}$$

A_P is the cylinder piston area, and D_P is the piston diameter.

When the flow restrictor is located downstream of the cylinder, the restriction builds up pressure to create the pressure drop required to "push" the fluid through it. Considering again that the fluid directly returns to the reservoir after the restrictor, the pressure in the rod end of the cylinder is equivalent to the pressure drop to net area. Now, the pressure in the back of the cylinder is the addition of the load-to-piston-area ratio and the pressure drop, as shown below:

$$P_R \approx \Delta P_{FCV} \rightarrow F_R \approx \Delta P_{FCV} {\cdot} A_R \rightarrow P_B \cong \frac{F_L + \Delta P_{FCV} \cdot A_R}{A_B} = \frac{F_L}{A_B} + \left(\frac{A_R}{A_B}\right) {\cdot} \Delta P_{FCV}$$

$$A_R = A_P - A_r = \pi \cdot \frac{{D_p}^2}{4} - \pi \cdot \frac{{D_r}^2}{4} = \pi \cdot \frac{{D_p}^2 - {D_r}^2}{4}$$

$$P_B \cong \frac{F_L}{A_B} + \left(\frac{A_P - A_r}{A_P}\right) \cdot \Delta P_{FCV} = \frac{F_L}{A_B} + \left(1 - \frac{A_r}{A_P}\right) \cdot \Delta P_{FCV}$$

1. ΔP_{FCV} = pressure drop across the flow restrictor valve,
2. A_r = cylinder rod area, and
3. D_r = rod diameter.

to obtain considering that the rod diameter is significantly smaller than the piston diameter:

$$D_r \ll D_p \rightarrow \frac{A_r}{A_p} \ll 1$$

$$P_B \cong \frac{F_L}{A_B} + \Delta P_{FCV}$$

From the analysis presented above, it can be inferred that the operating pressure in the cylinder—or in any other actuator, for that matter—is larger in the meter-out arrangement than in the meter-in arrangement. This translates into a larger operating pressure in the hydraulic circuit indeed in the pressure that the pump needs to provide. This is the major drawback of the meter-out arrangement; larger operating pressures mean stronger cylinders are required and more power consumed by the pump and lost when the pressure relief valve is open. The advantage of a meter-out arrangement is twofold; first, it guarantees that that side of the cylinder is always completely filled with hydraulic fluid, and second, it pressurizes that fluid returning back to the reservoir, allowing better control of the cylinder movement in any situation, whether loaded or unloaded or installed vertically or horizontally.

USING A REGENERATIVE CIRCUIT

A regenerative circuit is used when it is desirable to increase the extending speed of a DAC without the need of supplying additional flow, either using a second pump or one with variable displacement. The regenerative circuit can also be used when we want to simplify the type and number of valves required to operate the cylinder.

As discussed earlier in the chapter, the most common way to operate a DAC is to use a directional valve to change the direction of the fluid flow going to and coming from the cylinder when extending or extracting it. Extending a regular DAC requires the fluid flow to be sent to the back end of the cylinder, allowing the fluid in the rod end of the cylinder to go back to the reservoir. As already indicated, there is little or no pressure on the rod side of the typical hydraulic circuit at this time. The typical operation of a double-acting hydraulic cylinder is shown in Figure 8-10a. To retract the cylinder, the opposite is true: Fluid flow is directed to the rod end of the cylinder, allowing it to flow from the back end of the cylinder to the reservoir. Similarly, there would be little or no pressure on the back side of the hydraulic circuit.

In a regenerative circuit, the back and rod ends are connected when extending the cylinder. This translates into both ends of the cylinder having the same pressure. When the cylinder is extended, fluid flows from the pump to the back end of the cylinder, while the piston pushes fluid out of the rod end of the cylinder also to the back end, where it combines with the fluid coming from the pump. Hence, the fluid from the rod end helps extend the cylinder, thus creating a regenerating motion. To retract the cylinder, the regenerative circuit operates like the simple hydraulic circuit, pressurizing the rod side of the cylinder. The circuit is only regenerative when extending a cylinder. The cylinder in a regenerative circuit is retracted in the same manner as a regular DAC. The difference between a regular DAC and a regenerating DAC is the fact that there is pressure on the rod side of a regenerating cylinder in both instances, when it is extended and retracted. The operation of a double-acting hydraulic cylinder in a regenerative circuit is shown in Figure 8-10b.

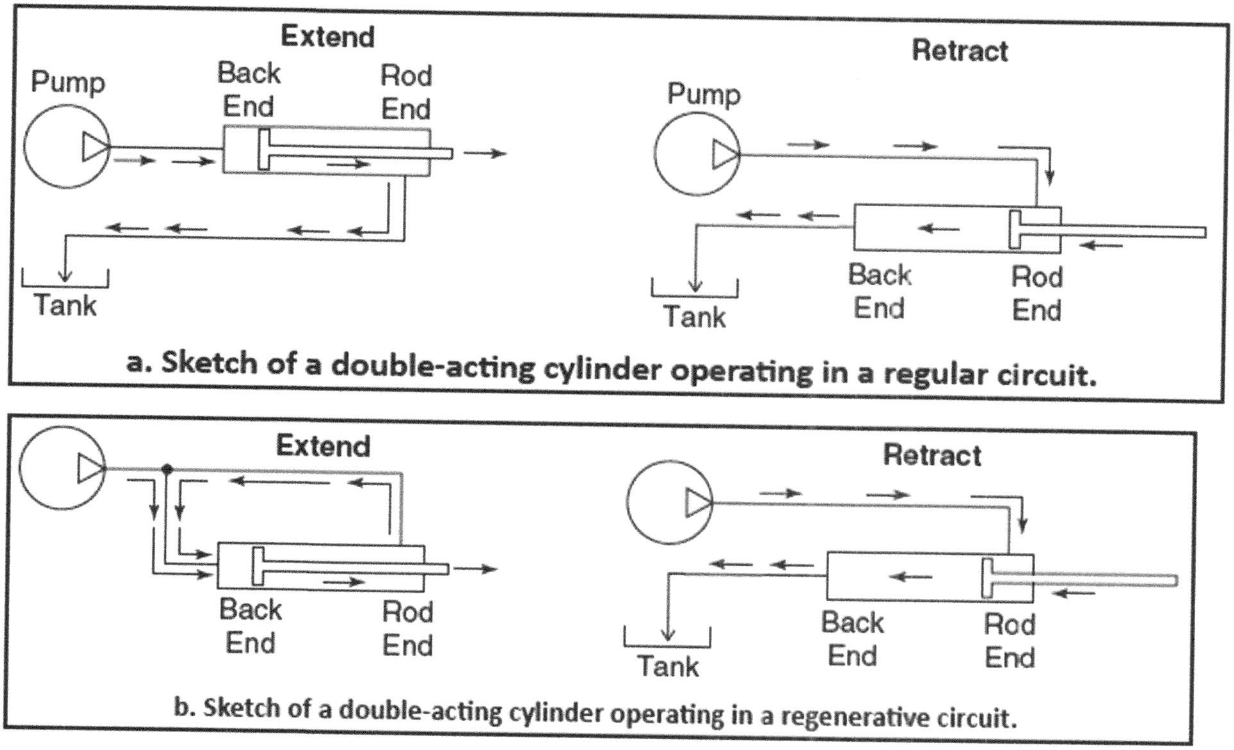

Figure 8-10 Comparing the operation of a double-acting cylinder in a regular circuit and a regenerative circuit.

Using a DAC in a regenerative circuit has a significant impact on the operating conditions of the cylinder and the hydraulic circuit during the extending stroke but no effect during the retracting stroke. As already indicated, the regenerative cylinder retracts like a regular cylinder, with the operating conditions being the same. Figure 8-11 shows a schematic with the operating conditions of a DAC in a regenerative circuit during the extending stroke.

Like in regular DACs, the forces acting in the cylinder are balanced (when moving at uniform speed and neglecting internal friction):

$$F_B - F_R = F_L$$

1. F_B = force exerted by the pressure on the piston in the back end of the cylinder,
2. F_R = force exerted by the pressure on the piston in the rod end of the cylinder, and
3. F_L = load driven by the cylinder.

Neglecting the losses in the lines connecting the back and rod ends of the cylinder, the pressure at both ends of the cylinders is the same. This pressure needs to be provided by the pump. The analysis of the forces leading to the estimation of this pressure is presented below:

$$F_B = P_B \cdot A_B$$

and

$$F_R = P_R - A_R$$

$$P_B \cdot A_B - P_R \cdot A_R = F_L$$

$$P_R = P_B = P_p \rightarrow P_p \cdot A_B = F_L + P_p \cdot A_R \rightarrow P_p \cdot (A_B - A_R) = F_L$$

$$P_p = \frac{F_L}{(A_B - A_R)}$$

$$A_B = A_p = \pi \cdot \frac{{D_p}^2}{4}$$

$$A_R = A_p - A_r = \pi \cdot \frac{{D_p}^2}{4} - \pi \cdot \frac{{D_r}^2}{4} = \pi \cdot \frac{{D_p}^2 - {D_r}^2}{4}$$

1. P_B = pressure on the piston at the back end of the cylinder,
2. A_B = cylinder effective area at the back end of the cylinder,
3. P_R = pressure on the piston at the rod end of the cylinder,
4. P_P = pressure provided by the pump,
5. A_R = cylinder effective area at the rod end of the cylinder,
6. A_P = cylinder piston area,
7. D_P = piston diameter,
8. A_P = cylinder rod area, and
9. D_r = rod diameter.

Hence, the minimum pressure to be provided by the pump is:

$$P_p = \frac{F_L}{\pi \cdot \frac{{D_r}^2}{4}}$$

The load that the regenerative cylinder can provide is given by:

$$F_L = P_p \times \pi \cdot \frac{{D_r}^2}{4}$$

The fact that the rod diameter of a cylinder is always significantly smaller than the piston diameter makes these findings relevant. The operating pressure of a regenerative cylinder is considerably larger when compared with that of a regular cylinder when driving the same load. Conversely, the load that a regenerative cylinder can drive is significantly smaller than the one provided by a regular cylinder when the supply pressure is the same.

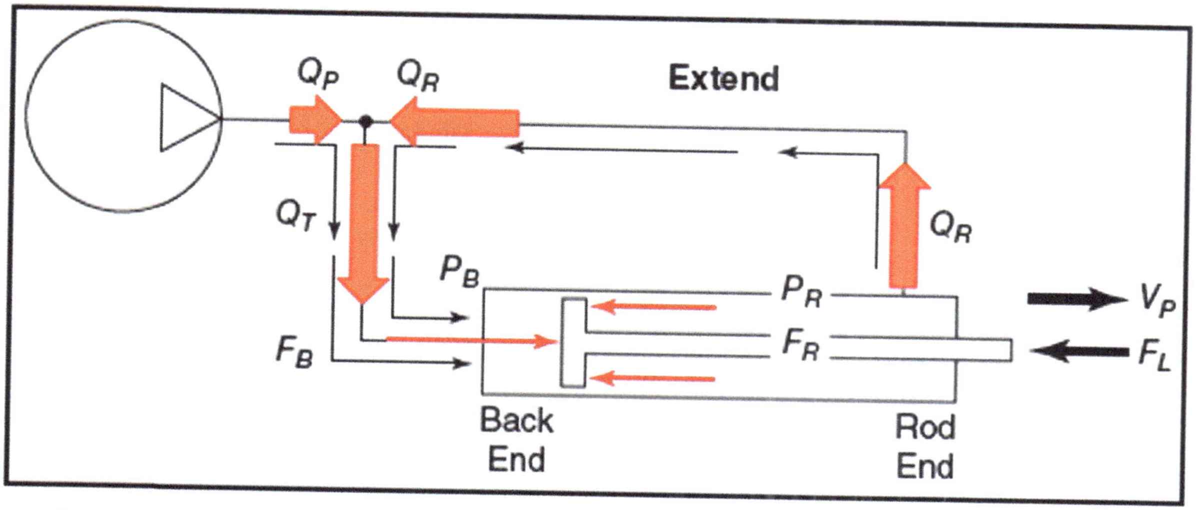

Figure 8-11 Operating conditions in a regenerative double-acting cylinder during the extending stroke.

Installing a regular DAC in a regenerative circuit also has an impact on the fluid flow distribution and the cylinder's extending speed. As was previously indicated, the fluid returning from the rod end is combined with the one provided by the pump and then supplied to the back end of a regenerative cylinder during the extending stroke. Using the principle of conservation of mass:

$$Q_T = Q_p + Q_R$$

1. Q_R = flow rate returning from the rod end of the regenerative cylinder,
2. Q_P = flow rate provided by the pump, and
3. Q_T = total flow rate supplied to the back end of the regenerative cylinder.

$$Q_R = V_R \cdot A_R$$

and

$$Q_T = V_B \cdot A_B$$

with

$$V_R = V_B = V_p$$

Upon substitution:

$$V_B \cdot A_B = Q_p + V_R \cdot A_R$$

$$V_p \cdot A_p = Q_p + V_p \cdot (A_p - A_r)$$

$$V_p = \frac{Q_p}{A_r}$$

The flow that a pump needs to supply for a given extending speed is:

$$Q_p = V_p \cdot A_r$$

Again, the difference in the effective areas of the regenerative cylinder when compared with that of a regular cylinder has significant consequences. First, the extending speed of a regenerative cylinder is considerably larger when compared with that of a regular cylinder for a given pump flow. Second, the flow that a pump needs to provide to a regenerative cylinder is significantly smaller than the one required by a regular cylinder when extending at the same speed.

Figure 8-12 presents the ISO drawing and simulated operation of a regenerative circuit. The regenerative feature is created by moving the feed of the rod-end cylinder upstream of the directional control valve (Figure 8-12a). In this regenerative arrangement, the cylinder will always retract when the hydraulic pump is on. As demonstrated, the flow a pump needs to provide to extend a regenerative cylinder is significantly lower than that of a regular double-acting cylinder. In Figure 8-12b, the pump provides only about 9 gpm (34 lpm); 23 gpm (87 lpm) are diverted to the reservoir through the pressure relief valve, while the total flow sent to the back end of the cylinder to extend it is roughly 80 gpm (300 lpm). About 70 gpm (265 lpm) are "regenerated" (i.e., provided by the rod end of the cylinder). It was also demonstrated that the operating pressures of a regenerative cylinder are significantly larger than those required in a regular DAC. In Figure 8-12b, we see that the estimated operating pressure in the given circuit is about 500 psi (3400 kPa), the setting of the pressure relief valve setting, when only about 70 psi (500 kPa) would be needed in a regular DAC for the same loading conditions. The retraction stroke of a regenerative cylinder is identical to that of a regular cylinder: All the flow coming from the pump goes to the rod end of the cylinder (see Figure 8-12c).

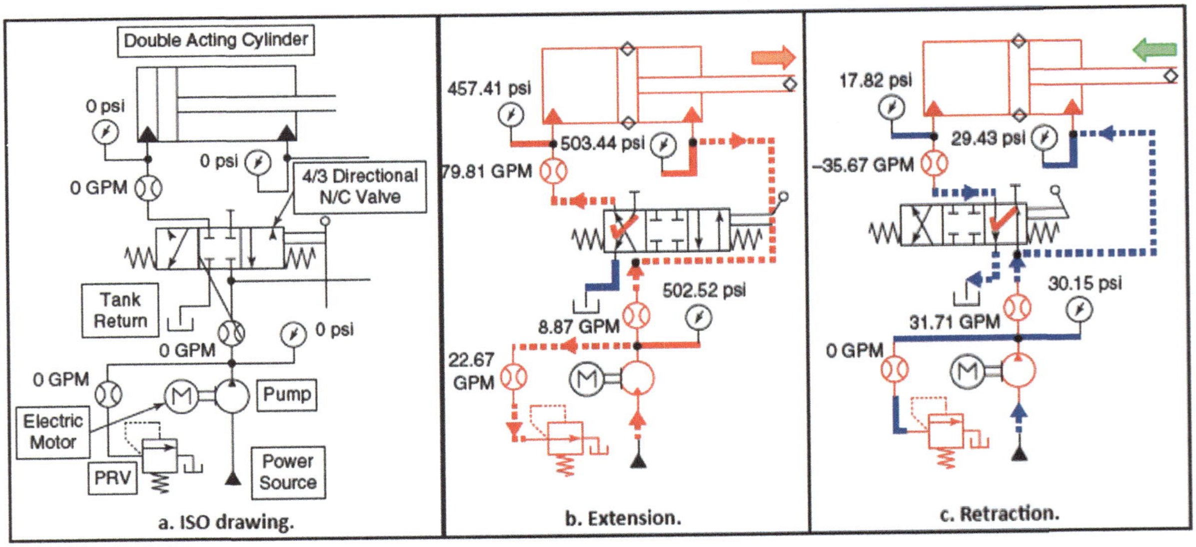

Figure 8-12 Rapid extension of a dual-acting cylinder by redirecting rod-end fluid return: Regenerative circuit (AS v.5.7E).

In addition to increasing the extending speed of a DAC, the regenerative circuit simplifies the circuit by reducing the complexity and number of valves required to operate the cylinder. Figure 8-13 compares three valve configurations: two for the regular DAC and one for actuating the cylinder in a regenerative circuit. As can be seen in the figure, the use of a regenerative circuit allows the system to be operated with a simple manifold, employing just two solenoid valves and a check valve, as opposed to some type of spool valve or a manifold with three or four valves needed to operate a regular DAC. The regenerative circuit is just a different way to operate and increase the extending speed of a DAC.

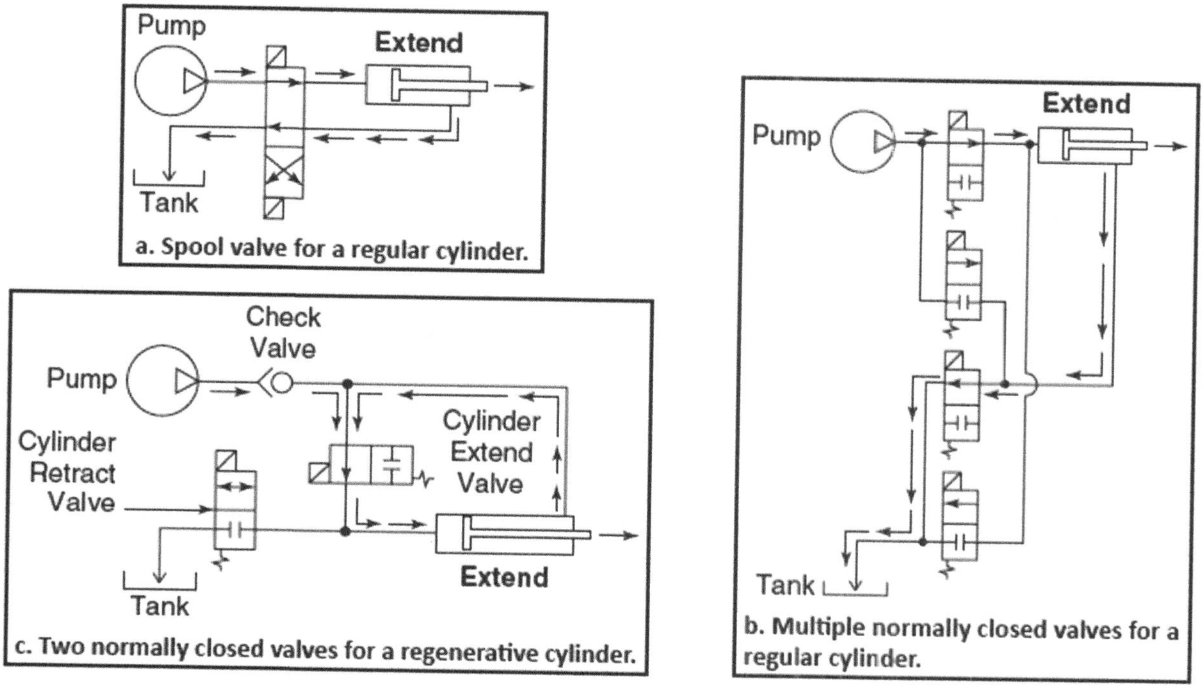

a. Spool valve for a regular cylinder.

b. Multiple normally closed valves for a regular cylinder.

c. Two normally closed valves for a regenerative cylinder.

Figure 8-13 Typical valve configurations required to operate a double-acting cylinder, regular and regenerative circuits.

Pump Unloading Circuit

Figure 8-14 shows the ISO diagram and the operation of what appears to be a typical DAC hydraulic circuit. In the circuit, an electric-driven hydraulic pump provides power to the cylinder through the directional valve. The flow returning from the cylinder goes back to the reservoir, and the circuit is protected with a pressure relief valve. What makes this particular circuit different from the regular DAC circuit is the addition of a nonreturn flow or check valve (see Figure 8-14a). The check valve is installed between the relief line branching point and the pilot line activating the pressure relief valve. What this check valve provides to the circuit is the ability to activate the pressure relief valve without the pump continuously providing the pressure necessary to keep the valve open. When idling or with the cylinder either fully extended or retracted and the directional valve still open, the increase in pressure activates the pressure relief valve to allow the excess flow to be diverted back to the reservoir. Once the pressure relief valve opens, the pressure at the pump discharge falls drastically, and the check valve closes, isolating the pump from the directional valve and the DAC or any other actuator (see Figure 8-14b). The extending and retracting stroke in this pump-unloading circuit are identical to that of a regular DAC, as the check valve remains open during the entire stroke (see figures 8-14c and 8-14d).

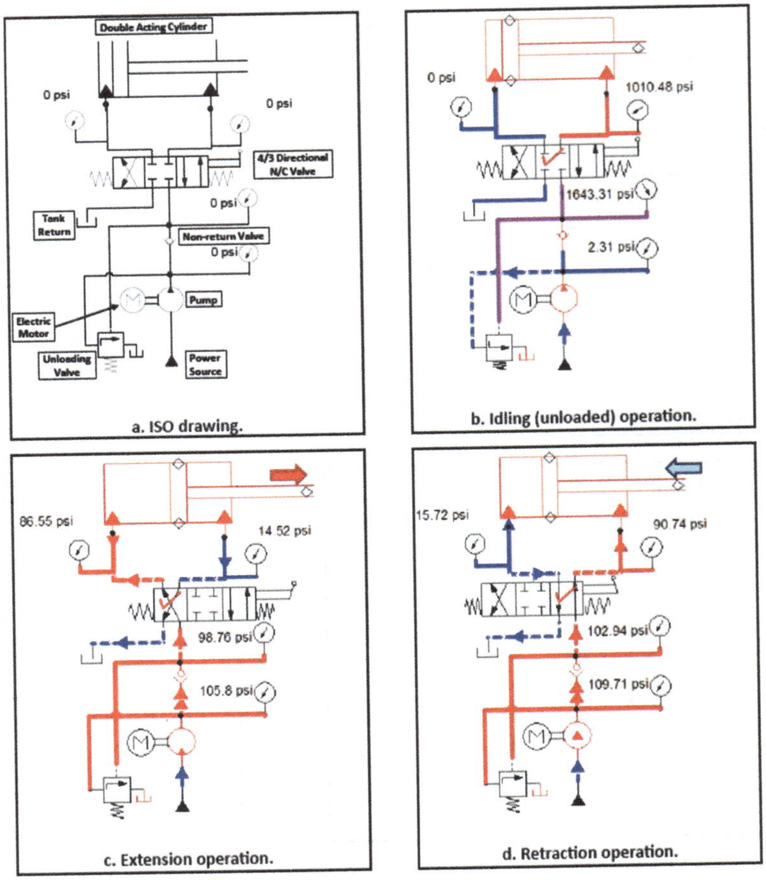

Figure 8-14 Depressurizing or unloading the pump in a hydraulic circuit by adding a pilot-fed pressure relief valve (AS v.5.7E).

Multiple Actuators: Operation and Control

Single- and Double-Acting Cylinders: Sequencing

Figure 8-15 compares two similar circuits with two hydraulic cylinders that operate in sequence. The distinction between the two circuits is the actuation. The cylinder sequence in the circuit in Figure 8-15a is both initiated and terminated manually, whereas the circuit sequence in Figure 8-15b is initiated manually but completed automatically. The actuation in both cases is initiated manually via push-button PB-1 (ladder diagram insert in both figures), extending the DAC. Similarly, a limit switch 1-LS activated by the DAC extends the single-acting cylinder (SAC). On the other hand, the retraction of both cylinders is initiated by a second push-button PB-2, whereas the retraction of both cylinders in the sequence of Figure 8-15b is initiated by a second limit switch 2-LS activated by the SAC.

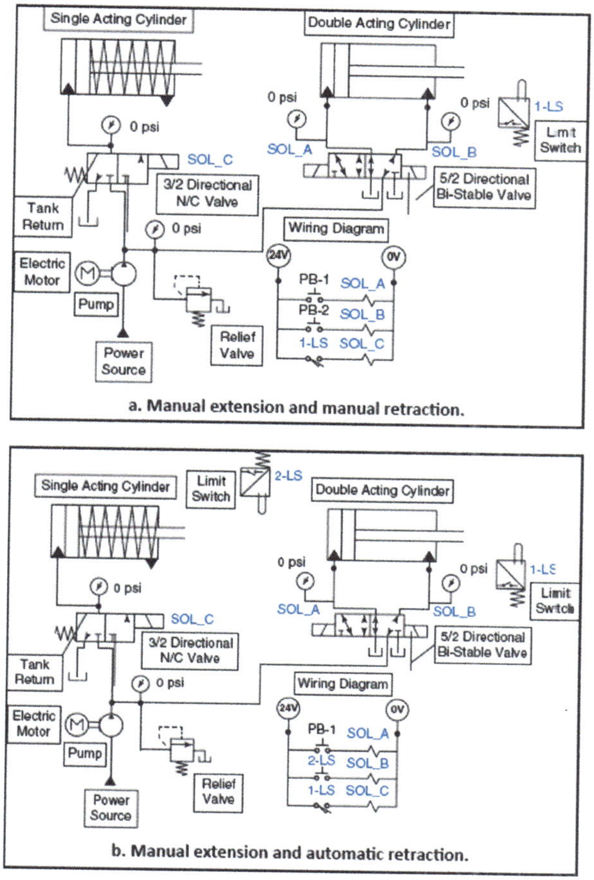

a. Manual extension and manual retraction.

b. Manual extension and automatic retraction.

Figure 8-15 Two-cylinder hydraulic circuits: Single-acting and double-acting cylinders.

Figure 8-16 compares the expected operation when extending the cylinders (via simulation) of the hydraulic circuits in Figure 8-15. Both cylinders extend under the same loading conditions. Once the extension sequence is initiated, the pressure difference between the back end and the rod end of the DAC reaches roughly 72 psi (500 kPa). Once the DAC fully extends, initiating the extension of the SAC, the pressure in the back end of the SAC is roughly 75 psi (520 kPa). The 3-psi (20-kPa) difference in pressure is associated with the force required to compress the spring at the stroke extension shown. The pressure values shown can be easily validated when it's known that both cylinders have an internal diameter of 6 inches (in; 150 millimeters [mm]), driving a load of 2,000 pounds-force (lbf; 9,000 newtons [N]).

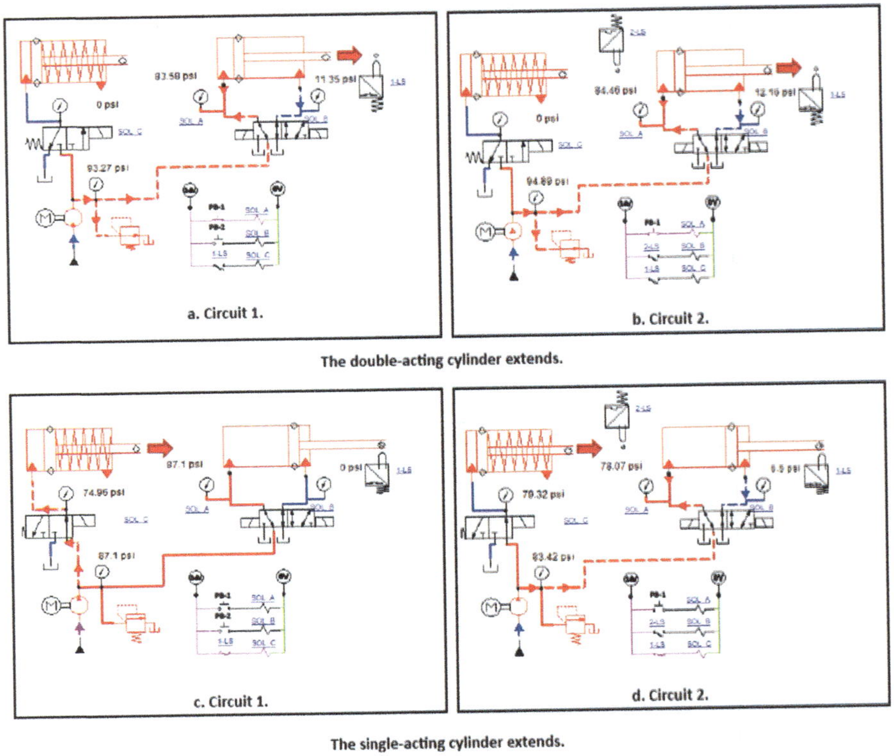

Figure 8-16 Two-cylinder hydraulic circuits' operation: Activating the extension stroke sequence (AS v.5.7E).

Figure 8-17 compares the expected operation when the cylinders retract (via simulation) of the hydraulic circuits in Figure 8-15. Both cylinders extend under the same loading conditions. Once the extension sequence is initiated, the pressure difference between the back end and the rod end of the DAC reaches roughly 72 psi (500 kPa). Once the DAC fully extends, initiating the extension of the SAC, the pressure in the back end of the SAC is roughly 75 psi (520 kPa). The 3-psi (20 kPa) difference in pressure is associated with the force required to compress the spring at the stroke extension shown. The pressure values shown can be easily validated, knowing that both cylinders have an internal diameter of 6 in (150 mm), driving a load of 2,000 lbf (9,000 N).

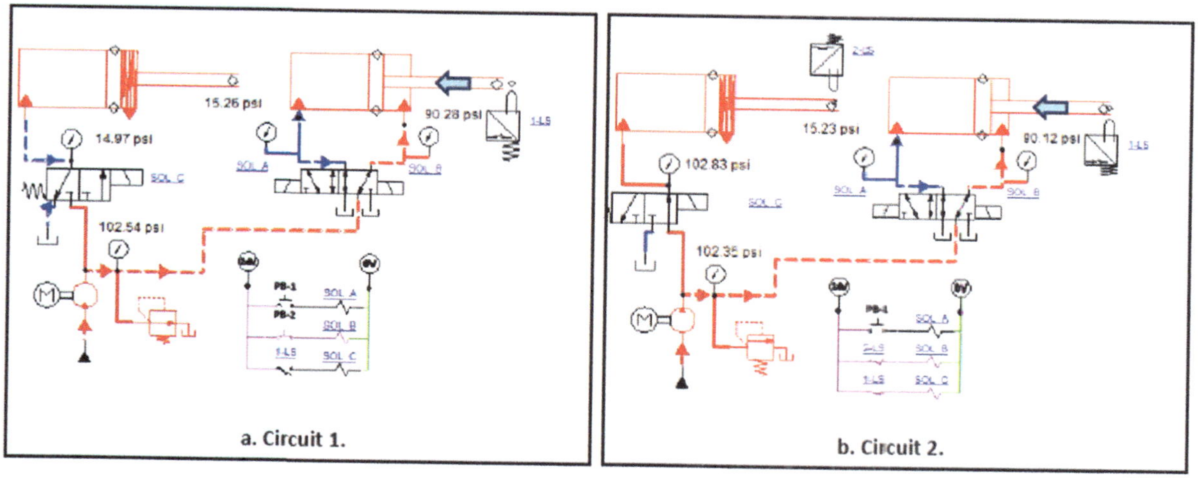

Figure 8-17 Two-cylinder hydraulic circuits' operation: Retraction stroke (AS v.5.7E). Circuit 1: Manual extension and manual retraction. Circuit 2: Manual extension and automatic retraction.

Double-Acting Cylinders: Parallel Arrangement

SPLITTING THE FLOW

Figure 8-18 presents a hydraulic circuit with two DACs arranged in parallel. A typical application of such an arrangement is portrayed in Figure 8-18b. Casting concrete columns and pillars requires very long molds that are normally closed, and open using hydraulic cylinders to move in and out of the mold walls. This application and many others require the cylinders to move simultaneously. Figure 8-18c shows the operation via simulation of two identical cylinders. Both cylinders drive equal loads, with supply lines with the same hydraulic characteristics. In these conditions, the extension speed of both cylinders is the same, as we can infer from the equal flow being supplied to each cylinder, half the flow supplied by the pump. Nevertheless, there is no way that exactly identical cylinders and/or supply lines can be manufactured. As with every other piece of machinery, hydraulic cylinders are manufactured within certain tolerances; therefore, the internal diameters (i.e., effective areas) and the friction at the seals will always be slightly different. Similarly, it is impossible—and in most cases impractical—to build a fluid distribution system that has equal friction losses between the system pump and each cylinder supply port. Figure 8-18d now shows the same cylinder arrangement, but with the cylinder driving uneven loads. In the case of uneven loading, the cylinder with the lower load extends first, in this case, the left cylinder. The heavily loaded cylinder (the cylinder on the right in Figure 8-18e) extends second, also taking in all the flow sent by the hydraulic pump.

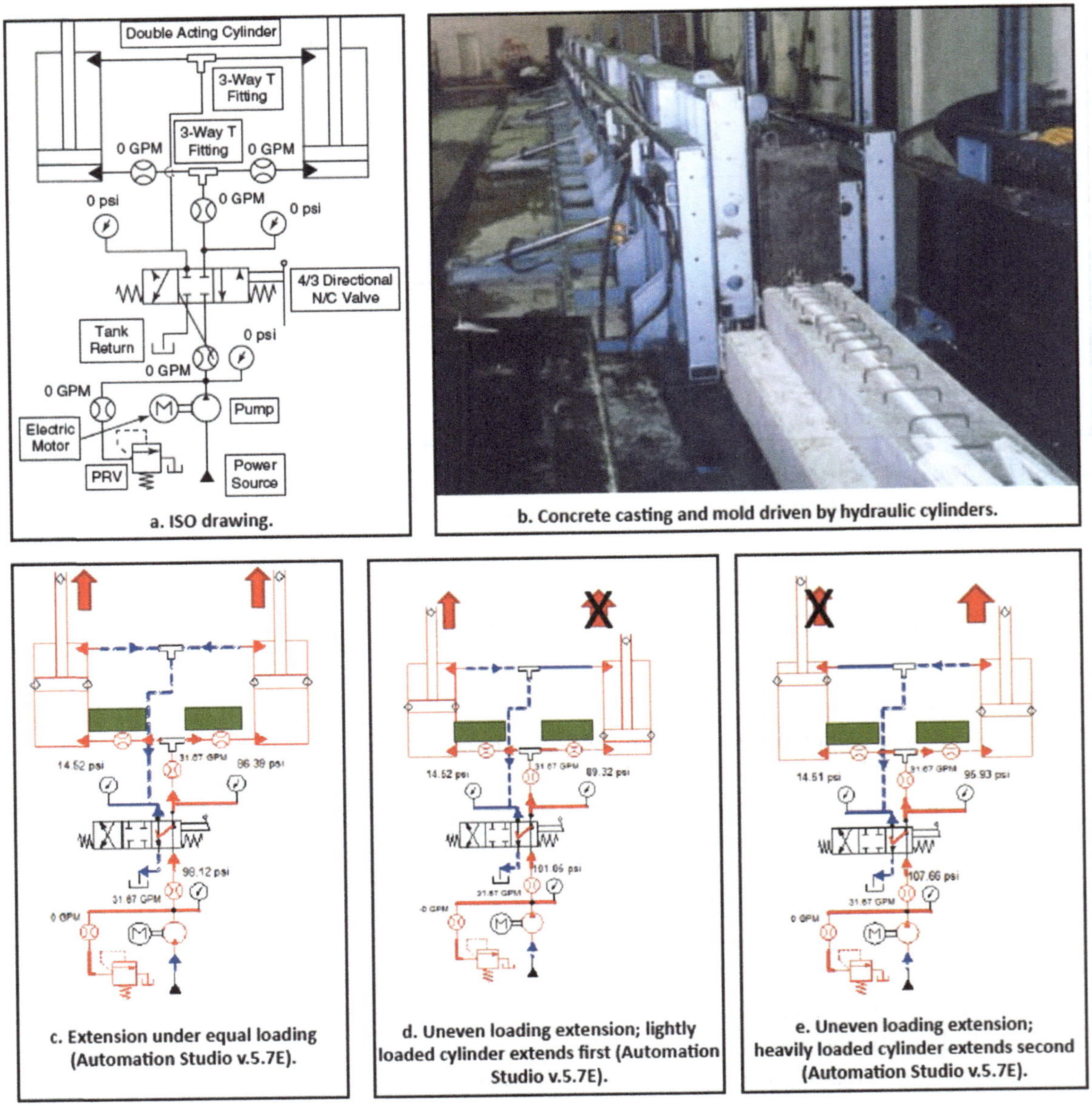

a. ISO drawing.

b. Concrete casting and mold driven by hydraulic cylinders.

c. Extension under equal loading (Automation Studio v.5.7E).

d. Uneven loading extension; lightly loaded cylinder extends first (Automation Studio v.5.7E).

e. Uneven loading extension; heavily loaded cylinder extends second (Automation Studio v.5.7E).

Figure 8-18 In-parallel arrangement, two-cylinder hydraulic circuits (identical double-acting cylinders).

FLOW DIVIDERS

Figure 8-19a presents another circuit with hydraulic cylinders connected in parallel, just like the arrangement previously displayed in Figure 8-18a. In this case, the flow splitter feeding the back-end ports of the cylinders has been replaced with a flow divider installed between the directional valve and the cylinder ports. The function of a flow divider is to guarantee that the flow rate supplied by the directional valve is split in flow rate fractions that do not change independently of the loading in the cylinders. One of the most common flow divider designs in the hydraulic industry is the rotary flow divider (Figure 8-19b). In Figure 8-19b, the main inlet supply port is shown together with two threaded relief valves. Figure 8-19b also shows that a rotary flow divider is a sandwich of external gear machines assembled in parallel. Figure 8-19d presents the internal arrangement of a similar flow divider but with four outlets. In the Figure, four pairs of gears are shown, one per outlet, mounted on a single shaft for a synchronous rotation. Figure 8-19d also shows the alignment sleeves (top) that locate each section for a concentric assembly and maintain close clearances. As we will discuss in the next chapter, external gear machines are of the positive-displacement type, meaning that they pump or take in a fixed amount of fluid per revolution. As such, linked identical gear pumps and motors will supply exactly the same flow to every component connected to them.

Figure 8-19c shows the operating conditions of the circuit in Figure 8-19a when the cylinders drive unequal loads. In the circuit, the load in the cylinder on the right is larger than the one on the left, as we can infer from the large pressure required to extend it. Nevertheless, the flow fed to each cylinder is exactly the same, guaranteeing the simultaneous extension within manufacturing tolerances. At uneven loading, the rotary flow divider distributes the input hydraulic power according to each actuator's requirement. The cylinder driving the heavier load requires more power.

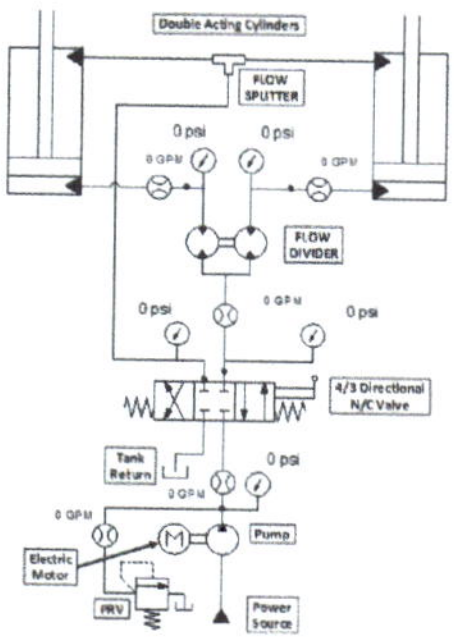

a. Circuit ISO drawing.

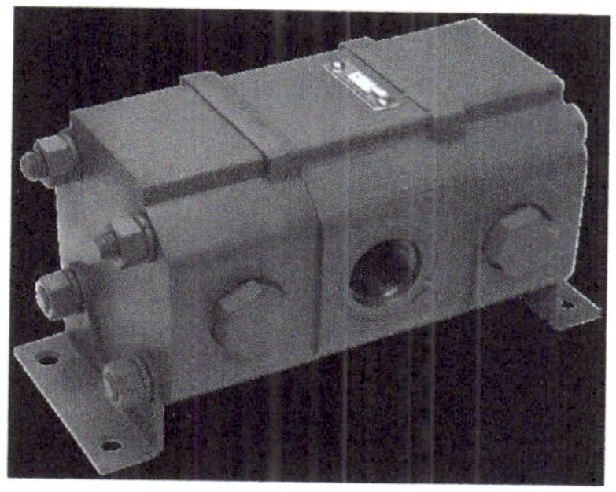

b. Rotary flow divider, external view. Courtesy of Viking.

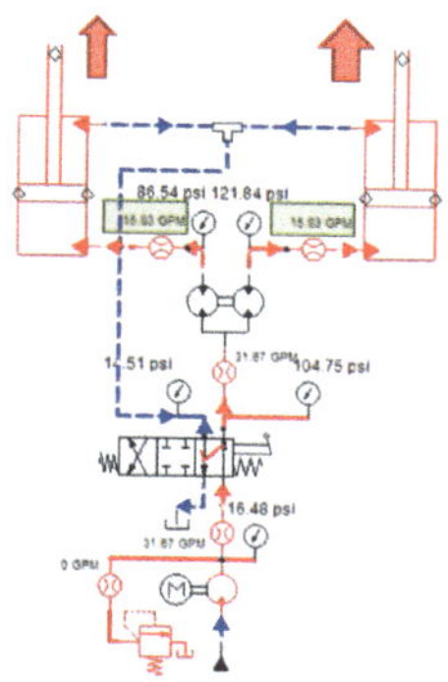

c. Circuit operation at uneven loading (AS v.5.7E).

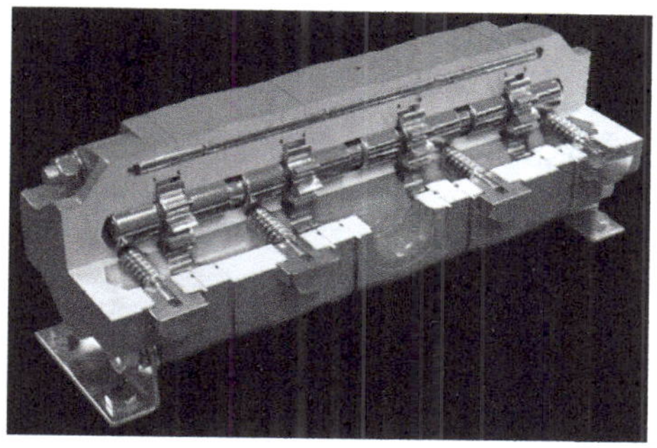

d. Gear flow divider, internal view. Courtesy of Viking.

Figure 8.19 Using a rotary flow divider to feed two identical cylinders in parallel to ensure synchronous extension.

The drawback to rotary flow dividers is they have moving parts. As with any type of machinery, an external gear flow divider will require bearings, seals, and built-in pressure relief valves. A flow divider design that has no moving parts is the humble spool type (see Figure 8-20). A humble spool valve functions as a flow divider by maintaining equal pressure drops through metering orifices situated in two spools that are linked by two "lugs" (Figure 8-20a). The spools are of the cartridge type and are typically inserted in the valve block where the inlet flow is split into two feeding ports (Figure 8-20a). The oil enters the divider through port 1, passing through both spools and out of ports 2 and 3 (see Figure 8-20b). Inside the valve body, the oil passes through the control orifices in the spools, and if the flow is equal, the spools remain in the central position (as shown in the second cross-section in Figure 8-20b). If there is a tendency for more oil to pass through one side than the other due to changes in the outlet pressure, then the pressure drop rises in that side, causing it to drag the other spool across and eventually restricting the outlet of the higher-flow side while keeping the lower-flow side open (third cross-section in Figure 8-20b). As soon as the pressure drop through the control orifice in both spools is equal, the assembly will maintain a metering position, keeping the flow from both outlets equal. Any change in the outlet pressures will cause the spool to move, accommodating the change by metering the oil through the path of least resistance. If the outlet of one port is blocked, all the inlet flow goes to the open port (shown in the last cross-section in Figure 8-20b).

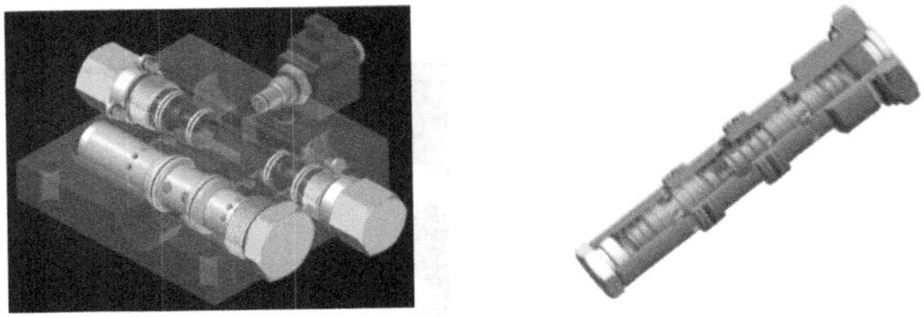

a. Isometric views. Courtesy of Industrial Hydraulics.

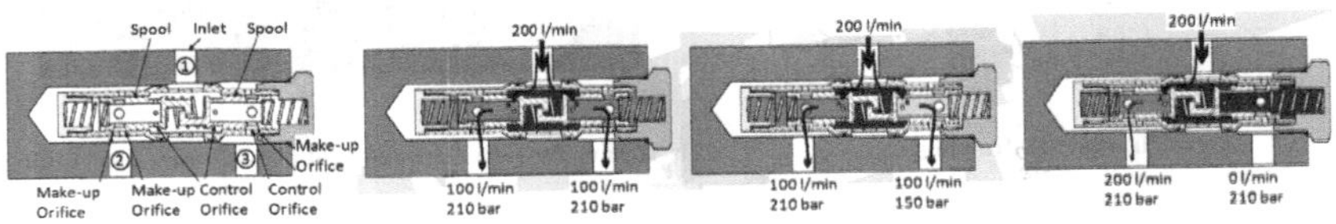

b. Cross-section with basic parts and operation.

Figure 8-20 Humble spool valve flow divider.

Mechanical Linkage

Figure 8-21 presents a third circuit, with hydraulic cylinders connected in parallel just like the two previous arrangements in figures 8-18 and 8-19. In this case, the cylinders are directly fed by flow splitters in both rod-end and back-end ports. The only difference between this new arrangement and the original circuit presented in Figure 8-18 is the mechanical link joining the two cylinders. The mechanical link is strong and rigid enough to sustain any uneven loading on the cylinders to the extreme that one cylinder might take the total load of the arrangement while the other is load free. The mechanical link ensures that both cylinders extend simultaneously, as indicated by the simulation shown in Figure 8-21b. One of the common examples of this application is the design of large excavators. Figure 8-21c presents a large excavator when digging a trench, with a close-up of the two cylinders that drive the excavator boom. In this application, the boom is the strong and rigid mechanical link.

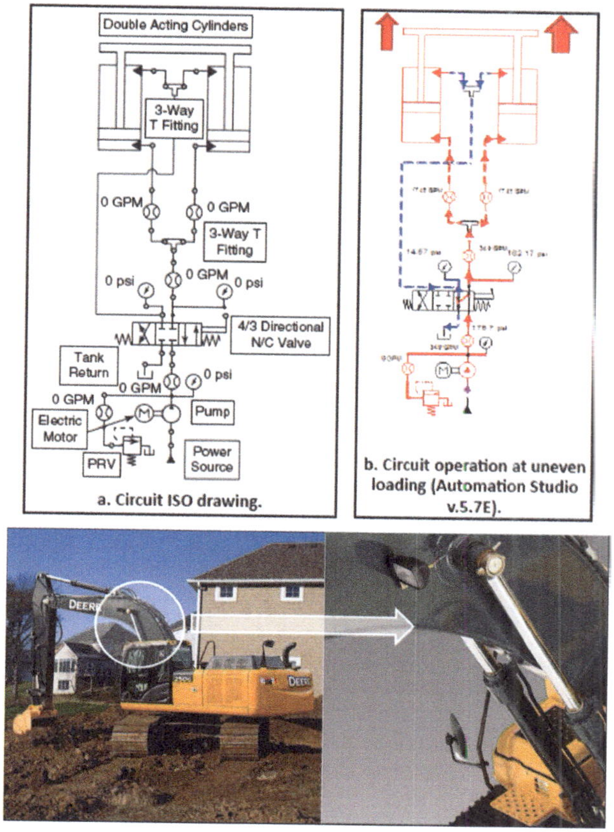

a. Circuit ISO drawing.

b. Circuit operation at uneven loading (Automation Studio v.5.7E).

c. Mechanically linked cylinders in a large excavator. Courtesy of John Deere.

Figure 8-21 Circuits with two identical cylinders in parallel and mechanically linked.

Pneumatic Circuits

Single-Actuator Operation and Control

SINGLE-ACTING CYLINDER: FLOW RESTRICTOR

Figure 8-22 shows the ISO diagram and operation of a single-acting pneumatic cylinder circuit with speed control. The speed control is attained by the addition of a meter-in flow restrictor valve; a pressure relief valve is added to protect the circuit (see Figure 8-22a). It should be noted that the circuit pressure relief valve is not generally needed in a pneumatic circuit because in practice, most systems have a pressurized storage tank that have one valve installed, as required by design standards.

The most significant difference between a pneumatic circuit and its hydraulic counterpart is the effect that the compressibility of the fluid used to transmit the power has in the operation and response of the main components of the system—the actuators in particular. Figures 8-22b–e compare the operating conditions of the circuit when initiating and completing the cylinder extending stroke with two different settings for the flow restrictor.

When initiating the sequence, the cylinder extends at roughly the same speed, flows of 3.1 cubic feet per minute (cfm) versus 2.14 cfm, independently of the flow restrictor opening, 100% open versus 10% open. The air passing through the directional valve is filling both the lines and the actuator at the minimum pressure required to push the load, roughly 70 psi (1,000 kPa). As the cylinder is filled, the extending speed increases (at constant load) with either restrictor setting, but it is significantly larger when the restrictor is fully open. In both cases, the pressure increases to levels that activate the added pressure relief valve, but with the restrictor fully open, the air rushing in completely fills the cylinder—excess flow is less than 1%—and extends it at very high speed when compared to the case with the restrictor open 10% only: 114.7 cfm versus 5.8 cfm. When the restrictor is open only 10%, the excess flow is as high as 145%.

The retracting stroke sequence for the cylinder in the circuit under analysis (Figure 8-22a) is presented in figures 8-22f–h. If the push-button valve remains activated after the cylinder reaches the end of stroke, the pressure increases until the pressure relief valve opens to divert all the flow, as seen in Figure 8-22f. Once the push button is released, the cylinder does not retract immediately as it would occur with a hydraulic cylinder. The compressed air filling the cylinder and lines will expand and will do so for a period of time. The time required for the pressure in the cylinder to fall low enough to allow the compressed return spring to push back the cylinder piston. It should be noted that the flow restrictor does not affect the retracting speed, as the orifice is bypassed through a check valve (one-way flow restrictor).

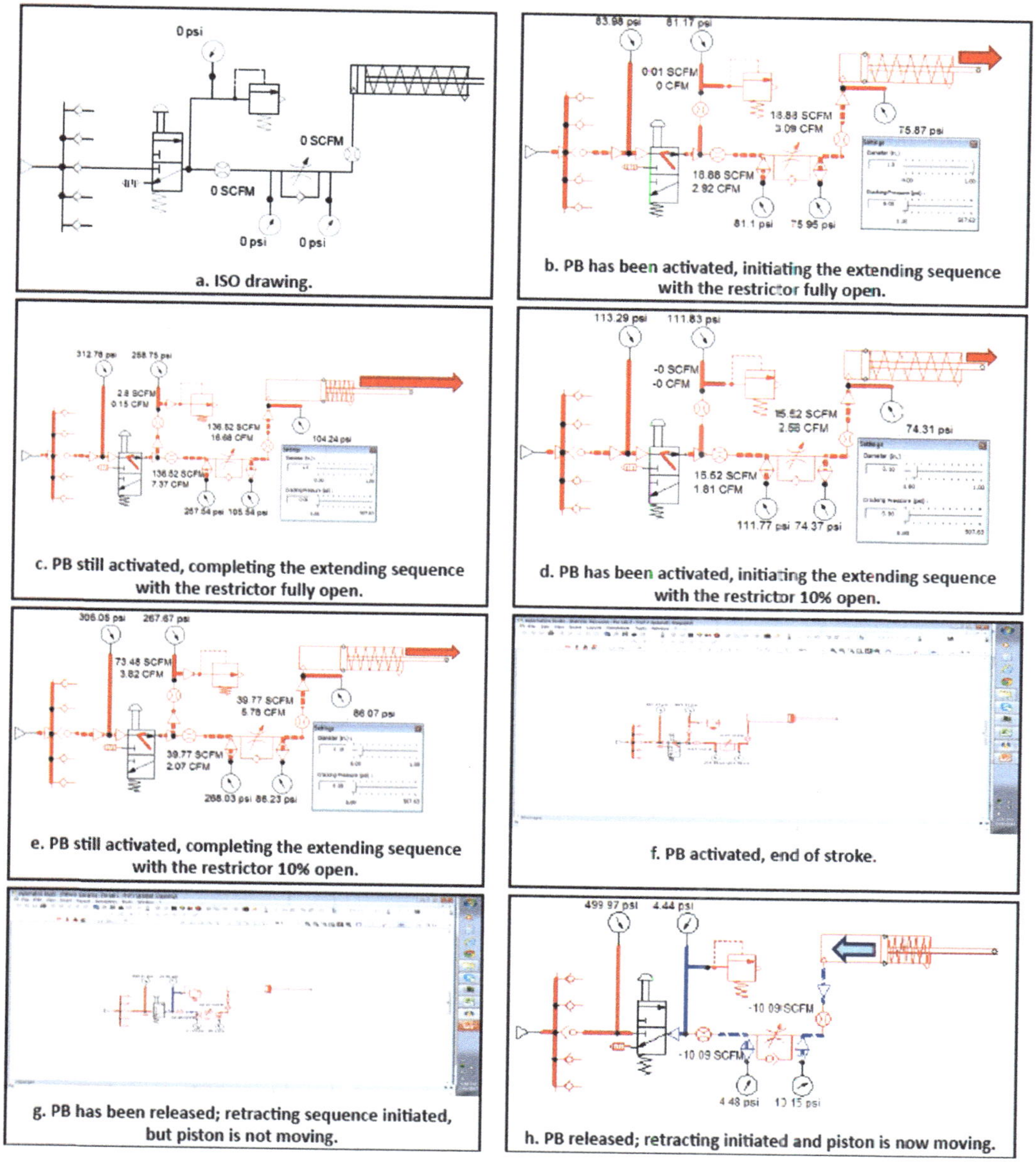

Figure 8-22 Single-acting cylinder pneumatic circuit with a meter-in flow restrictor for speed control (AS v.5.7E).

Double-Acting Cylinders

REMOTE AND SAFE OPERATION: OR LOGIC CONTROL

Figure 8-23 compares two double-acting pneumatic cylinder circuits with logic control. Figure 8-23a presents the ISO diagram of the circuit with an OR logic control strategy. Two normally closed, push button–activated valves can independently be used to extend the cylinder. Figures 8-23b and 8-23c present, respectively, the circuit in the idling condition and during the extending stroke, once either push button is activated and held. This push-button valve arrangement allows the circuit to be activated from separate locations; for instance, a push button might be located near the machinery, whereas the other is located remotely in a control room. In contrast, Figure 8-23d presents the ISO diagram of the same circuit but with an AND logic control strategy. The same normally closed, push-button valves now must be simultaneously activated to extend the cylinder. Figure 8-23e demonstrates the circuit condition when only one PB is activated; the cylinder remains retracted just like the idling condition. It is only when both push-button valves are activated (Figure 8-23d) that the cylinder extends. This push-button valve arrangement ensures safety by requiring the operator to use both hands to activate the circuit.

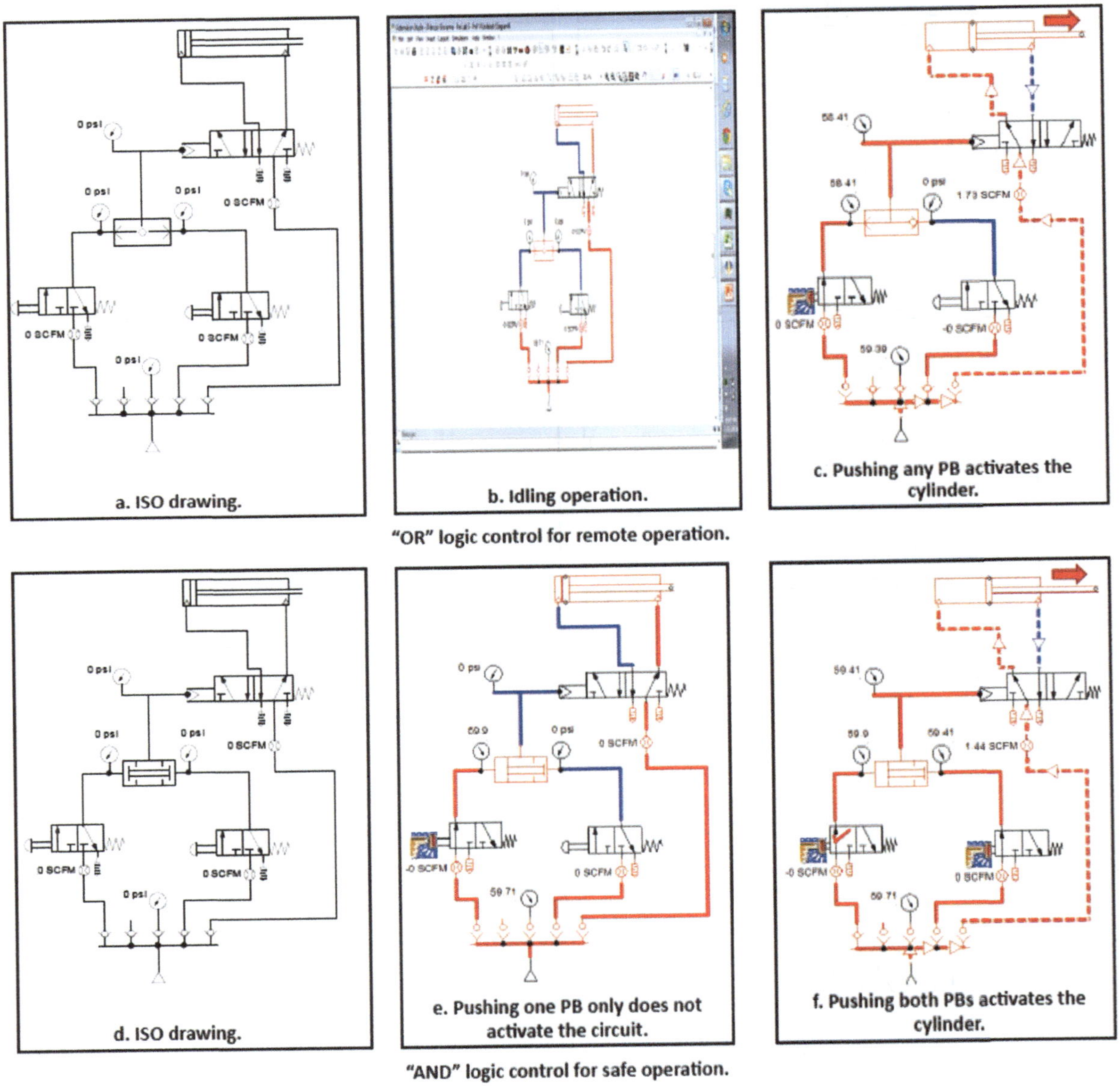

Figure 8-23 Double-acting cylinder pneumatic circuits with logic control for remote and safe operation (AS v.5.7E).

SAFE RECIPROCATION: AND LOGIC CONTROL

In Figure 8-24, we see a double-acting pneumatic cylinder circuit with logic control. Two normally closed, push button–activated valves are installed in series with two other normally closed valves activated by rollers. The rollers are activated by a cam attached to the cylinder rod that senses the beginning and ending of each stroke. The valve arrangements activate the pilots of the bistable valve that powers the DAC (see the ISO diagram in Figure 8-24a). Figure 8-24b shows the cylinder's extending sequence; pushing the MS button activates the bistable to extend the cylinder only if the inner roller valve is activated, indicating that the cylinder is fully retracted and can safely initiate the extending stroke. Figure 8-24c shows the cylinder at the end of the stroke. Once the cylinder is fully extended, it activates the outer roller valve, but the cylinder remains extended because the ME button has not been activated. Once the ME button is pushed, the bistable valve is piloted to retract the cylinder, as the outer roller valve activation indicates the cylinder was fully extended and could safely be retracted (see Figure 8-24d). As discussed in Chapter 7, the arrangement of the two normally closed valves in series constitute a logic strategy in which both valves need to be activated to produce the desired outcome. In this case, the cylinder is extended only when the circuit senses it was fully retracted. Conversely, the cylinder is retracted only when the circuit senses it was fully extended. The valve arrangements ensure the safe and complete extension or retraction of the cylinder in each stroke.

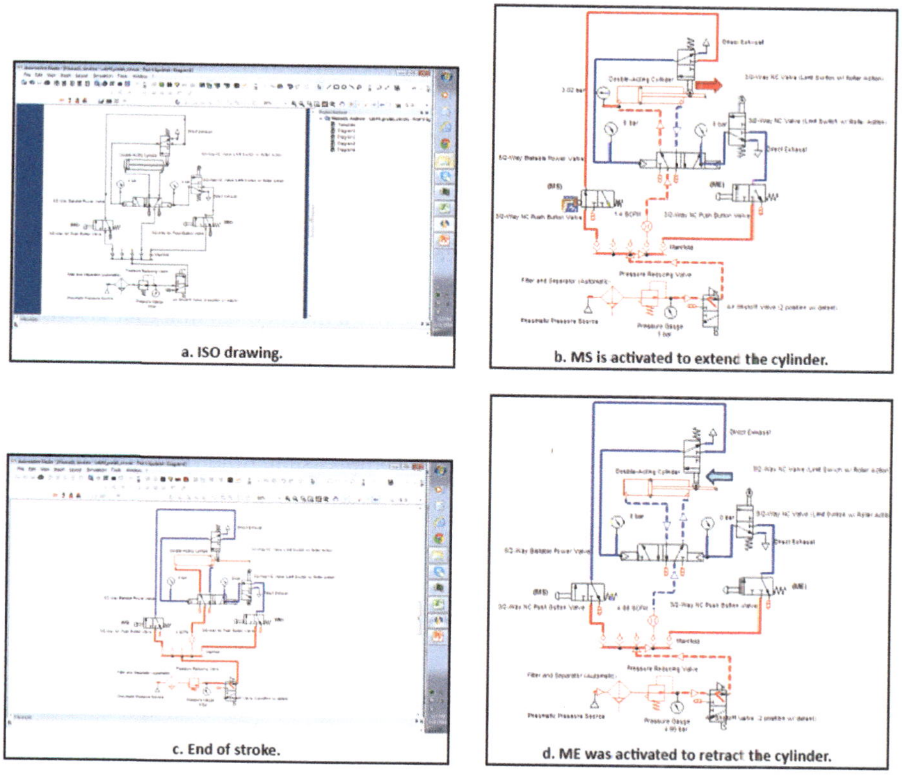

a. ISO drawing.

b. MS is activated to extend the cylinder.

c. End of stroke.

d. ME was activated to retract the cylinder.

Figure 8-24 Double-acting pneumatic cylinder circuit with AND logic control (AS v.5.7E).

Summary

While an SAC can carry load only in one direction, a DAC can exert loads in both opposite directions. After the SAC has carried on its work in one stroke using fluid power, it must return to the original position using other mechanisms—most commonly, a spring. On the other hand, a DAC is driven by the fluid from the circuit in both directions. Depending on the return mechanism, the designer must be careful in selecting an SAC circuit, as it most likely will require larger operating pressures for equal loading when compared with a DAC.

In some cases, it is advisable or even required to be able to adjust an actuator's operating speed. The most common method of regulating the speed of an actuator is using a flow restrictor. In every circuit, the flow restrictor can be located either upstream or downstream from the actuator. The main advantage of regulating the flow upstream—meter-in regulation—is that the actuator will be subjected to lower pressures when compared to regulating the flow downstream—meter-out regulation. One advantage of a meter-out arrangement is guaranteeing that the side of the cylinder returning the fluid back to the reservoir is completely filled with hydraulic fluid.

Using a regenerative circuit to activate a DAC is an alternative to using flow restriction to change or regulate its speed. In a regenerative circuit, fluid from the rod end is directed to the back end of the cylinder, reducing significantly the amount of fluid that the pump in the circuit needs to provide or, conversely, increasing significantly the extending speed. In addition to increasing the extending speed of a DAC, the regenerative circuit simplifies the circuit by reducing the complexity and the number of valves required to operate the cylinder. The major drawback of using a regenerative circuit is dealing with larger operating pressures in the cylinder and the circuit and smaller load-carrying capacity when compared with a regular circuit.

Synchronizing cylinders to operate in parallel requires adding components to the circuit that can guarantee the cylinders extend and retract together. Feeding the cylinder simultaneously, splitting the flow with tees does not guarantee the simultaneous operation. There are no two identical cylinders; hence, any minimal difference in diameter, friction at the seals, installation, and anchoring will cause uneven loading, so the cylinder with slightly less loading will always extend first. The two most common methods to synchronize cylinders are to use flow dividers to feed each cylinder with exactly the same flow or to use a mechanical link that is strong and rigid enough to sustain any uneven loading on the cylinders.

The most significant difference between a pneumatic circuit and its hydraulic counterpart is the effect that the compressibility of the fluid used to transmit the power has in the operation and response of the main components of the system—the actuators, in particular. Using a compressible fluid translates into actuator hesitation, flow variation, lag, and sluggish response, among other characteristics. The main advantage of a pneumatic circuit is the cleanliness that inherently offers and the vast supply of fluid in the case of a system using atmospheric air.

Exercises

1. A DAC is hooked up in a regenerative circuit (Figure 8-10b). The operating pressure in the circuit is 1,000 psi. The piston area is 25 square inches (in^2), and the rod area is 7 in^2. The pump actual flow is 20 gpm. Find the following:
    a. Cylinder speed and load capability in the extending direction
    b. Cylinder speed, and load capability in the retracting direction
2. The lifter shown in Figure 8-25 has two suction cups. Each cup has a 6-in lip outside diameter and a 5-in lip inside diameter. The operating vacuum pressure of the system is 10 psi. How heavy of an object can

be lifted with the vacuum lifter? Hint: Recall the definition of vacuum pressure.

Figure 8-25

3. The hydraulic circuit below shows two DACs connected in series. The rod-end of cylinder 01 supplies the oil to the back end of cylinder 02. Cylinder 01 has an internal diameter of 4 in and a 2-in rod. Related to this circuit, find the following:
    a. The diameter of cylinder 02 if both cylinders need to extend simultaneously (i.e., at the same speed).
    b. With both cylinders extending, the load on each cylinder is 10,000 lbf. If cylinder 02 has a diameter of 3.5 in and a 1.5-in rod, what would the minimum pressure that the pump needs to provide? Hint: Neglect any friction and pressure losses in pipes and valves; assume that the pressure relief valve remains closed and that the fluid returns to an open tank.

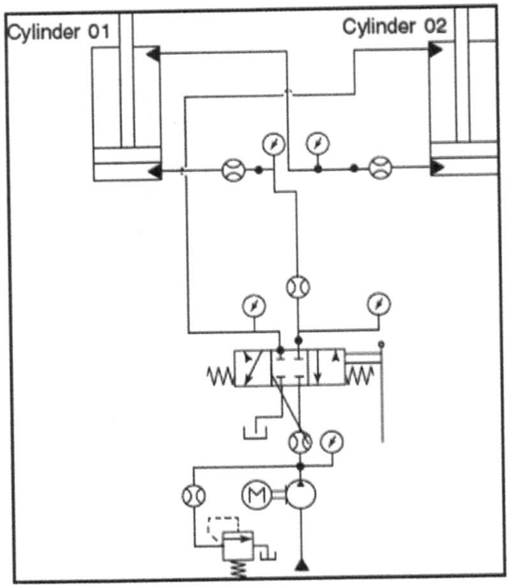

IMG 8.1

4. Label each component and describe the operation of the circuit below, generating a phase diagram.

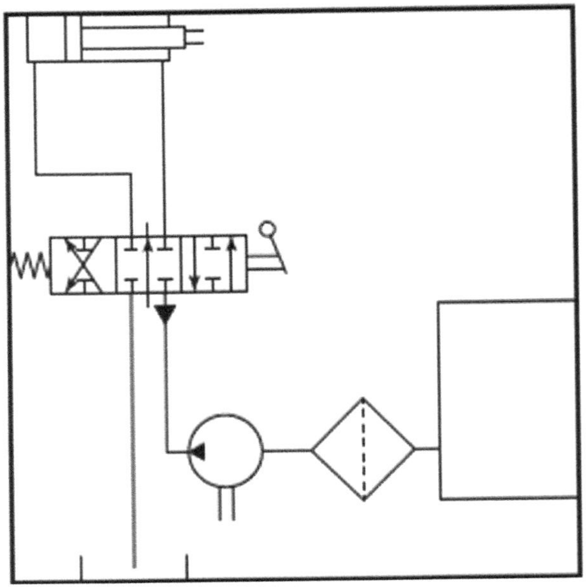

IMG 8.2

5. Label each component and describe the operation of the circuit below using a phase diagram. Explain the sequence of operation, referring to the phase diagram.

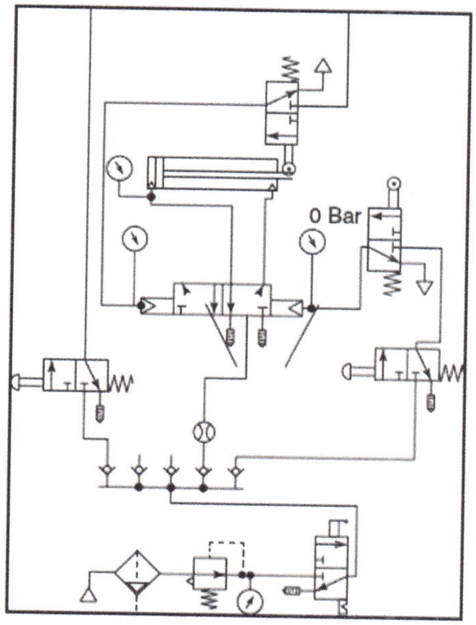

IMG 8.3

6. Describe the operation of the circuit below and generate a phase diagram. To start the sequence, the PB1 push-button is pressed.

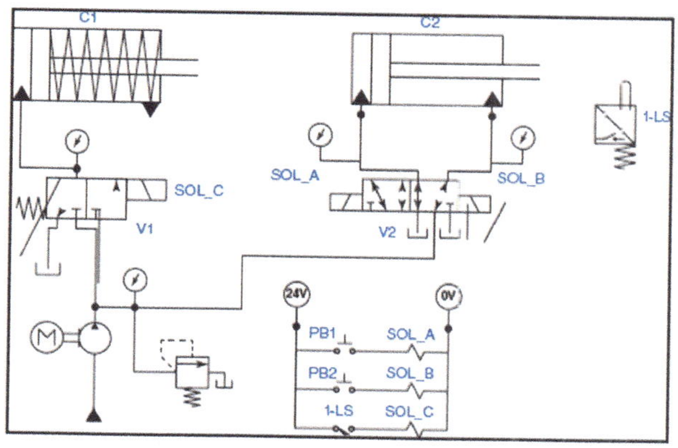

IMG 8.4

7. For the meter-in system shown below, we have the following data: flow control valve capacity coefficient (Cv) = 1.2 gpm/psi 1/2; cylinder piston area = 3.14 in^2; cylinder load = 5,000 lbf; specific gravity of oil = 0.95; pressure relief valve setting = 1,500 psi. Determine the cylinder speed.

   Qfcv = Cv * (deltaP/SG)1/2

8. How heavy of an object can be lifted with a suction cup that has a 6-in lip outside diameter and a 5-in lip inside diameter with a vacuum suction pressure of 10 psi (i.e., a 10-psi vacuum;

remember what a vacuum is)?

9. You are asked by your boss to determine whether a new vacuum cup design for placement on a car windshield will work in place of the traditional hand placement. The windshield area is 4.2 square feet, with a glass density of 2.51 grams per cubic meter with a weight of 30 lbf. The system must have a safety factor of 3. The vacuum cup is designed to have an outer diameter of 5 in and an inner diameter of 4.85 in and is operating at a vacuum of 3 psi. Will this work? Justify your answer by showing calculations. (Atmospheric pressure is 14.7 psi; absolute, 0 psi gage pressure.)

Figure Credits

Fig. 8-1: Tiesse, http://commons.wikimedia.org/wiki/File:Cric_005.jpg, 2009.
Fig. 8-6a: Galen Frysinger, "Erie Canal lock," http://www.galenfrysinger.com/erie_canal_lock__33.htm. Copyright © by Galen Frysinger.
Fig. 8-6b: Galen Frysinger, "Erie Canal lock (close-up)," http://www.galenfrysinger.com/erie_canal_lock__33.htm. Copyright © by Galen Frysinger.
Fig. 8-6: All images Copyright © by Tug 44.
Fig. 8-9: Copyright © by Duk (CC BY-SA 3.0) at http://commons.wikimedia.org/wiki/File:Hydraulic_circuit_directional_control.svg.
Fig. 8-10: Copyright © 2006 by HWH Corporation.
Fig. 8-11: Copyright © 2006 by HWH Corporation.
Fig. 8-13: Copyright © 2006 by HWH Corporation.
Fig. 8-18b: Copyright © by J. McCoy Equipment.
Fig. 8-19b: Viking Pump Inc, "Rotary flow divider, external view.," http://vp.salesmrc.com/pdfs/Sec_341.6.pdf, pp. 1. Copyright © 2010 by IDEX Corporation.
Fig. 8-19d: Viking Pump Inc, "Gear Flow Divider, Internal View," http://vp.salesmrc.com/pdfs/Sec_341.6.pdf, pp. 3. Copyright © 2010 by IDEX Corporation.
Fig. 8-20a1: Eaton, "Diff-Lock Valve Including By-pass Solenoid Sector Valves," http://www.eaton.com/ecm/groups/public/@pub/@eaton/@hyd/documents/content/pll_2269.pdf, pp. 283. Copyright © 2007 by Eaton.
Fig. 8-20a2: Eaton, "2CFD40," http://www.eaton.com/ecm/groups/public/@pub/@eaton/@hyd/documents/content/pll_2269.pdf, pp. 281. Copyright © 2007 by Eaton.
Fig. 8-21c1: Deere & Company, "Mechanically Linked Cylinders in a Large Excavator," https://www.deere.com/en_US/docs/construction/excavators/250g_lc/dkax250290g.pdf. Copyright © by Deere & Company.
Fig. 8-21c2: Deere & Company, "Mechanically Linked Cylinders in a Large Excavator (close-up)," https://www.deere.com/en_US/docs/construction/excavators/250g_lc/dkax250290g.pdf. Copyright © by Deere & Company.
Fig. 8-25: Herbert Reichart, https://commons.wikimedia.org/wiki/File:Little_vakuumlifter.jpg, 2009.
IMG 8-1: Copyright © by Larry Villasmil and Robert Garrick with Automation Studio. Reprinted with permission.
IMG 8-2: Copyright © by Larry Villasmil and Robert Garrick with Automation Studio. Reprinted with permission.
IMG 8-3: Copyright © by Larry Villasmil and Robert Garrick with Automation Studio. Reprinted with permission.
IMG 8-4: Copyright © by Larry Villasmil and Robert Garrick with Automation Studio. Reprinted with permission.

CHAPTER 9A

Fluid Power Actuators: Hydraulic Machinery

Overview and Learning Objectives

Fluid power actuators and machinery are at the heart of any fluid power system. Actuators typically provide mechanical power in the form of linear or rotary motion. An actuator converts the fluid power provided by either a hydraulic or a pneumatic system into the motion required to perform work, from lifting a heavy load to opening or closing a large control valve. Now, a pump or a compressor typically provides the fluid power in the system. In this chapter, we present simultaneously the principle of operation and the most typical designs of actuators and machinery (i.e., hydraulic pumps and motors). The similarity of pumps and motors used in the fluid power industry is conducive to this unique approach, as most of them are of the positive displacement type.

Pumps and motors play the most important role in a hydraulic system. A machine converts one form of energy to another. In fluid power, the main purpose is to convert mechanical energy into hydraulic energy, transmit that fluid energy, deliver it to where it is needed or required, and convert that hydraulic energy back to mechanical energy. The pump is the fluid machine that provides the hydraulic energy to the fluid. The motor, or actuator, is the fluid machine that converts the hydraulic energy provided to the fluid back into effective mechanical work.

Figure 9-1 shows the most basic fluid power machine, a piston within a cylinder that displaces back and forth. Acting as a fluid pump, the piston is mechanically pushed to squeeze the liquid out of the cylinder. Acting as a fluid motor, the piston is pushed by the liquid being forced into the cylinder so that the piston exerts a mechanical force.

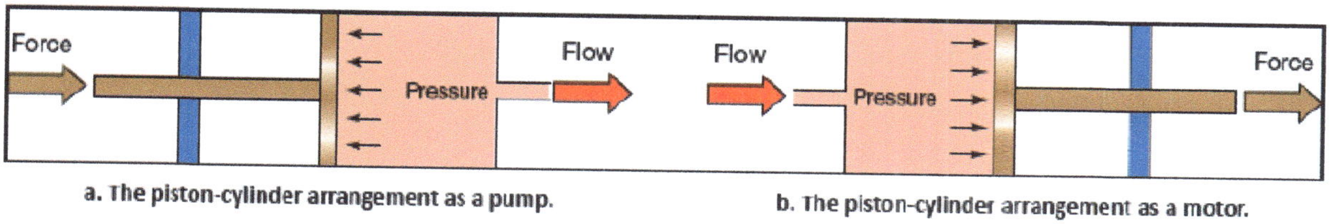

Figure 9-1 A piston-cylinder arrangement, the most basic fluid power machine.

After completing this chapter, students will be able to answer the following questions:

1. What is a positive-displacement machine? How does it differ from a dynamic machine?
2. How does fluid power machinery actually "pump" the fluid or get "motorized" by the pressurized fluid?
3. How can the flow rate delivered or handled by a fluid power machine be estimated (positive-displacement type), including its volumetric efficiency?
4. How do piston and gear fluid power machines operate?

5. How can the pressure-torque-size relationship for fluid power machines be evaluated, including the determination of the mechanical and overall efficiency?
6. What is the power required to drive a hydraulic machine?
7. What is the power extracted from a hydraulic motor or actuator?
8. How can the performance of hydraulic machines be compared?
9. What are the main construction features of hydraulic machines?

General Classification

It is frequently thought that the pump generates the pressure required by a hydraulic system. In reality, the pump provides the flow required by the system. As in any other fluid flow system, the pressure in the hydraulic system is determined by the load on the motors and actuators connected to the system and the losses in the feeding lines. The design of the pump determines the method by which the mechanical energy provided to the pump is converted into hydraulic energy as well as its effectiveness. There are mainly two ways in which a pump provides or delivers fluid to the hydraulic system. One way is to actually "trap" or "take" the fluid from the suction side of the pump and "push" or "squeeze" it into the discharge line. This type of pump is traditionally referred to as a ***positive-displacement pump***. An alternate way of providing fluid to the hydraulic system is to accelerate the fluid by increasing its kinetic energy, which would be subsequently converted to pressure energy. This alternate pumping method is generated by a machine commonly referred to as a ***dynamic pump***.

The construction features of the pumps and motors used in the fluid power industry are relatively similar. As such, the general classification of fluid power machinery is described by concentrating mostly in pumps. Figure 9-2 presents common examples of positive displacement pumps. Figure 9-2a presents the lubricating pump of a Toyota engine. The pump is an internal gear pump in which the gear axes of rotation are offset. The pump draws oil from the crankcase when the gear teeth separate as they rotate; the oil drawn is then trapped between adjacent teeth and pushed out when the gear teeth progressively intermesh. Finally, figures 9-2b and 9-2c present a schematic and a picture of a typical water well pump. In this case, the pump lever is connected to a piston that moves up and down inside a vertical cylinder. When the piston is pushed down by the lever, the water trapped below in the cylinder moves up though the check valve located in the piston. When the piston is lifted by the lever, it pushes out the water but also lifts water from the well to refill the cylinder.

a. An internal gear pump.

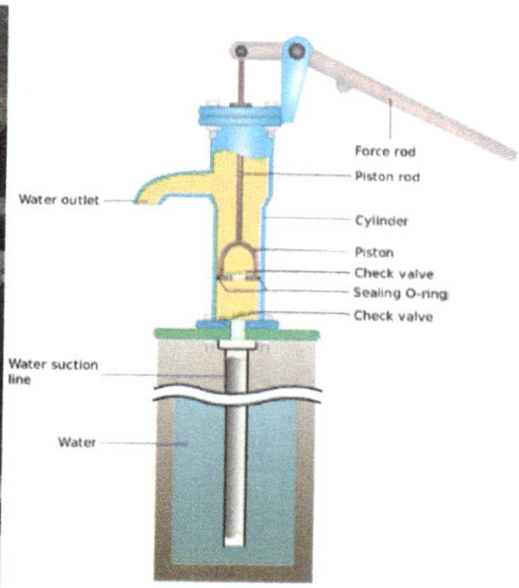

b. Schematic of a water well piston pump.

c. A water well piston pump.

Figure 9-2 Typical examples of everyday positive-displacement pumps.

Figure 9-3 presents common examples of dynamic *pumps*. Figure 9-3a displays a small drain pump installed in a laboratory where there is no sewage piping. The pump is located below the sink so that the water drained is collected in the pump casing, where a rotating impeller accelerates it toward the periphery, where it is expelled toward the pump discharge via a casing volute. In contrast, Figure 9-3b shows a large multistage (multiple impellers operating in series) centrifugal pump installed on an offshore platform. This pump arrangement injects filtered lake water at very high pressure into oil reservoirs. The sectional view of a single-stage, horizontally mounted centrifugal pump can be seen in Figure 9-3c. The cutout in the figure exposes the main component of a dynamic or centrifugal pump, the impeller (101). The impeller works in conjunction with the casing (100) to generate the pressure head required by the hydraulic system. Figure 9-3c also shows the main shaft (122), the support bearings (112, 168), and the mechanical seal chamber (184M). Figure 9-3d shows a schematic of an open section of the casing of a similar single-stage pump. The figure shows the impeller, including the suction eye, where the fluid is drawn and the vanes, including the one that imparts the kinetic energy to the fluid and propels it outward. Figure 9-3d shows that the impeller rotates counterclockwise, and in doing so, it propels the liquid outwardly but in the same direction that will be collected in the volute casing. In the figures, it can be seen that the volute casing is shaped as a spiral with a progressively increasing cross-section, and the liquid will slow down, converting kinetic energy back into pressure.

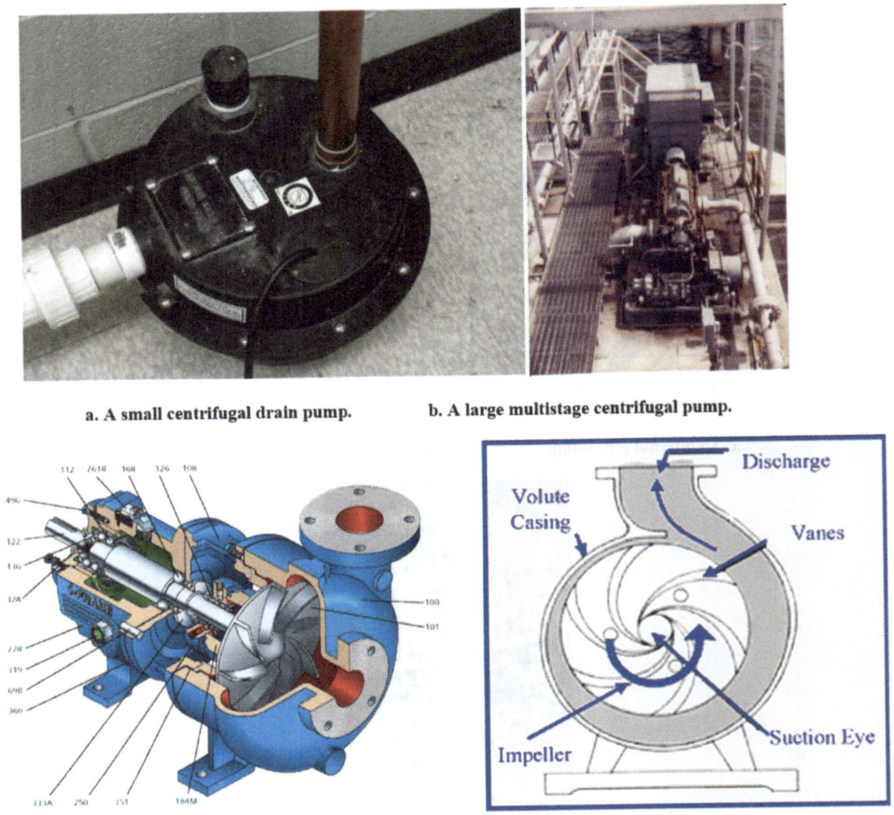

a. A small centrifugal drain pump.

b. A large multistage centrifugal pump.

c. Sectional view of a single-stage centrifugal pump.

d. Main parts of a single-stage pump.

Figure 9-3 Typical examples of everyday dynamic pumps.

Both positive-displacement and dynamic pumps are typically classified by their design construction. Figure 9-4 presents a simplified classification tree for positive-displacement pumps. The primary distinction among designs is the movement of the pumping element. In a reciprocating pump or motor, a piston, plunger, or cup moves back and forth alternately. In a rotary machine, pumping elements of various types rotate continuously. Similarly, Figure 9-4b shows a simplified classification tree for dynamic (centrifugal) pumps. The primary distinction among designs is the location of the impeller relative to the shaft support. Medium- to small-process pumps have typically an overhung impeller, whereas large and high-pressure pumps have one or multiple impellers located between bearings. Although most centrifugal pumps are of horizontal shaft design, vertical pumps are common in applications where the actual pump installation is above the liquid level, such as in water wells, or when the pump suction pressure is relatively low, such as in cryogenic applications.

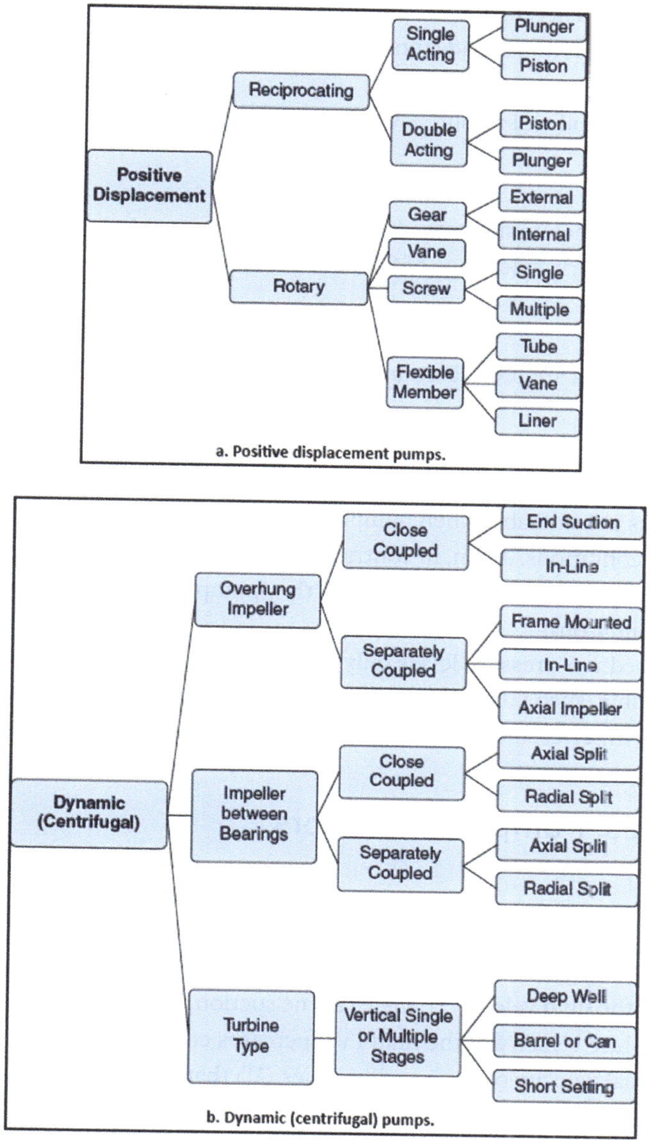

Figure 9-4 Pump classification trees.

Most, if not all, fluid power systems require providing pressurized fluid at a variable rate. A positive-displacement pump delivers a fixed amount of fluid when operating at constant speed. In these pumps, the rotor has or creates very tight chambers that alternately connect to the suction, where its volume of fluid is drawn, trapped, and moved toward the discharge. Because the chambers are separated from each other, there is no connection or possibility of a clear passage from the outlet to the inlet. This characteristic is the most desirable feature in a fluid power machine. A positive-displacement pump is able to provide the required volumetric flow independently of the pressure in the hydraulic system. In contrast, dynamic pumps have open passages between the discharge and the suction, so they cannot provide constant volumetric flow; most importantly, they have inherent limits in the pressure they can overcome. Therefore, fluid power systems are exclusively fitted with positive-displacement machines. These pumps have the following advantages over their dynamic counterparts:

- Very high-pressure capability. Pressure levels up to 12,000 pounds per square inch (psi; 83 gigapascals) are not uncommon.
- Relatively small size.
- High overall efficiency maintained over the design range of operating conditions.
- Ability to operate in a very wide range of flow and pressure conditions, therefore maintaining high overall efficiency.

In contrast, dynamic pumps are preferable and used in processes that:

- Require a very high flow when compared with positive-displacement pumps.
- Handle low-viscosity fluids. In contrast with positive-displacement pumps, performance in dynamic pumps deteriorates considerably when pumping viscous fluids.
- Operate at steady conditions. Again, in contrast with positive-displacement pumps, performance in dynamic pumps deteriorates considerably when they pump at a flow rate that deviates from the impeller design conditions.
- Demand low- to medium-pressure levels only. Although pressure levels up to 5,000 psi (34 megapascals [MPa]) are achievable, such pumps require very high rotational speed and/or an arrangement of multiple impellers in series, increasing the size, complexity, and costs.

Hydraulic Machines: Pumps and Motors

Pumping and Motoring Theory

A dynamic machine imparts energy to the fluid by accelerating it and increasing its kinetic energy. Figures 9-5a and 9-5b present schematics of the fluid velocity triangles at the suction and discharge of a single impeller of a typical centrifugal pump. The absolute velocity of the fluid (V) increases continuously from inlet to discharge, primarily by the action of the tangential velocity of the impeller vanes (U) that increases linearly with the radial location on the impeller. This increase in velocity—indeed, kinetic energy—is converted to pressure in the pump volute. In a positive-displacement machine, the pumping element draws fluid from the inlet, traps the fluid when the machine is completely filled, and then delivers the fixed amount of fluid toward the discharge. As emphasized earlier, there is no connection or the possibility of a clear passage between the outlet and inlet in a positive-displacement pump. Figure 9-5c shows the schematic of a plunger pump. The plunger moves back and forth alternately inside a cylinder that has inlet and suction check valves. When the plunger moves back, it creates a reduction in pressure within the cylinder. Once the pressure inside the cylinder falls below the prevailing pressure in the inlet, the inlet check valve opens, allowing the fluid being in the inlet to fill the cylinder. Once the plunger is fully retracted, the cylinder is completely filled, and the suction check valve closes. Then, the plunger starts to move forward, pushing the fluid trapped in the cylinder, increasing its pressure up to the value prevalent on the discharge side, and opening the check valve. The retraction and extension of the plunger occurs continuously, hence creating what is typical of reciprocating positive-displacement pumps: pulsating flow. A plunger can only either draw or deliver fluid at once; therefore, a single-plunger pump will draw or deliver fluid only half the time.

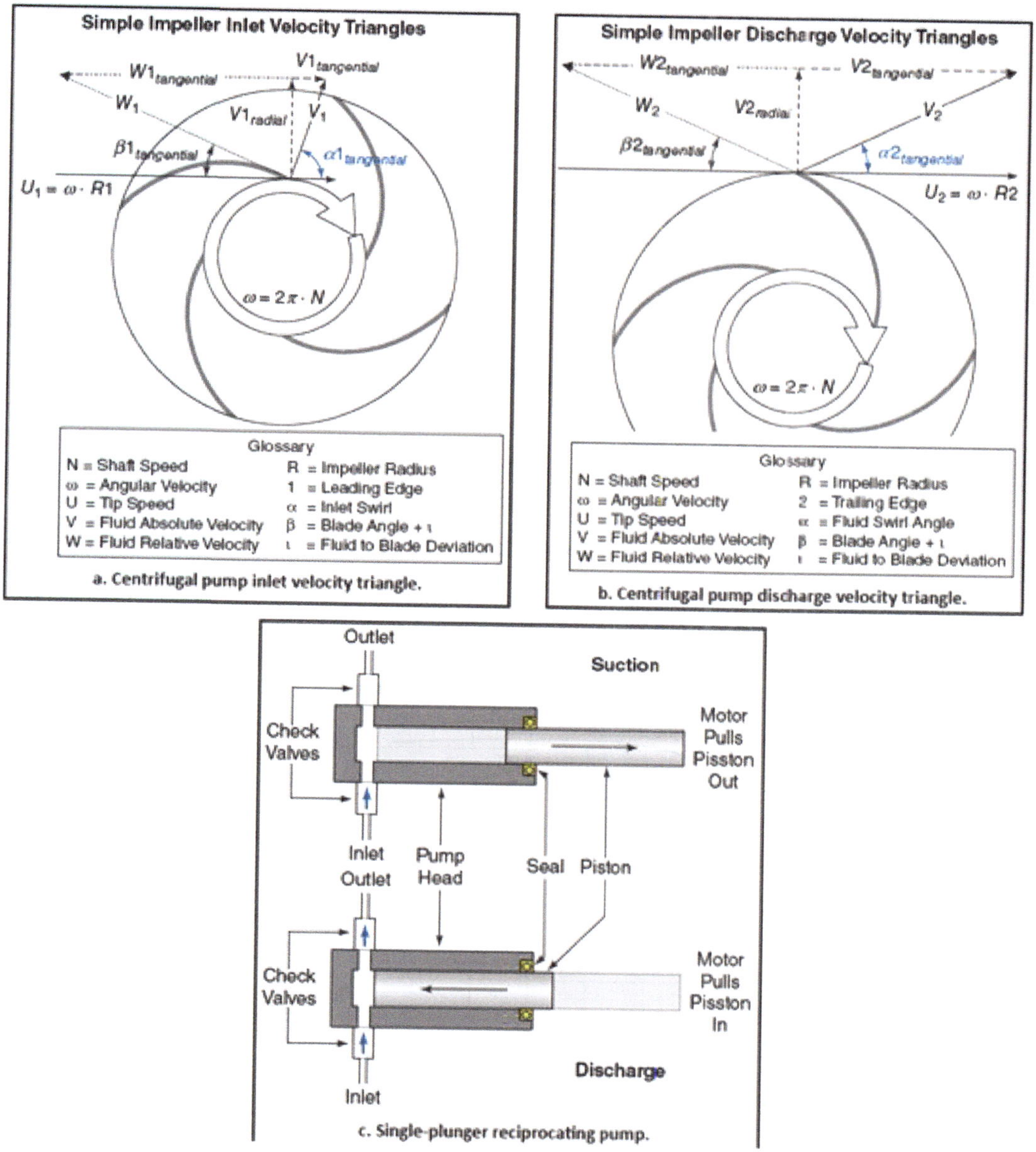

a. Centrifugal pump inlet velocity triangle.

b. Centrifugal pump discharge velocity triangle.

c. Single-plunger reciprocating pump.

Figure 9-5 Principle of operation of centrifugal and reciprocating pumps.

Another way to contrast the difference in the way these two kinds of pumps is comparing typical performance curves. Figure 9-6a displays a typical performance curve of a single-stage overhung impeller centrifugal pump operating at constant speed, 3,560 revolutions per minute (rpm). In Figure 9-6a, the manufacturer includes the efficiency map (purple lines), power curves (brown lines), net-positive-suction-head-required curves (red lines), and pressure head (black lines) versus the flow rate of the same impeller design cut at four different outlet diameters. The pressure-head-versus-flow-rate curves represent the maximum head, or the pressure that the pump can generate when delivering a particular flow in any of the four impeller diameters, as shown in Figure 9-6a. As shown in the figure, this pump—and, typically, most dynamic pumps—generates the maximum head or pressure at zero flow.

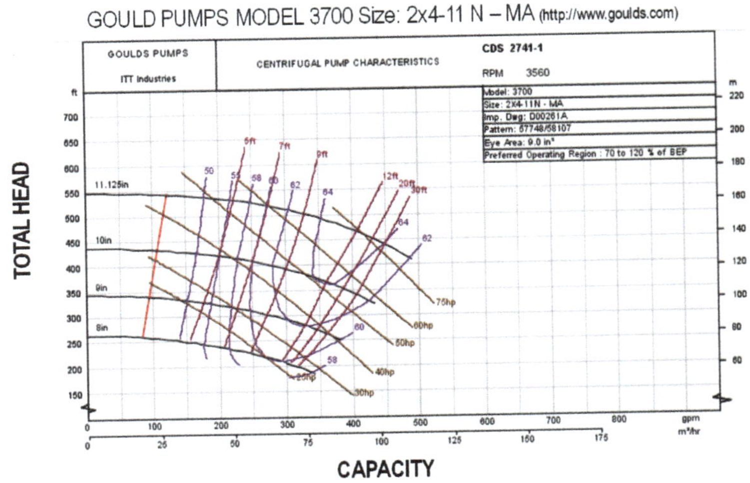

a. Single-stage centrifugal pump.

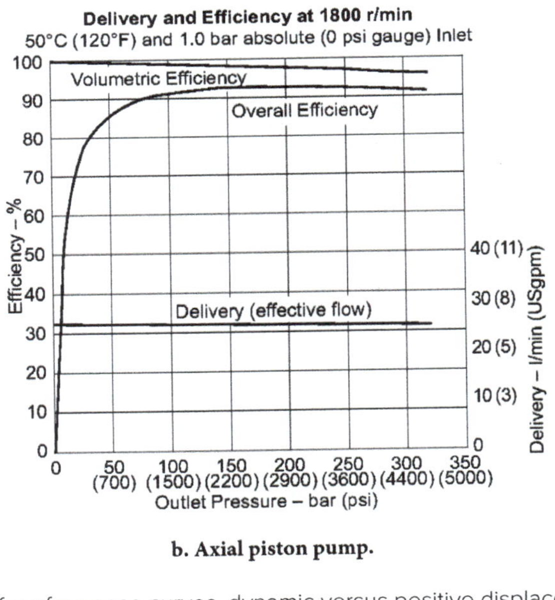

b. Axial piston pump.

Figure 9-6 Comparison of performance curves, dynamic versus positive displacement pumps.

Most manufacturers also include the preferred operating flow rate range in terms of the flow at the best efficiency point, as shown in Figure 9-6a. As was just discussed for Figure 9-6a, there is a lot of information on a performance curve of a dynamic pump. In contrast, positive-displacement pumps have more straightforward performance curves (see Figure 9-6b). Operating at constant speed, any positive-displacement pump delivers a mostly constant flow rate. In Figure 9-6b, the actual or effective flow being delivered by the pump appears as a horizontal line. The flow rate decreases only very slightly as the discharge pressure is increased. The flow reduction shown in the pump performance curve of Figure 9-6b is equivalent to only 4% of the theoretical flow when the pump operates at nearly the maximum pressure of 4,400 psi (30.4 MPa) and the maximum rotational speed of 1,800 rpm. For practical purposes, this variation of 4% in the flow delivered by a positive displacement in the design pressure range is negligible when compared with that of the centrifugal pump—120%—when both operate at constant speed. Figure 9-6b also shows the curves of volumetric and overall efficiency. Volumetric efficiency is always high at any pressure. Overall efficiency is also high except at low pressures.

Positive-Displacement Machines (Pumps and Motors)

In this section, we focus on estimating the volume displaced or taken by a positive-displacement machine pump. The size and type of pump or motor determine the volume that each can trap or displace in a particular period of time. The great majority of pump and motor manufacturers express this volume per cycle of operation. In most machines, a cycle is one rotation or a single revolution of the machine's rotor. Consequently, in the following subsections, we proceed to calculate the theoretical volume that a positive-displacement machine traps or displaces per one revolution of the pumping elements for the most popular machine designs.

Piston Type

Due to the high-pressure requirements, most piston machines used in the fluid power industry are the **plunger type**. For clarification, we must mention that all manufacturers in this industry refer to this design as a piston type. Nevertheless, it is important to note that where large-positive displacement machines are used (the oil industry is an example), there exists a clear distinction between plunger and piston pumps. A plunger machine can only be single action, but a piston machine can be dual action, as both chambers within the cylinder, in front and behind, can be used to trap or displace fluid (see *Figure 9-7*). As plunger design pumps are referred to as piston pumps in the fluid power industry, such nomenclature will be used throughout the rest of the book. Typical single (action) piston pumps are shown in Figure 9-8.

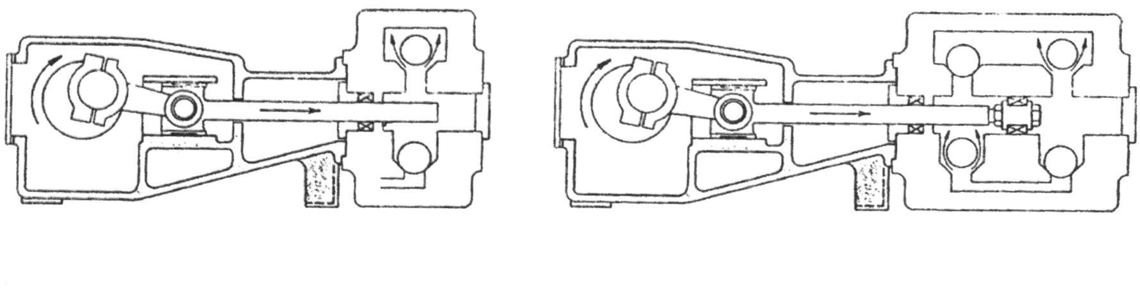

a. Single-action design plunger-type pump. **b. Dual-action design piston-type pump.**

Figure 9-7 Plunger versus piston pump design (using in the oil, food, and other general process industries).

To estimate the volume displaced by a piston pump, the analysis turns to the schematics shown in Figure 9-8. They represent typical piston machines, a pump like the ones described in Figure 9-9a, or an actuator such as a hydraulic cylinder (see Figure 9-8b).

Once the pump has been filled with fluid, a force is required to push forward the piston and overcome the pressure prevailing on the discharge side of the pump. As we discussed in Chapter 3, this force must be equal to:

$$Force = pressure \times Area$$

The required area is the one displaced by the piston, although technically it is the cross-section of the cylinder, as the piston itself is slightly smaller than the cylinder (there is a sealed clearance between the piston and the cylinder walls to allow free movement of the piston). The area of the piston *AP* is estimated as:

$$A_p = \frac{\pi}{4} D_p^2$$

where D_P is the diameter of the piston. Then, for any piston with a given diameter, the volume displaced by the piston would be directly proportional to the distance traveled by it. The maximum distance that the piston can travel is the cylinder stroke *S*. Therefore, the volume displaced is estimated as:

$$V_D = A_P \times S$$

The units of *VD* would be:

$$V_D = in^2 \times in = in^3$$

$$V_D = cm^2 \times cm = cm^3$$

So, it is clear that *VD* is measured in units of volume. Now, in checking typical manufacturer catalogs of hydraulic pumps and motors, one will find that all manufacturers express *VD* in the following units:

$$V_D = {in^3}/{rev} or V_D = {cm^3}/{rev}$$

How is this apparent discrepancy reconciled? Simply recall that the piston of a single-action pump extends, displacing fluid only once per one cycle or a single revolution of the driving mechanism if this mechanism rotates rather than reciprocates. It is thus implied that the distance traveled by the piston during a pumping extending cycle is the stroke, *S*.

The next step is to determine the volumetric flow delivered by a single-action pump. Recalling the principle of conservation of mass or the continuity equation from Chapter 5, we would have:

$$Q = V \times A$$

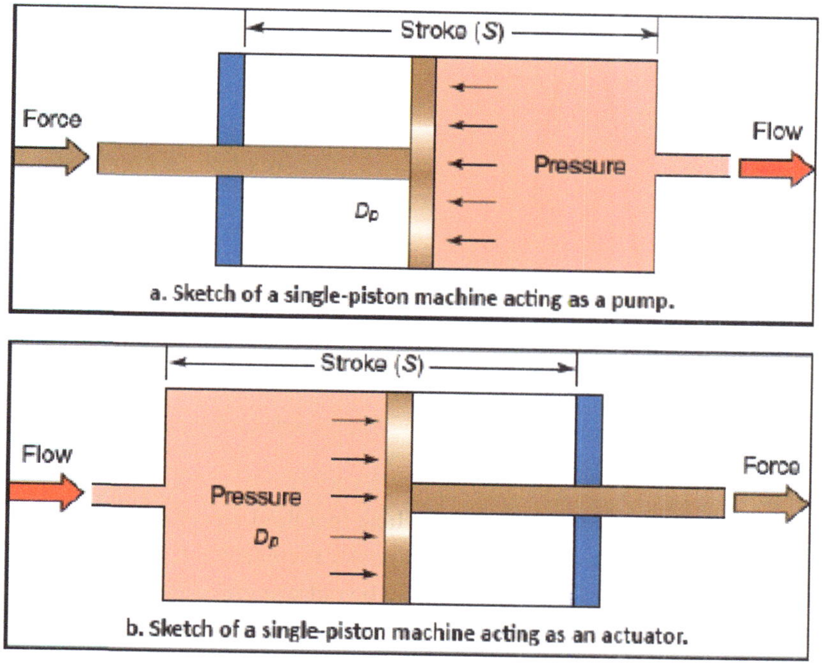

Figure 9-8 Single (action) piston machine schematics

V represents the velocity of the piston, and *A* represents its cross-sectional area. If one is interested in the average flow rate delivered by the pump during the extension, the piston velocity value should be the average speed, i.e.:

$$V = \frac{S}{t_{ext}}$$

In this equation, *t*ext can be interpreted as the period of time that the pump takes to complete a pumping cycle. Nevertheless, if one is interested in the actual flow that the pump delivers to the system, the **instantaneous piston velocity** is needed. The pump's instantaneous piston velocity is determined by the action mechanism. Most hydraulic machines are driven by a rotating shaft, so additional mechanisms are required to convert this rotary or circular motion into reciprocating or linear motion. In these cases, the instantaneous velocity of the piston is typically a sinusoidal function. In a single-action pump with one piston, the flow rate will be maximum midway through the stroke and zero at the beginning and end of the stroke. The flow rate over time of a single-action pump with one piston operating at constant speed looks like the plot in Figure 9-9.

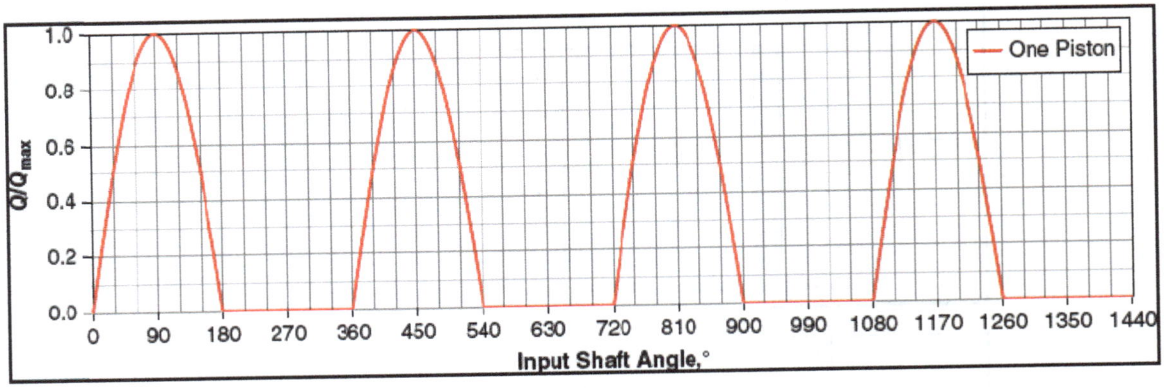

Figure 9-9 Normalized instantaneous flow rate of a single-action, single-piston pump.

Now, thinking of yourself as a user of a piston machine in a fluid power system, what would be the consequences of installing a single-piston machine in the fluid power system? First, in a pump, the flow delivered will be intermittent, and in a motor, the output torque will also be pulsating. Second, an intermittent flow will generate pressure fluctuations in supply lines and hoses leading to potential premature failures, not mentioning jerky operation and noise. How did manufacturers handle this problem? They simply arranged asynchronously multiple pistons in parallel. The intermittent flows of the pistons overlap, so the magnitude of the flow pulsations is considerably reduced although not eliminated in its entirety (see Figure 9-10). The most significant improvement is that any multiple-piston pump or motor will provide or take fluid continuously, in contrast with a single-piston machine, in which there is no flow at all half of the time. It is clear that the flow variation reduces with the number of pistons. In ideal conditions, the flow fluctuation in a three-piston pump is 13.4%, whereas in the most common piston pump design—eight pistons—the fluctuation reduces to 7.6%.

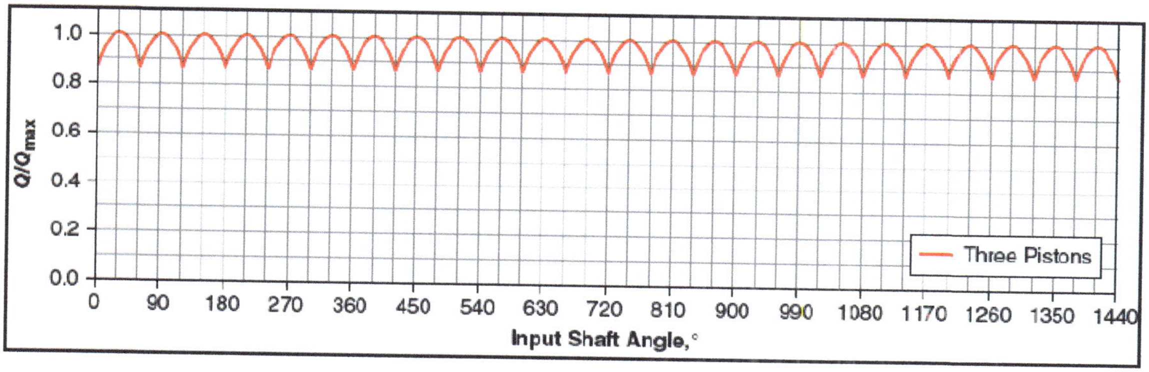

a. Triple-piston pump.

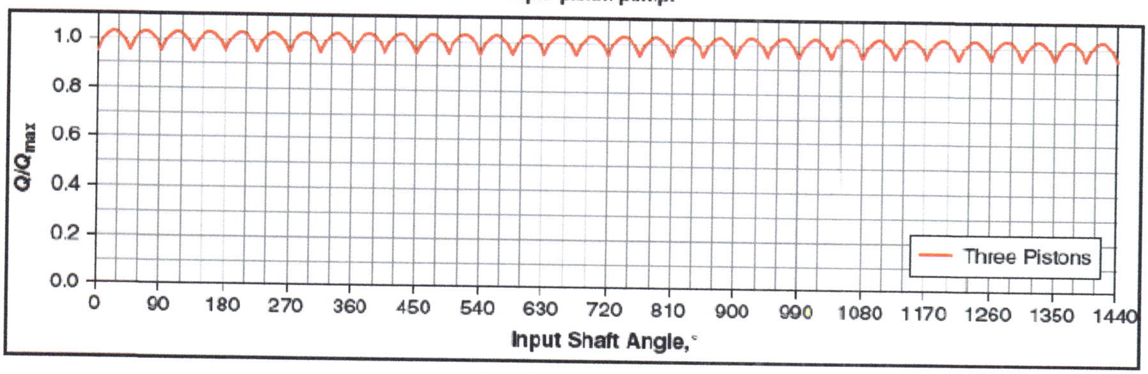

b. Eight-piston pump.

Figure 9-10 Normalized instantaneous flow rate of a single-action, multiple-piston pump.

IDEAL FLOW RATE IN A PISTON-TYPE MACHINE

To estimate the pump average flow rate, we can also consider the period of a complete pumping cycle. A single-piston pump pushes fluid out during the extending stroke but takes in fluid during the retracting stroke without pumping out any fluid. As most fluid power machines are driven by rotating machinery, the extending-retracting cycle corresponds to the time period that the input shaft takes to complete one revolution. The average velocity is related to this time as follows:

$$V = \frac{S}{t_{cycle}}$$

with *tycle* given by:

$$t_{cycle} = \frac{1}{N}$$

where N is the frequency or rotational speed at which the machine operates. The average linear piston velocity becomes:

$$V = \frac{S}{t_{cycle}} = \frac{S}{\frac{1}{N}}$$

$$V = S \times N$$

If we denote NP as the number of pistons, upon substitution, the ideal flow rate of a piston machine operating at constant rotational speed is given by:

$$Q = V \times A = S \times N \times N_P \times A_P$$

Upon rearrangement:

$$Q = N_P \times A_P \times S \times N$$

$$Q = V_D \times N$$

with:

$$V_D = N_P \times A_P \times S$$

where VD represents the volume displaced by all pistons of the machine in one revolution. This term is typically called ***volumetric displacement***. This (ideal flow rate) equation indicates that the average volumetric flow provided by the pump is directly proportional to the size of the pump and its operating speed.

As a designer of fluid power equipment and machinery, one should consider that fluid power systems are mostly used in mobile and construction equipment, such as excavators and cranes. Loading and flow demands by this type of equipment is extremely variable. What would be the approach to supply or take flow with a piston machine that operates continuously in such a system? Looking at the flow equation, let's evaluate the possibilities. One option would be varying the rotational speed *N*. Such an approach requires either variable speed drivers and speed controls. If the speed is fixed, flow control valves are needed to divert the excess flow leading to inefficiency and frequent stops and goes with waste of power. An alternative could be modifying the pump (or motor) itself, altering the volumetric displacement (V_D). The cylinder bores and pistons are physical components of the machine; their diameters are fixed. Therefore, the area displaced by the pistons is already set, defined by the size of the machine. Similarly, the number of pistons—also a physical characteristic of the machine—is already established, too. What about the stroke? Is the distance traveled by the piston within each cylinder during each cycle fixed? Certainly, the stroke is the maximum distance that the piston can move within each cylinder, and indeed it is determined by the physical length of the cylinder block. Nevertheless, what if the stroke of each piston could be adjusted by reducing the distance they travel on each revolution? The possibility of modifying the stroke of each piston would create the flexibility of adjusting the pumped or admitted flow, even if the machine is operating at constant speed.

IDEAL FLOW RATE IN A BENT-AXIS PISTON-TYPE MACHINE

In a bent-axis machine, the cylinder block (and machined bores) containing the cylinders is tilted at a certain angle relative to the input shaft. It is this tilt that converts the rotary motion of the shaft into the reciprocating motion needed to retract and extend each piston. Figure 9-11 depicts two fixed-displacement bent-axis machines: a pump and a motor. Figure 9-11a shows a Rexroth AA2FO axial-piston fixed pump for use in mobile and stationary applications for hydrostatic drives. Figure 9-11b shows the arrangement and major components of a Vickers SE104D bent-axis motor. Like the AA2FO pump, the Vickers SE104D bent-axis motor drive shaft is displaced by an angle or bent relative to the cylinder block and piston's center line. The basic operation of a bent-axis hydraulic motor is simply the reverse of a bent-axis hydraulic pump. In the motor, the hydraulic fluid enters the machine through a valve block that can feed either port, inlet, or outlet by commuting the flow in a nonrotating valve plate. From the inlet port, the fluid pressurizes the cylinder bores until the resulting forces from the piston rods overcome the resisting torque of the load driven by the motor shaft. The internals of a piston-type hydraulic motor are shown in Figure 9-11b.

a. Pump.

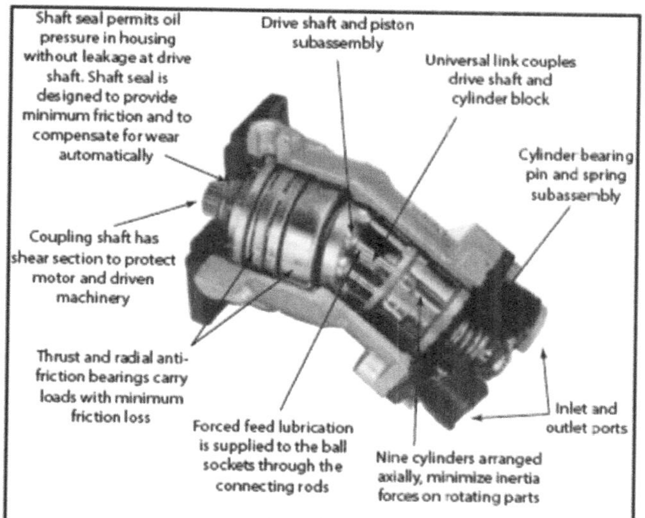

b. Motor internals.

Figure 9-11 Fixed-displacement bent-axis piston machines.

A sketch of the disk drive shaft, piston subassembly, and cylinder block is presented in Figure 9-12a. The sketch presents one single piston at the top of the disk rotation, with piston fully retracted, and at the bottom of the disk rotation, with the piston fully extended. The overlaid triangle is shown enlarged in Figure 9-12b. Using basic geometry and corresponding and alternate interior angles, it can be seen that the maximum distance traveled by the piston or stroke S is related to the diameter of the circle containing the bores in the cylinder DB by the same cylinder block to input shaft angle θ or bent-axis angle. Therefore, the stroke of each piston can be related to the bent-axis angle using basic trigonometry:

$$S = D_B \times \tan\theta$$

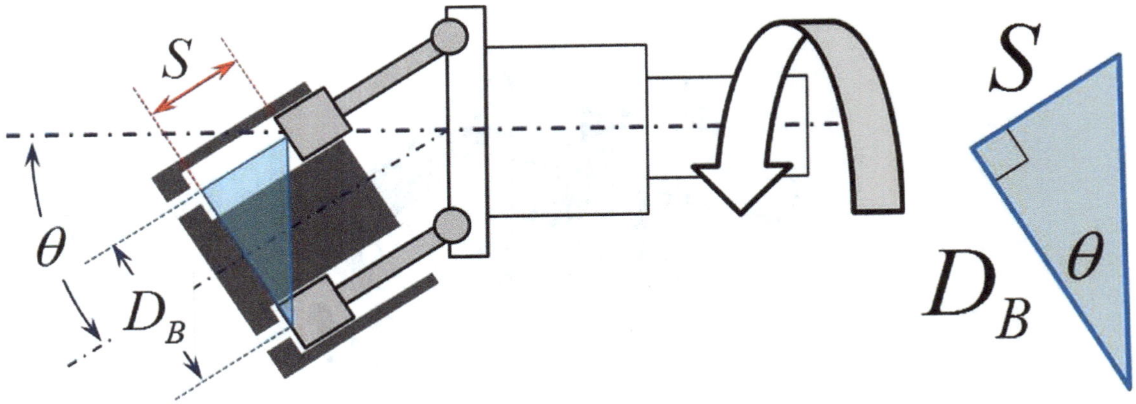

a. Drive shaft and cylinder/piston assembly. **b. Stroke and bent-axis angle.**

Figure 9-12 Volumetric displacement analysis of a fixed-displacement bent-axis piston machine.

The volume displaced by a single piston can be expressed as:

$$V_P = A_P \times D_B \times \tan\theta$$

The total volume displaced by the pump would be given by:

$$V_D = N_P \times A_P \times D_B \times \tan\theta$$

and the volumetric flow rate:

$$Q = V_D \times N$$

where:

- Θ = bent-axis angle,
- DB = cylinder bore circle diameter,
- S = piston stroke,
- AP = cross-sectional area of the pistons' bore, and
- w = width of the slotted rotor and vanes.

Figures 9-13a and 9-13b present a Parker V14 large-frame, variable-displacement, bent-axis motor. In a variable-displacement machine, the volume of fluid displaced by each piston in a pump or the volume required to extend each piston in a motor is modified by adjusting the stroke. Figure 9-13c shows the relationship between the piston stroke and the bent-axis angle. When comparing the machines in Figures 9-12 and 9-13, it can be concluded that a piston pump and a piston motor are similar.

a. Motor, external view.

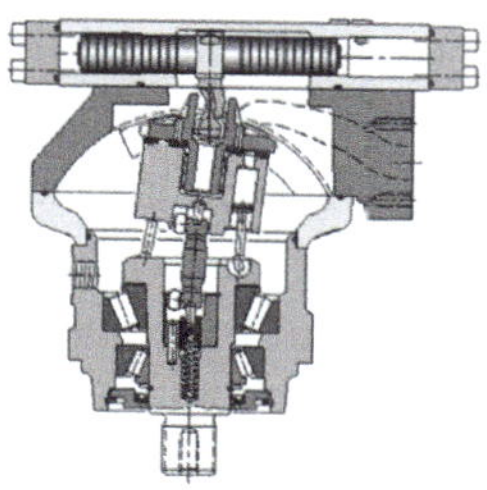

b. Motor cutout.

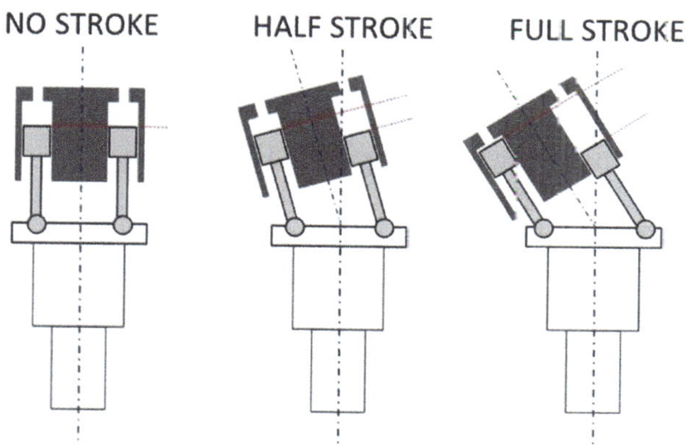

c. Varying the stroke in a bent-axis machine.

Figure 9-13 Variable displacement bent-axis piston machine.

AXIAL-PISTON OR SWASH PLATE–TYPE MACHINE

In an **axial-piston** design machine, the cylinder block axis is parallel to the input shaft center line. In these machines, the angle required to convert rotary motion into linear motion is provided by a shoe plate that follows a tilted swash plate. The swash plate can be either fixed or movable. A fixed or stationary swash plate creates a constant stroke on each piston (e.g., a fixed volumetric displacement machine). A movable or adjustable swash plate creates a variable stroke on each piston (e.g., a variable volumetric displacement machine). Although swash plates can feasibly act as motors, most manufacturers design and sell swash plate machines to be operated as pumps. The volumetric displacement and volumetric flow rate of an axial piston machine can also be calculated with the formulas developed previously for bent-axis piston pumps.

Gear Type

Machines of the gear type are the most widely used in general hydraulic systems. When compared with other designs, gear-type machines are relatively simple in construction, reliable in operation, and cost-effective. The trade-off is a slight reduction in the overall efficiency, particularly in high-pressure applications. Typically, a gear machine, either a pump or a motor, requires the meshing of two geared rotors.

An **external gear** machine essentially consists of a pair of meshing gears supported in parallel bearing blocks or bushings. Figures 9-14a and 9-14b compare side by side an exploded view of a Rexroth F Series external gear pump and a cutout of a Rexroth F Series external gear motor. The main components of both external gear pumps and external gear motors are the same, and the internal layout is similar. The case or housing holds the bearing blocks or bushings that support the two rotors. Each individual rotor has a gear that meshes or mates with a gear in the second rotor to form a pair of meshing gears. The case or housing turns into an enclosure with front and end covers. At one end, the input or output shaft protrudes from the front cover where it is sealed by a shaft seal ring.

An alternate gear machine configuration is that where the gears are meshed internally, as shown in Figure 9-14c. Figure 9-14c presents an **internal gear pump,** which has only one geared rotor driving a second gear or idler that rotates within. The main components of an internal gear pump are labeled in Figure 9-14c. The main gear rotor (1) drives the idler gear (2) and is mounted on the shaft (3). The pumping enclosure is formed between the casing (4), head (5), and bracket (6). Figure 9-14c also shows an internal pressure relief valve (7), designed to protect the pump from overloading. Depending on the gear design, the meshing of the internal gears might require the use of a crescent for proper operation, as shown in Figure 9-14d.

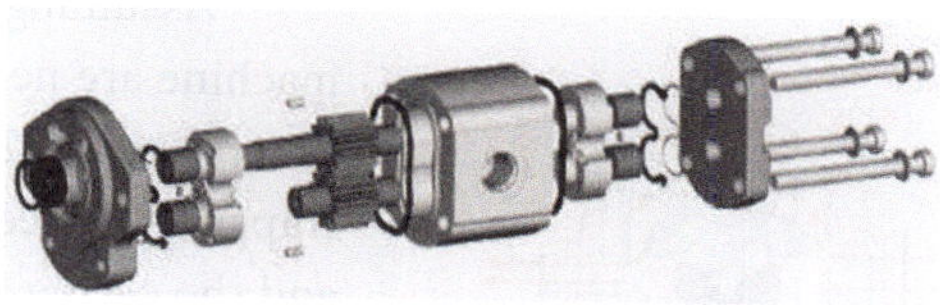

a. External gear pump.

b. External gear motor.

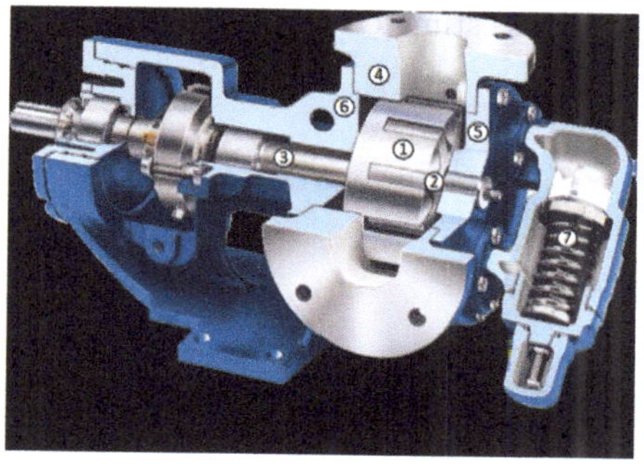

c. Internal gear pump with crescent.

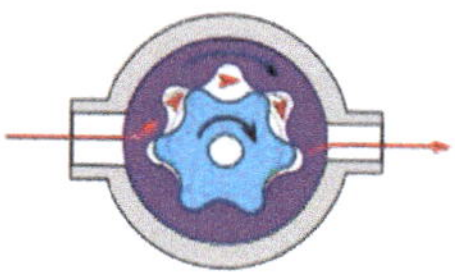

d. Internal gear pump without crescent.

Figure 9-14 Gear machines.

Gear pumps and motors are true rotary machines. The rotational motion of the gears directly creates the pumping action, in contrast with piston machines, which need a motion conversion mechanism. To estimate the volume trapped by a gear pump or the one required to displace a gear motor, the analysis turns to the schematics shown in Figure 9-15. The schematic represents an external gear machine, where each mating gear has six teeth. Gear machines require very tight clearances in both between the teeth tip and the casing and between the gear rotors and the end plates to prevent and minimize the leakage from high-pressure to low-pressure zones.

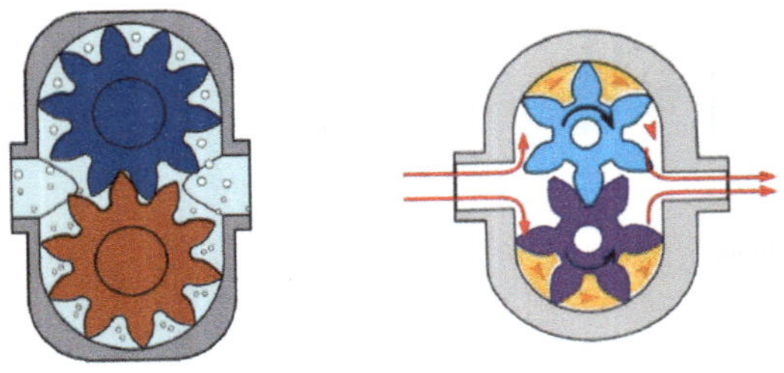

Figure 9-15 Gear machine operation.

Assuming that internal clearances in the gear machine are negligible, it should be relatively easy to visualize that the motoring or pumped fluid is being trapped between adjacent teeth of each gear rotor and the casing and pushed by the gear teeth, as highlighted in Figure 15-b. As the gear rotates, the fluid is discharged or pumped into the outlet port. The gears are continuously intermeshed; the teeth of one gear rotor are always in contact with the teeth of the other gear rotor. There is not a free path for the fluid to go from the outlet to the inlet or vice versa. Therefore, the volumetric displacement of a gear machine corresponds to the volume of fluid that can be trapped between the gear teeth and the casing. Assuming a perfect intermeshing (i.e., the tooth of one gear fully fills the space between two adjacent teeth of the other gear), the total space between the teeth of one gear can be estimated as:

$$A_{1G} = \frac{1}{2} \times \left[\frac{\pi}{4}(D_O^2 - D_I^2)\right]$$

where:

DO = gear external or tip diameter, and

DI = gear internal or root diameter.

Defining *w* as the width of each gear and considering that a gear machine requires two gears, the volumetric displacement of a gear machine can be estimated as:

$$V_D = (2 \times A_{1G}) \times w$$

$$V_D = \left[\frac{\pi}{4}(D_O^2 - D_I^2)\right] \times w$$

The amount of fluid trapped by a gear machine is essentially the volume of the pockets created in the space between the gear teeth and the casing. The gear interteeth space does not change; therefore, the volume calculated with the formula above corresponds to the only value of volumetric displacement a gear machine can have. In other words, gear machines are inherently of the fixed-volumetric-displacement type. As with piston pumps, the implied units of *VD* are units of volume per revolution of the input shaft, as

$$(A_{1G} \times w)$$

represents the volume of fluid displaced or trapped by one gear in one revolution.

Performance: Torque, Power and Efficiency: Volumetric, Mechanical, and Overall

Torque Analysis

As it has been shown, positive displacement machines in general can be used interchangeably as either pumps or motors in the fluid power industry. Most manufacturers of this type of machinery sell similar, equivalent, or even the exact same design of equipment for both services. As such, the torque analysis is done in a hydraulic motor, the machine that converts fluid power into mechanical power (i.e., its purpose is to specifically generate torque at a particular rotational speed). Of the three machines studied so far—piston, gear, and vane machines—the one that is conducive to a simpler analysis to derive the theoretical torque developed by hydraulic motors is the gear machine. Figure 9-16 shows a gear motor sketch with the pressurized zones and the forces exerted by the prevailing pressure on the teeth of the gear rotors. Both sketches will be the base of the torque analysis.

In Figure 9-26a, the red-highlighted zone corresponds to the supply side of the motor, where high-pressure fluid is provided to actually run it. Conversely, the blue-highlighted zone corresponds to the outlet side of the motor, where the prevailing pressure is relatively low. To facilitate the analysis, it will be assumed that the discharge side is directly connected back to an open supply tank open to the atmosphere. With no back pressure on the discharge side of the motor, the torque analysis requires only determining the forces created by the pressure prevailing on the supply side.

The forces created by the supply pressure in a gear motor are presented in Figure 9-16b. As discussed in previous chapters, pressure creates forces that are perpendicular to the surface where they are exerted. As the teeth have curved surfaces, it is advisable to split the actual force at each point into the radial and tangential components. The tangential force components are the ones contributing to the torque developed by the motor, as the radial components are directed toward each gear rotor axis of rotation. Turning again to Figure 9-16b, tangential forces are generated only on the teeth surfaces. As presented in the rotors sketch, tangential forces generated on two adjacent teeth on each rotor cancel each other. In the intermeshing region, one tooth on each gear has partial surfaces exposed to the pressure defined by the point of contact. The point of contact changes continuously during the gear intermeshing; therefore, the teeth surface area exposed to pressure also varies. Nevertheless, the tangential forces on these opposing gear teeth cancel each other out on average, as the point of contact varies symmetrically. Thus, the conclusion is that only one tooth contributes to the torque developed by a gear motor. As the intermeshing varies constantly, the tooth that generates the torque alternates from one gear to the other.

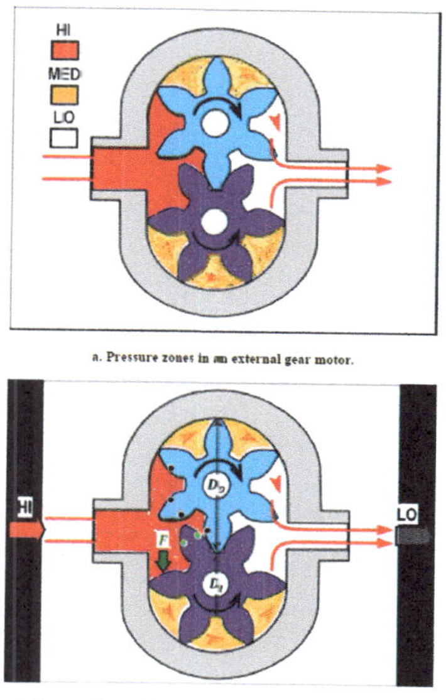

Figure 9-16 Origin of the torque exerted by a gear motor.

Recalling the pressure-force relationship, the force on any area exerted by any given pressure value is given by:

$$F = P \times A$$

The projected area on the tooth that creates a tangential force on the gear rotor is given by:

$$A_T = \left[\frac{1}{2}(D_O - D_I)\right] \times w$$

where:

- DO = tip diameter of the gear rotor,
- DI = root diameter of the gear rotor, and
- *w* = width of the gear rotor.

The torque developed by the tangential force is given by:

$$T = F_T \times r_{F_T} = (P \times A_T) \times r_{F_T}$$

As the tangential force varies linearly with the rotor radius, the effective force acting on the tooth can be concentrated at half the height of the tooth. Consequently, the acting radius for this force can be estimated as half the average of the tip and root diameter of the gear:

$$r_{F_T} = \frac{1}{2} \times \left[\frac{1}{2}(D_O + D_I)\right]$$

Therefore:

$$T_T = \left\{P \times \left[\frac{1}{2}(D_O - D_I) \times w\right]\right\} \times \left[\frac{1}{4}(D_O + D_I)\right]$$

$$T_T = P \times \left[\frac{1}{8}(D_O^2 - D_I^2)\right] \times w$$

The theoretical formula to estimate the volumetric displacement of a gear machine can be rearranged as:

$$V_D = \left[\frac{\pi}{4}(D_O^2 - D_I^2)\right] \times w = \frac{2\pi}{1}\left[\frac{1}{8}(D_O^2 - D_I^2)\right] \times w$$

$$\frac{V_D}{2\pi} = \left[\frac{1}{8}(D_O^2 - D_I^2)\right] \times w$$

Upon substitution of the rearranged volumetric displacement formula into the theoretical developed torque equation:

$$T_T = P \times \frac{V_D}{2\pi}$$

The above equation indicates that the torque developed by a gear motor is directly proportional to the supply pressure and its volumetric displacement (i.e., its size). Any application that would require variations in torque would need either a supply system that can adjust the pressure fed to the motor or several gear motors with different volumetric displacements. The analysis of the next section that relates mechanical power to fluid power (hydraulic) will reveal that the torque formula derived for a gear motor is valid for all positive displacement machines.

This simple formula only estimates the effective or average theoretical torque developed by a gear motor in one revolution. The instantaneous torque that the gear motor develops fluctuates between maximum and minimum values determined by the shape of the teeth and the variation on the net teeth area exposed to the supply pressure as the point of contact between rotors moves during the gear intermeshing. Deriving the theoretical torque formulas for other positive displacement machines is not as straightforward as in most cases; more than one element trapping fluid contributes to the torque developed in the case of a motor or the torque required to operate a pump.

Power Analysis

It has been already established that fluid power machinery converts energy and power from one form to another. A pump converts mechanical power into fluid power. A motor converts fluid power into mechanical power. In an ideal conversion process, all the energy from the power source is converted to usable energy on the power consumption side. On that premise, the definition of mechanical power and fluid (hydraulic) power can be combined to validate the formula derived to calculate the theoretical torque in a gear motor. This analysis is also fundamental for the discussion about estimating the efficiencies of positive displacement machines.

Recalling the mechanical and fluid power formulas:

$$P_{O_{M_T}} = T \times \omega = T_T \times \omega$$

$$P_{O_{H_T}} = P \times Q = P \times Q_T$$

where:

- TT = theoretical torque required or exerted by the machine,
- ω = angular speed,
- P = pressure in the discharge or supply system, and
- QT = theoretical volumetric flow supplied or required by the machine.

In an ideal conversion process, mechanical and fluid powers are the same; therefore:

$$P_{O_{M_T}} = P_{O_{H_T}}$$

$$T_T \times \omega = P \times Q_T$$

Rewriting angular speed ω and volumetric flow rate Q as:

$$\omega = 2\pi \times N = 2\pi \times \frac{1}{t_{period}}$$

$$Q_T = \frac{V_{T_{period}}}{t_{period}}$$

Establishing the time that any positive displacement machine takes to complete an operational cycle as the period for the analysis, the actual interval corresponds to the time needed to complete one revolution of the pumping or motoring elements. In that case, we can rewrite the power equality equation as:

$$T_T \times 2\pi \times \frac{1}{t_{period}} = P \times \frac{V_{T_{Period}}}{t_{period}}$$

$$T_T \times 2\pi = P \times V_{T_{cycle}}$$

The theoretical volume of one cycle or one revolution of a positive-displacement machine is simply the volumetric displacement; therefore:

$$T_T = P \times \frac{V_D}{2\pi}$$

The power conversion analysis confirms that the equation derived to calculate the theoretical torque developed for a gear motor is applicable to any positive displacement machine. The theoretical torque of a positive displacement machine is directly proportional to the supply pressure and its size.

Volumetric Efficiency

In fluid power systems, positive displacement machines face significant changes in the operational conditions. They have to be able to withstand large and sudden variations in pressure, volumetric flow, and changes in temperature. Therefore, although the rotating components are designed to operate with very tight clearances within the machine casing, the high-pressure levels involved create small amounts of backflow or leakage through such clearances. The pressure difference between ports of the machine and the size and extension of the internal clearances determines the amount of leakage, reducing the effective amount of fluid that the machine can actually handle.

Figure 9-17a redraws the gear motor sketch indicating the pressurized zones. As the amount of leakage is driven by pressure, the leakage flows from zones of high pressure, the motor supply, to zones of low pressure, the motor exhaust. In a gear motor and gear machines in general, the leakage primarily occurs between the teeth tip, endplates, and the casing. In a motor, the leakage does not contribute to the torque developed by it. Therefore, to develop the torque estimated above, a larger flow rate must be supplied to the motor to compensate for the leakage. As both theoretical flow and leakage go in the same direction, the relationship between them can be expressed as:

$$Q_A = Q_T + Q_L$$

where:

- QA = actual flow rate supplied to the motor,
- QT = theoretical or effective flow rate that contributes to develop the torque in the motor, and
- QL = backflow or leakage.

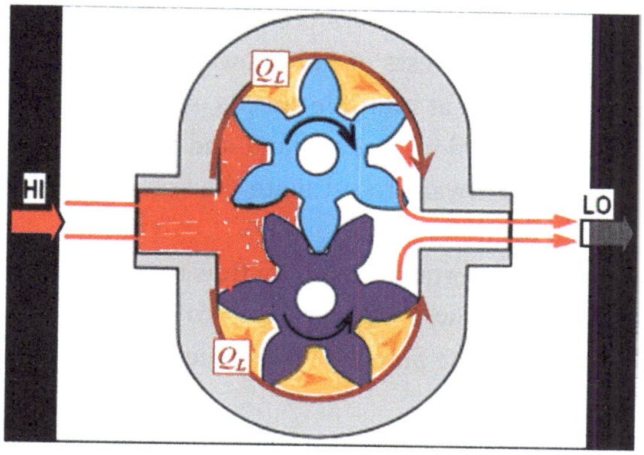

a. Pressure zones and leakage in an external gear motor.

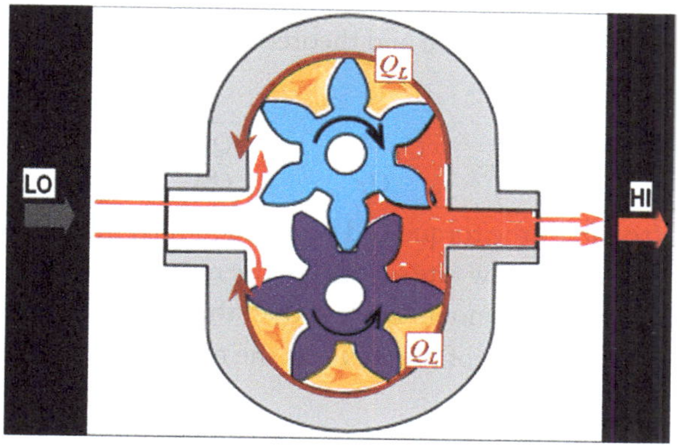

b. Pressure zones and leakage in an external gear pump.

Figure 9-17 Simplified leakage path in gear machines.

Figure 9-17b inverts the pressurized zones of the gear motor sketched in Figure 9-17a to represent an identical gear machine, operating now as a pump. As the amount of internal leakage is driven by pressure independently of the machine purpose, the internal leakage now flows in opposite direction to the theoretical flow rate that the pump pushes toward the discharge. As a result, to estimate the actual flow rate that the pump can really provide to the system, it is necessary to subtract the internal leakage from the flow rate that the pump could theoretically deliver. Hence, in a pump, the relationship between the actual flow rate, theoretical flow rate, and leakage can be expressed as:

$$Q_A = Q_T - Q_L$$

where:

- QA = actual flow rate delivered by the pump,
- QT = theoretical flow rate that the pump could deliver under ideal conditions, and
- QL = backflow or leakage.

The volumetric efficiency data that manufacturers present on either pumps or motors are generally estimated with actual testing. During performance tests of fluid power machinery, manufacturers measure multiple variables and among them the actual flow rate at multiple speeds, several ranges of pressures, and with fluids of different viscosity. With the measured actual flow rate at hand and the theoretical flow determined by the size, type, and the operating speed, it is straightforward to estimate the volumetric efficiency of the machine once its purpose is established.

In a pump, the objective is to provide a certain amount of flow to the fluid power system (i.e., the actual flow rate that the system needs). To provide that flow, the pump needs to generate a slightly larger amount to compensate for leakage. The volumetric efficiency of the pump is simply the actual to theoretical flow ratio:

$$\eta_{V_P} = \frac{Q_A}{Q_T}$$

Upon substitution of the definitions of actual and theoretical flow rates for a pump:

$$\eta_{V_P} = \frac{Q_T - Q_L}{Q_T} = 1 - \frac{Q_L}{Q_T}$$

In a motor, the objective is to satisfy the torque and power requirements of an external system providing the motor with the necessary flow—the actual flow rate going into the motor. Providing the torque and power needed requires feeding the motor a slightly larger amount of flow than the minimum or theoretically required to compensate for leakage. The volumetric efficiency of the motor is the opposite of that of a pump (i.e., the theoretical to actual flow ratio):

$$\eta_{V_M} = \frac{Q_T}{Q_A}$$

Upon substitution of the definitions of actual and theoretical flow rates for a motor:

$$\eta_{V_M} = \frac{Q_T}{Q_T + Q_L} = \frac{1}{1 + \frac{Q_L}{Q_T}}$$

Examination of both formulas reveals that the volumetric efficiency of both pumps and motors reduces as the internal leakage increases. As expected, the maximum efficiency is 100% ($\eta = 1.0$)— achievable only in ideal conditions.

Mechanical Efficiency

Like all rotating machinery, positive displacement machines also have energy and power losses associated with internal friction. In a positive displacement machine, most of this internal friction is associated with the fluid viscosity. The very tight clearances present in the casing, bearings, and seals cause hydrodynamic shearing of any viscous fluid, leading to losses in the form of heat. As with volumetric efficiency, significant changes in operational conditions such as pressure and operational speed also affect the amount of viscous internal friction. Figure 9-18 sketches a gear machine operating as a pump and as a motor indicating the pressurized zones. As viscous or mechanical friction and losses are primarily due to the rotation of the main rotor and pumping elements, internal friction always opposes rotation.

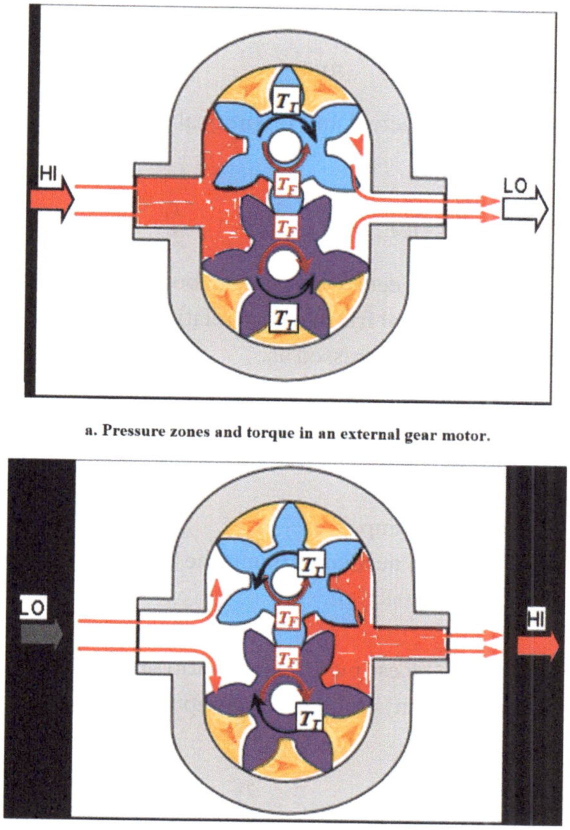

a. Pressure zones and torque in an external gear motor.

b. Pressure zones and torque in an external gear pump.

Figure 9-18 Simplified torque path in gear machines.

In a motor, the actual or output torque is the hydraulic or theoretical torque that the motor can develop under the given conditions but is reduced by the amount of torque required to overcome the losses associated with the internal friction. The relationship between the actual torque, hydraulic or theoretical torque, and friction torque can be expressed as:

$$T_A = T_T - T_F$$

where:

- *TA* = actual torque developed by the motor,
- *TT* = theoretical or hydraulic torque that the motor develops in ideal conditions, and
- *TF* = torque lost to internal viscous friction.

The motor's mechanical efficiency can be estimated as the ratio of the actual torque output at the motor shaft and the theoretical torque the motor would generate in ideal conditions (i.e., with no friction):

$$\eta_{M_M} = \frac{T_A}{T_T}$$

Upon substitution of the definitions of actual and theoretical torques for a motor:

$$\eta_{M_M} = \frac{T_T - T_F}{T_T} = 1 - \frac{T_F}{T_T}$$

In a pump, the input or actual torque needs to overcome both the hydraulic torque required to pump the fluid and the torque associated with the internal friction. The relationship between the actual torque, hydraulic or theoretical torque, and friction torque can be expressed as:

$$T_A = T_T + T_F$$

where:

- TA = actual torque supplied to the pump,
- TT = theoretical or hydraulic torque needed to pump the fluid toward the discharge, and
- TF = torque lost to internal viscous friction.

The pump mechanical efficiency can be estimated as the ratio of the theoretical torque required to drive the pump in ideal conditions (i.e., with no friction) and the actual input torque at the pump shaft:

$$\eta_{M_P} = \frac{T_T}{T_A}$$

Upon substitution of the definitions of actual and theoretical torques for a pump:

$$\eta_{M_P} = \frac{T_T}{T_T + T_F} = \frac{1}{1 + \frac{T_F}{T_T}}$$

As discussed regarding volumetric efficiency, most manufacturers conduct testing for both pumps and motors to provide actual values for mechanical efficiency. Adding a dynamometer or using calibrated electrical motors to test fluid power machinery allows manufacturers to measure actual torque values. Comparing actual torque values with those calculated with the theoretical formula allows the estimation of the mechanical efficiency of the machine once its purpose is established.

Examination of both formulas reveals that the mechanical efficiency of both pumps and motors reduces as the internal friction torque increases. As expected, the maximum efficiency is 100% (η = 1.0)—achievable only in ideal conditions.

Overall Efficiency

The ***overall efficiency*** of any energy or power conversion process is essentially the ratio between the amount of useful energy or output of a system to the energy supplied or input to the same system. In positive-displacement machines, the overall efficiency amounts to the ratio between mechanical power and hydraulic power values. What follows is the derivation of the overall efficiency formulas for both pumps and motors. The analysis demonstrates that the overall efficiency of a fluid power machine is simply the product of the mechanical and volumetric efficiencies.

As the purpose of a pump is to convert mechanical power to hydraulic power, the overall efficiency can be expressed as:

$$\eta_{O_P} = \frac{P_{O_H}}{P_{O_M}} = \frac{P \times Q_A}{T_A \times \omega}$$

With:

$$T_A = \frac{T_T}{\eta_{M_P}}$$

$$T_T = P \times \frac{V_D}{2\pi}$$

$$\omega = 2\pi \times N$$

$$Q_T = V_D \times N$$

To get:

$$\eta_{O_P} = \frac{P \times Q_A}{\frac{\left(P \times \frac{V_D}{2\pi}\right)}{\eta_{M_P}} \times (2\pi \times N)} = \frac{Q_A}{V_D \times N} \times \eta_{M_P} = \frac{Q_A}{Q_T} \times \eta_{M_P}$$

$$\eta_{O_P} = \eta_{V_P} \times \eta_{M_P}$$

Similarly, as the purpose of a motor is to convert hydraulic power to mechanical power, the overall efficiency can be expressed as:

$$\eta_{O_M} = \frac{P_{O_M}}{P_{O_H}} = \frac{T_A \times \omega}{P \times Q_A}$$

With:

$$T_A = \eta_{M_M} \times T_T$$

$$T_T = P \times \frac{V_D}{2\pi}$$

$$\omega = 2\pi \times N$$

$$Q_T = V_D \times N$$

To get:

$$\eta_{O_M} = \frac{\left[\eta_{M_M} \times \left(P \times \frac{V_D}{2\pi}\right)\right] \times (2\pi \times N)}{P \times Q_A} = \frac{V_D \times N}{Q_A} \times \eta_{M_M} = \frac{Q_T}{Q_A} \times \eta_{M_M}$$

$$\eta_{O_M} = \eta_{V_M} \times \eta_{M_M}$$

Comparing both efficiency formulas, it is found that the overall efficiency for both pumps and motors is the product of the volumetric and mechanical efficiencies:

$$\eta_O = \eta_V \times \eta_M$$

The overall efficiency of a positive-displacement machine is a good indicator of the integrity of the machine. When brand new, the machine performance in terms of overall efficiency is mostly defined by the machine design. When in use and after hours of operation, the machine efficiency deteriorates progressively. A simple predictive maintenance tool is to monitor periodically the overall efficiency. Minimum or acceptable efficiency levels can be established, and any sudden drop in efficiency values could be an indication of internal damage in the machine. Machinery performance evaluations via overall efficiency is very common in the industry, as it is relatively easy to determine with simple measurements in the field and more reliable than other predictive maintenance technologies, such as determining the remaining bearing life of a machine via vibration signatures and temperature trends.

Key Equations

Piston Machine

$$V_D = N_P \times A_P \times S$$

$$A_P = \frac{\pi}{4} D_P^2$$

where:

1. *DP* = diameter of the piston,
2. *S* = cylinder stroke, and
3. *NP* = number of pistons.

Units of *VD* would be:

$$V_D = {}^{in^3}/_{rev}$$

Or:

$$V_D = {}^{cm^3}/_{rev}$$

$$Q = V_D \times N$$

where N is the frequency or rotational speed at which the machine operates.

Bent-Axis or Swash Plate Machine

$$V_D = N_P \times A_P \times D_B \times \tan\theta$$

where DB is the diameter of the circle containing the bores in the cylinder, and

$$\theta$$

is the same cylinder block to input shaft angle or bent-axis angle.

Gear Machine

$$V_D = \left[\frac{\pi}{4}(D_O^2 - D_I^2)\right] \times w$$

where:

- DO = gear external or tip diameter,
- DI = gear internal or root diameter, and
- w = width of each gear.

Torque Analysis

$$T_T = P \times \frac{V_D}{2\pi}$$

Power Analysis

The theoretical mechanical and fluid power formulas are as follows:

$$P_{O_{M_T}} = T \times \omega = T_T \times \omega$$

$$P_{O_{H_T}} = P \times Q = P \times Q_T$$

where:

- TT = theoretical torque required or exerted by the machine,
- ω = angular speed,
- P = pressure in the discharge or supply system, and
- QT = theoretical volumetric flow supplied or required by the machine.

Volumetric Efficiency

Volumetric efficiency of the pump:

where:

- QA = actual flow rate delivered by the pump,
- QT = theoretical flow rate that the pump could deliver under ideal conditions, and
- QL = backflow.

$$\eta V_p = \frac{Q_A}{Q_T}$$

$$\eta V_P = 1 - \frac{Q_L}{V_D \times N}$$

Volumetric efficiency of the motor:

$$\eta V_M = \frac{Q_T}{Q_A}$$

$$Q_A = Q_T + Q_L$$

where:

- QA = actual flow rate supplied to the motor,
- QT = theoretical or effective flow rate that contributes to develop the torque in the motor, and
- QL = internal leakage.

$$\eta V_M = \frac{Q_T}{Q_T + Q_L} = \frac{1}{1 + \frac{Q_L}{Q_T}}$$

$$\eta V_M = \frac{1}{1 + \frac{Q_L}{V_D \times N}}$$

Mechanical Efficiency

Mechanical efficiency of the motor:

$$\eta_{M_M} = \frac{T_A}{T_T}$$

$$T_A = T_T - T_F$$

where:

- *TA* = actual torque developed by the motor,
- *TT* = theoretical or hydraulic torque that the motor develops in ideal conditions, and
- *TF* = torque lost to internal viscous friction.

$$\eta_{M_M} = \frac{T_T - T_F}{T_T} = 1 - \frac{T_F}{T_T}$$

$$\eta_{M_M} = 1 - \frac{T_F}{\frac{1}{2\pi} P \times V_D}$$

Mechanical efficiency of the pump:

$$\eta_{M_P} = \frac{T_T}{T_A}$$

$$T_A = T_T + T_F$$

where:

- *TA* = actual torque supplied to the pump,
- *TT* = theoretical or hydraulic torque needed to pump the fluid toward the discharge, and
- *TF* = torque lost to internal viscous friction.

$$\eta_{M_P} = \frac{T_T}{T_T + T_F} = \frac{1}{1 + \frac{T_F}{T_T}}$$

$$\eta_{M_P} = \frac{1}{1 + \frac{T_F}{\frac{1}{2\pi} P \times V_D}}$$

Overall Efficiency

A pump converts mechanical power into hydraulic power, so the overall efficiency is defined as:

$$\eta_{O_P} = \frac{P_{O_H}}{P_{O_M}} = \frac{P \times Q_A}{T_A \times \omega}$$

With:

$$T_A = \frac{T_T}{\eta_{M_P}}$$

$$T_T = P \times \frac{V_D}{2\pi}$$

$$\omega = 2\pi \times N$$

$$Q_T = V_D \times N$$

To get:

$$\eta_{O_P} = \frac{P \times Q_A}{\frac{\left(P \times \frac{V_D}{2\pi}\right)}{\eta_{M_P}} \times (2\pi \times N)} = \frac{Q_A}{V_D \times N} \times \eta_{M_P} = \frac{Q_A}{Q_T} \times \eta_{M_P}$$

$$\eta_{O_P} = \eta_{V_P} \times \eta_{M_P}$$

A motor converts hydraulic power into mechanical power, so the overall efficiency is defined as:

$$\eta_{O_M} = \frac{P_{O_M}}{P_{O_H}} = \frac{T_A \times \omega}{P \times Q_A}$$

With:

$$T_A = \eta_{M_M} \times T_T$$

$$T_T = P \times \frac{V_D}{2\pi}$$

$$\omega = 2\pi \times N$$

$$Q_T = V_D \times N$$

To get:

$$\eta_{O_M} = \frac{\left[\eta_{M_M} \times \left(P \times \frac{V_D}{2\pi}\right)\right] \times (2\pi \times N)}{P \times Q_A} = \frac{V_D \times N}{Q_A} \times \eta_{M_M} = \frac{Q_T}{Q_A} \times \eta_{M_M}$$

$$\eta_{O_M} = \eta_{V_M} \times \eta_{M_M}$$

Exercises

1. What is a positive-displacement machine? How does it differ from a dynamic machine?
2. Find the theoretical volumetric flow rate in liters per minute (lpm) of an axial piston pump that operates at 1,200 revolutions per minute (rpm). The pump has nine pistons that are 15 millimeters (mm) in diameter. The pistons are arranged on a 130-mm-diameter piston circle. The offset angle is 10 degrees.
3. What is the theoretical volumetric flow rate in units of gallons per minute (gpm) of an axial piston pump that operates at 2,000 rpm? The pump has nine pistons that are 0.5 inches (in) in diameter. The pistons are arranged on a 6-in-diameter piston circle. The swash plate offset angle is 9 degrees.
4. Find the swash plate offset angle for an axial pump that has a theoretical volumetric flow rate of 35 gpm at 3,300 rpm. The pump has nine 5/8-in-diameter pistons arranged on a 6-in piston circle diameter.
5. What is the offset angle for an axial piston pump that has a theoretical volumetric flow rate of 0.002 cubic meters per second (cm^3/s) at 3,600 rpm? The pump has seven 16-mm-diameter pistons arranged on a 130-mm piston circle.
6. Internal combustion engines (ICEs) have multiple moving parts. To reduce wear, most ICEs rely on a lubrication system pressurized by a positive-displacement pump. Typically, this pump is an external gear pump like the one shown in Figure 9-19. Each rotor has a teeth tip or outside diameter of 51 mm and a teeth root or inside diameter of 25 mm. The height or width of each rotor is 50 mm.
    a. Determine the volumetric displacement of the gear pump (volume of oil displaced by this pump in one full revolution). **Answer should be provided in liters per revolution.**
    b. Calculate the **theoretical** flow rate delivered by the pump when operating at 1,500 rpm. **Answer should be provided in lpm.**

Figure 9-19

7. You are using an Eaton Series 26 pump with a volume displacement of 30.6 cm^3 per revolution from the chart below. It operates at 3,000 rpm. Calculate the theoretical flow rate in lpm. Compare this result with the listed minimum output flow of 84.4 lpm. Why are these numbers different?

Table 9.1: Eaton Series 26 pump general specifications and performance data.

Displacement cm^3/r (in^3/r)	22.5 (1.37)	24.3 (1.48)	25.2 (1.54)	27.7 (1.69)	29.0 (1.77)	30.6 (1.87(
Max intermittent pressure bar (PSI)	241 [3500]	241 [3500]	241 [3500]	241 [3500]	235 [3400]	224 [3250]
Rated speed (RPM)	3000	3000	3000	3000	3000	3000
Minimum output low at 207 bar (300 PSI) and rated speed LPM (GMP)	62,1 [16.4]	67.0 [17.7]	69.7 [18.4]	76.5 [20.2]	79.9 [21.1]	84.4 [22.3]
Input power at 207 bar (300 PSI) and rated speed and cont. Pressure kW (HP)	27.3 [36.6]	30.5 [40.9]	31.0 [41.6]	33.4 [44.8]	35.4 [47.4]	37.5 [50.1]
The performance data in the table above and the following graphs were collected using a mineral base oil with a viscosity of 133 SUS at 49 degrees C [120 degress F]Source: Eaton, 2016.						

8. A hydraulic motor receives a flow rate of 60 lpm at a pressure of 9,000 kilopascals (kPa). The motor speed is 800 rpm. If the motor has a power loss of 3 kilowatts, find the motor actual output torque (in newton-meters [Nm]) and overall efficiency (in percentages).
9. A gear pump has an 82.6-mm outside diameter, a 57.2-mm inside diameter, and a 25.4-mm width. If the actual pump flow rate is 0.0018 m^3/s at 1,800 rpm and rated pressure, what is the volumetric efficiency? (Solve in percentages.)
10. You are using a Parker TJ series 0045 motor per Figure 9-20. The motor has a geometric displacement (V_d) of 41 cm^3 per revolution at 1024 rpm and 140 bar pressure. What is the maximum torque output of the motor at 140 bar pressure? How does this answer compare with the 71 Nm on the chart. Why are these numbers different?
11. Find the offset angle (in degrees) for a bent-axis axial piston pump that actually delivers 20 gpm at 3,000 rpm. The volumetric efficiency is 95%. The pump has nine pistons in total; each piston is 5/8 in in diameter arranged on a 5-in piston circle diameter.
12. A pump operates at 2,500 psi and 1,000 rpm. It theoretically delivers 5 gpm. It requires 9 hp (actual power) to drive the pump. Determine the overall efficiency of the pump (in percentages).
13. You are evaluating a large displacement sleeve bushing motor for an application. The product is a Permco 460 Series unit with the following specifications: V_d 7.2 in^3 per revolution (2-in internal gear width) and an operating pressure of 3,500 psi that operates at 2,400 rpm.

360/460 SERIES

MOTOR PERFORMANCE CHART

SPEED [rpm]	GEAR WIDTH [in] 1		1 1/4		1 1/2		1 3/4		2		2 1/4		2 1/2	
	Flow (GPM)	Torque (in-lb)	Flow (GPM)	Torque (in-lb)	Flow (GPM)	Torque (in-lb)	Flow (GPM)	Torque (in-lb)	Flow (GPM)	Torque (in-lb)	Flow (GPM)	Torque (in-lb)	Flow (GPM)	Torque (in-lb)
900	18.5	1907	22.2	2393	26	2879	30.2	3365	34.4	3850	37.7	3982	40.9	4114
1200	23.3	1918	28.1	2398	33	2879	38.5	3361	44.1	3843	48.4	3970	52.7	4097
1500	27.9	1910	34.1	2385	40.2	2861	46.9	3339	53.6	3817	59.1	3935	64.5	4053
1800	32.7	1901	40.1	2377	47.5	2853	55.4	3321	63.3	3789	70	3900	76.8	4011
2100	37.6	1890	46.1	2362	54.6	2834	63.6	3294	72.6	3753	80.7	3867	88.8	3981
2400	42.5	1881	52.3	2349	62	2817	72.2	3274	82.3	3731	91.7	3848	101.1	3964
PSI	3500										3250		3000	

Figure 9-20

14. The flow listed is 82.3 gpm (specified [actual] flow) and 3,731 inch-pounds of (specified [actual] torque). What are the motor volumetric and mechanical efficiencies? **Answers should be provided in percentages.**
15. You are using a Parker TJ series 0045 motor per Figure 9-21. What is the maximum torque output of the motor at 140 bar pressure? How does this answer compare with the 71 Nm on the chart? Why are these numbers different?

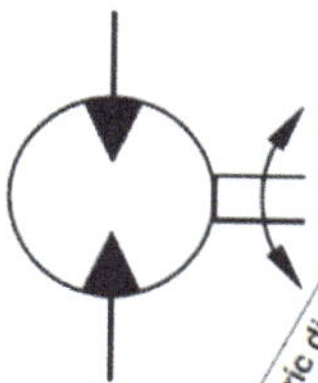

Motor Series TJ	Geometric displacement / Geom. Schluckvolumen / Cylindrée / Despazamientos — cm³/rev, in³/rev	Max. speed @ Max. intermittent flow / Max. Drehzahl Intermittierender Betrieb: / Vitisse de rotation maxi / Velocidad maxima a caudal intermitente maximo — rev/min	Max. oil flow / Max. Schluckstrom / Débit d'huile maxi / Caudal Maximo de Aceite — cont / int* l/min g/min	Max differential pressure / Max. Druckgefälle / Chute de pression maxi / Presion diferencial maxima — cont / int* bar psid	Max. supply pressure / Max. Eingangsdruck / Pression maxi entrée / Presion maxima de alimentacion — max bar psig	Max. torque / Max. Drehmoment / Couple maxi / Torque Maximo — cont / int* Nm lb-in	Max. performance / Max. Leistungabgabe / Puissance de sortie maxi / Maximo rendimieto — max KW HP	Min. starting torque / Min. Anlaufmoment / Couple min. fourni au dé manrage / Torque minimo de arranque — cont / int* Nm lb-in
TJ 0045	41 2.5	1024	34 42 9 11	140 190 2000 2750	200 2900	71 99 624 876	10.4 13.9	46 64 41[illegible] 565

Figure 9-21

16. A hydrostatic transmission, which is a closed-loop hydraulic circuit like the one shown in Figure xx, operates at 1,750 psi. The hydraulic pump has the following characteristics: $V_d = 5\ in^3$, $\eta_V = 88\%$, $\eta_m = 92\%$, and running at 1,200 rpm. If a typical reversible hydraulic motor has a volumetric efficiency of 90% and a mechanical efficiency of 95%, determine the following:

a. What is the minimum recommended size (V_d) for the hydraulic motor if it needs to run at 400 rpm?
b. If this hydraulic motor is installed, what is the actual output torque?

Reference

Eaton. 2016, October. Technical Catalog Gear Pumps and Motors, Series 26. Publication E-PUGE-MC001-E3.

Figure Credits

Fig. 9-2a: Ll1324, http://commons.wikimedia.org/wiki/File:Oil_Pump_from_Toyota_Coaster_Bus.JPG, 2012.
Fig. 9-2b: Copyright © by Manco Capac (CC BY-SA 3.0) at http://commons.wikimedia.org/wiki/File:Hand_pump-en.svg.
Fig. 9-2c: Copyright © by Tomasz Kuran (CC BY-SA 3.0) at http://commons.wikimedia.org/wiki/File:Rabe_pump01.jpg.
Fig. 9-3c: ITT Corporation | Goulds Pumps, "Goulds CV 3196 i-FRAME pump," http://www.gouldspumps.com/Products/CV-3196-i-FRAME/. Copyright © by ITT Corporation.
Fig. 9-3d: ITT Corporation | Goulds Pumps, "Main Parts of a Single Stage Pump," http://www.gouldspumps.com/ittgp/medialibrary/goulds/website/Products/CV-3196-i-FRAME/CV3196_iFrame_Web.pdf?ext=.pdf. Copyright © by ITT Corporation.
Fig. 9-5a: Copyright © by Mkoronowski (CC BY-SA 3.0) at http://commons.wikimedia.org/wiki/File:Impeller_inlet_velocity_triangles.PNG.
Fig. 9-5b: Copyright © by Mkoronowski (CC BY-SA 3.0) at http://commons.wikimedia.org/wiki/File:Impeller_discharge_velocity_triangles.PNG.
Fig. 9-5c: Copyright © by H Padleckas (CC BY-SA 2.5).
Fig. 9-6a: ITT Corporation | Goulds Pumps, "Single-Stage Centrifugal Pump," . Copyright © by ITT Corporation.
Fig. 9-6b: Copyright © by Eaton.
Fig. 9-7a: Copyright © 2015 by Hydraulic Institute. Reprinted with permission.
Fig. 9-7b: Copyright © 2015 by Hydraulic Institute. Reprinted with permission.
Fig. 9-11a: Copyright © by Bosch Rexroth Corporation.
Fig. 9-11b: Copyright © by Eaton.
Fig. 9-13a: Copyright © by Parker Hannifin Corporation.
Fig. 9-13b: Copyright © by Parker Hannifin Corporation.
Fig. 9-14a: Copyright © by Bosch Rexroth Corporation.
Fig. 9-14b: Copyright © by Bosch Rexroth Corporation.
Fig. 9-14c: Copyright © by Eaton.
Fig. 9-14d: Copyright © by Duk (CC BY-SA 3.0) at http://commons.wikimedia.org/wiki/File:Gear_pump_2.png.
Fig. 9-15a: Jahobr, http://commons.wikimedia.org/wiki/File:Gear_pump_animation.gif, 2011.
Fig. 9-15b: Copyright © by Duk; Adapted by Larry Villasmil and Robert Garrick (CC BY-SA 3.0) at http://commons.wikimedia.org/wiki/File:Gear_pump.png.
Fig. 9.16a: Copyright © by Duk; Adapted by Larry Villasmil and Robert Garrick (CC BY-SA 3.0) at http://commons.wikimedia.org/wiki/File:Gear_pump.png.
Fig. 9-16b: Copyright © by Duk; Adapted by Larry Villasmil and Robert Garrick (CC BY-SA 3.0) at http://commons.wikimedia.org/wiki/File:Gear_pump.png.
Fig. 9-17a: Copyright © by Duk; Adapted by Larry Villasmil and Robert Garrick (CC BY-SA 3.0) at http://commons.wikimedia.org/wiki/File:Gear_pump.png.
Fig. 9-17b: Copyright © by Duk; Adapted by Larry Villasmil and Robert Garrick (CC BY-SA 3.0) at http://commons.wikimedia.org/wiki/File:Gear_pump.png.
Fig. 9-18a: Copyright © by Duk; Adapted by Larry Villasmil and Robert Garrick (CC BY-SA 3.0) at http://commons.wikimedia.org/wiki/File:Gear_pump.png.
Fig. 9-18b: Copyright © by Duk; Adapted by Larry Villasmil and Robert Garrick (CC BY-SA 3.0) at http://commons.wikimedia.org/wiki/File:Gear_pump.png.
Table 9-1: Copyright © 2016 by Eaton.
Fig. 9-20: Copyright © by Permco, Inc.
Fig. 9-21: Copyright © by Parker Hannifin Corporation.

CHAPTER 9B

Fluid Power Actuators: Pneumatic Machinery

Overview and Learning Objectives

Compressors and air motors play the most important role in a pneumatic system. The compressor is the fluid machine that provides the pneumatic energy to the fluid. The motor, or actuator, is the fluid machine that converts the pneumatic energy provided to the fluid into effective mechanical work.

After completing this chapter, students should be able to do the following:

- Explain the compression and motoring features of a pneumatic machine.
- Describe the main construction features of pneumatic machines.
- Discuss how a piston compressor operates.
- Define what actual flow rate of a compressor is.
- Estimate the theoretical power required to drive a compressor.
- Estimate the theoretical power extracted from an air motor.
- Compute the performance of a pneumatic machine.

Pneumatic Machines: Compressors and Motors

Compression and Motoring Theory

Just as with hydraulic machines, compressors and pneumatic motors primarily differentiate by the way they impart or convert fluid power. Fluid power applications typically require low flow and high pressure; therefore, most fluid power applications involve positive displacement machines. In contrast, dynamic compressors and motors are of practical use when the operational conditions demand high flow and low-to-moderate pressure, although high pressures are achievable.

The most significant difference between hydraulic and pneumatic machines is that the latter must be designed to handle the fluid compressibility. Gases are readily compressible: as the pressure changes, so do the density and volume of the gas. In a **dynamic compressor,** the impellers and rotor vanes that increase the gas velocity are optimized to operate with a fluid that continuously changes density. In multistage compressors, the impeller on a centrifugal compressor or the rotor vanes in an axial compressor are different in each stage. Figure 9-22a presents a back-to-back multistage centrifugal compressor. The impeller radii's and impeller vanes' thicknesses reduce from stage to stage as the gas is compressed, occupying less volume. In Figure 9-22a, the gas enters axially in the most left impeller and leaves radially on the third impeller coming back in on the rightest impeller, flowing in the opposite direction. The arrangement portrayed in Figure 9-22a is referred to as back-to-back. The back-to-back impeller arrangement is designed to minimize the axial thrust and facilitate intercooling when high efficiencies are required. A centrifugal compressor is capable of generating pressure in the order of thousands of pounds per square inch (psi) while handling flows in the order of thousands of cubic feet per minute (cfm). Figure 9-22b presents a multistage axial compressor, where the gas flows from the bottom (closest to the viewer) to the top (far from the viewer). An axial compressor is comprised of rotating and static aerodynamic blades with a rotor that typically has a constant radius while the blades reduce in height at each stage, in an arrangement that normally exceeds 10 stages. Figure 9-22b shows how two mechanics check the clearance between the rotor blades and the casing to ensure that design specifications are met. In contrast with centrifugal compressors, axial compressors generate pressures close to only 100 psi but handle much larger flows in the order of hundreds of thousands of cubic feet per minute. Dynamic machines can also be used to extract energy from a gaseous stream.

a. Back-to-back multistage centrifugal compressor. **b. Axial compressor.**

Figure 9-22 Pneumatic machines, dynamic.

The principles of operation and the design of a **pneumatic positive-displacement machine** are very similar to those used in hydraulic applications. The pumping elements draw gas from the inlet, trap the gas when the machine is completely filled, and deliver it toward the discharge. The key feature remains eliminating or minimizing the presence of clear passages between the outlet and inlet in a compressor or vice versa in a motor. Nevertheless, the gas compressibility limits significantly the capacity of a positive-displacement machine. As gases are readily compressible and expandable, the progressive decrease in volume occurring toward the outlet of a compressor does not guarantee that the gas will be delivered to the discharge lines. Inside any compressor, the gas will be compressed until the internal pressure exceeds the prevailing pressure in the system. It is at this point that gas will start being delivered. If the pressure in the system is larger than the pressure that the compressor can create during the compression stroke or rotation, no gas will be delivered to the system. If no gas is delivered to the discharge, the gas trapped inside the compressor will expand during the admission stroke or rotation. The pressure inside the compressor will not be reduced below the pressure prevailing in the suction lines, and no gas will be admitted in the inlet of the compressor. This limitation is more evident in reciprocating machines. In rotary machines, the rotating elements create multiple pockets, so the trapped gas can effectively be pushed toward the discharge and will flow back to the inlet only through the internal clearances.

Positive-Displacement Machines (Compressors and Motors)

The focus of this section is a discussion on the effective volume displaced or taken by a positive-displacement pneumatic machine, either a compressor or a motor. As discussed previously, the size and type of the machine determine the volume that can trap or displace in a particular period of time. In contrast with hydraulic machines, the great majority of compressor and motor manufacturers do not disclose the theoretical volume per cycle of operation. The information available related to flow capabilities of a pneumatic machine are the effective or actual volumetric flow rate and the volumetric efficiency.

Piston Type

On pneumatic applications, all machines are of the true piston type. Plunger-type machines are not used due to practical size limitations, as plungers would need to be excessively large to displace the volumes required in typical pneumatic applications. Piston machines can be of dual action, as both chambers within the cylinder—in front and behind the piston—can be used to trap or displace fluid (see Figure 9-23). A single-acting piston is typically directly driven by a crankshaft and a connecting rod that is connected to the piston by a pin. In a double-acting machine, the piston is driven by a tensioned rod, which is driven by a crosshead that reciprocates in a slide under the action of a connecting rod and the crankshaft (see Figure 9-23). Pneumatic reciprocating machines that handle air are typically single-acting. The leakage that passes through the piston rings would simply go to the crankcase and back to the atmosphere with no harm, as the crankcase is typically vented. In contrast, double-acting machines like the one shown in Figure 9-23 are required when the gases under compression are hazardous, corrosive, or poisonous. A double-acting piston design allows the addition of a specially designated vented spaces or compartments between the crosshead and the compressor cylinder (see Figure 9-23). The leakage from the cylinder into these spaces can be treated and/or vented to a safe location.

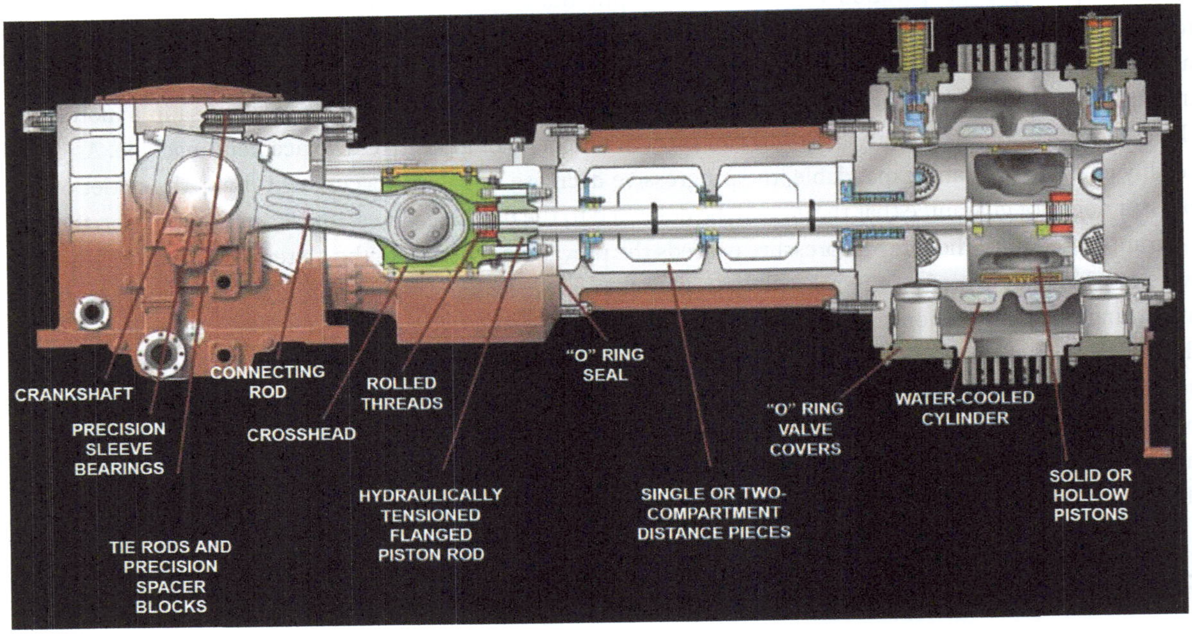

Figure 9-23 Double-acting piston compressor.

In a single-acting reciprocating compressor, one cycle comprises four events. When the piston moves forward, the cylinder volume reduces, compressing and discharging the gas. When the piston moves backward, the cylinder volume increases, re-expanding the gas not delivered and drawing in new gas. Figure 9-24 offers a schematic of how a single-acting reciprocating compressor works. With the piston at the bottom dead center (i.e., completely retracted), the cylinder is completely filled with gas, and the compression process starts (1). As the piston moves forward, the cylinder volume reduces, increasing the pressure.

When the cylinder pressure reaches a value slightly above the discharge pressure (2), the discharge valve opens, and the gas is pushed out while the piston completes the rest of the stroke. At the top dead center—the piston fully extended—the discharge valve closes (3). To avoid mechanical damage, a dead space must exist between the piston face and the cylinder head; therefore, some pressurized gas remains in the cylinder.

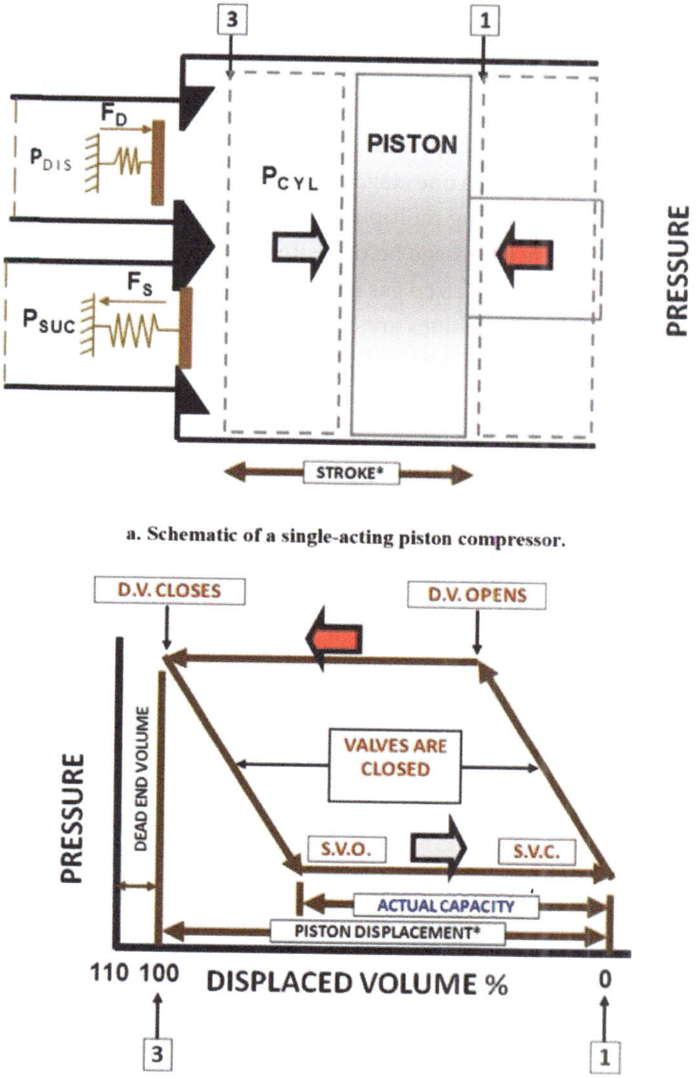

a. Schematic of a single-acting piston compressor.

b. Pressure versus piston displacement, simplified diagram.

Figure 9-24 How a reciprocating compressor works.

This trapped gas expands as the piston moves backward during the retracting stroke. The trapped gas expansion continues until the cylinder pressure reaches suction pressure (4) and the suction valve opens, allowing fresh gas to fill the cylinder. One the piston reaches the dead bottom center (1), a new compression cycle initiates.

The pressure-versus-piston-displacement diagram in Figure 9-24b presents the correlation between the position of the piston and internal cylinder pressure. If there were no dead space at the top of the cylinder, all the gas drawn in from the suction would be delivered once the discharge valve opens. Similarly, during the piston retraction, there would not be gas trapped at the top, and the cylinder pressure would have immediately reduced below suction pressure, allowing fresh gas to fill completely the cylinder. In these conditions, the volumetric efficiency of the compressor would be 100%. In reality, the need for dead space reduces the volumetric efficiency. The volumetric efficiency of a reciprocating compressor is approximated by the percentage of the stroke for which the suction valve remains open during the complete retracting stroke.

For practical purposes, the dead-end volume limits the discharge to suction pressure ratio that a single-stage compressor can effectively overcome. When high discharge pressures are required, several cylinders are arranged in series. The pressurized gas discharged from one stage is taken in by the following stage. In this way, the pressure ratio required by the process is spread among multiple stages, increasing the volumetric efficiency. In addition, cooling the gas stream on the discharge of a stage before entering the next stage also increases the efficiency, as the actual volume occupied by the hot pressurized gas reduces when it is being cooled.

Although multistage reciprocating machines are capable of handling very large pressure ratios, most pneumatic machines used in the fluid power industry operate at moderate discharge pressures and pressure ratios. Pneumatic compressors handling air are typically of one or two stages and designed for pressures in the order of tens or few hundreds of pounds per square inch (see Figure 9-25). Figure 9-25 presents a two-stage, 5-horsepower (hp) compressor capable of delivering 16.8 cubic feet per minute at a maximum pressure of 175 psi.

Figure 9-25 Two-stage reciprocating compressors.

Vane Type

Like hydraulic machines, vane compressors and motors are relatively simple in design, having very few moving parts while being somewhat easy to manufacture. Pneumatic vane machines, as the name implies, have vanes or paddles that are always in contact with the casing and self-adjust for wear (see Figure 9-26). The compression or motoring action in a pneumatic vane machine is similar to that of a hydraulic vane machine. A paddled rotor (Figure 9-26a) is eccentrically mounted in a circular or cycloidal housing. The eccentricity brings the rotor close to the housing wall so that a crescent-shaped cavity is formed. The inlet of a vane compressor extends from the point at which pockets between vanes are the smallest to where they are the largest so that the increasing volume draws fluids from the suction line (Figure 9-26b). The outlet of a vane compressor extends from the point where pockets between vanes are the largest to where they are the smallest so the reducing volumes push fluids toward the discharge line. In a vane motor, the opposite is true. The inlet corresponds to the housing section where the pocket volumes increase as the pressurized fluid pushes the vanes, creating torque and spinning the rotor. At the outlet, the pocket volumes reduce so that the vanes in this housing section push out the low-pressure fluid. The pockets between vanes are truly formed when the entire the rotor is enclosed, sealed within the casing by two side plates.

Figures 9-26c and 9-26d depict an Atlas Copco LZL Series Vane motor and its basic components. As Atlas Copco explains, these motors are designed to give outstanding starting and low speed performance. This is achieved by using a multiple vanes (1) and by vane/cylinder sealing obtained through a combination of pressurized air and interconnecting pins (2). These vane motors in particular are supplied with replaceable/removable internal restrictors in the connection ports, to control/limit the maximum speed. These motors have a wide speed and torque range and need lubricated air for proper operation.

a. Vane compressor rotor.

b. Natural gas vane compressor.

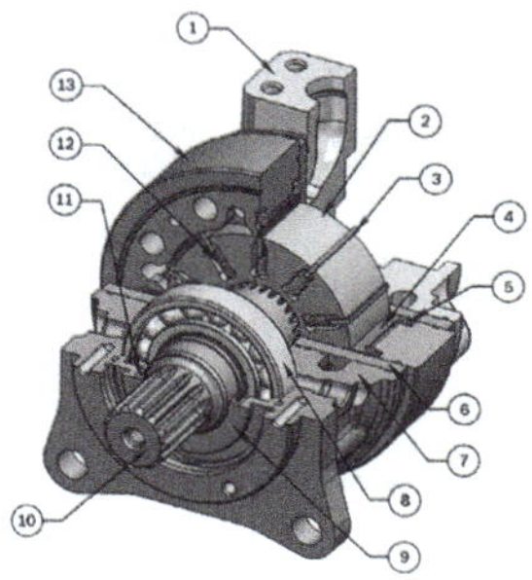

Figure 9-26 Pneumatic vane machines.

Screw Type

Screw machines (Figure 9-27a) are comparable with vane and gear machines in terms of their simplicity of construction, reliability, and cost-effectiveness. The trade-off is the challenge of ensuring minimal clearances between the screw helical threads and between the screws themselves and the casing.

Typically, screw machines have two rotors, male and female, that intermesh and rotate in parallel as shown in Figure 9-27b. The rotor intermeshing can be created in two ways. One way is to design the rotors so one screw drives the order like in a gear-type machine (see Figure 9-28c). In this case, the female rotor is driven by direct contact from the male rotor, which is itself powered by the compressor driver. To avoid damage on the screws' surfaces in contact when the treads intermesh, oil or any other viscous fluid is injected to create a lubricating film that also cools and seals the rotors and the casing. This design is normally referred to as an oil-flooded machine. Figure 9-27c shows a single-stage Ingersoll Rand Nirvana oil-flooded screw compressor and its main components. This compressor is driven by an integral electric motor, directly mounted on the casing.

The second screw meshing option is to add timing gears to the rotors so that they intermesh without contact (Figure 9-27d). This design is normally referred to as a dry screw machine. Pneumatic screw machines are typically designed for compression applications.

a. Typical screw compressor.

b. Intermeshed screw compressor rotor.

Figure 9.27a-b

c. Oil-flooded screw compressor.

d. Dry screw compressor.

Figure 9-27c-d Pneumatic Screw Machines

Figure 9-27d shows a two-stage Ingersoll Rand Nirvana dry screw compressor and its main components. It should be noted that an oil-flooded or wet compressor is more efficient as the internal clearances are much tighter and the fluid itself acts as a sealant. Nevertheless, dry compressor manufacturers such as Ingersoll Rand also use flooding in dry compressor designs as a way to cool down the compressed fluid, which in turn also increases the efficiency of the entire compression process.

In a screw machine, the inlet exposes one end of the rotor to the suction or supply line. The outlet of the screw machine exposes the opposite end of the rotor to the discharge or exhaust line. The compression or motoring action in a pneumatic screw machine is the result of elongated and twisted pockets created by the pairing of the helix intermeshing of the multiple thread screws that rotate within the casing at a tight clearance (see Figure 9-28). Pneumatic screw machines are typically designed for compression applications. Screw machines are seldom used as pneumatic motors.

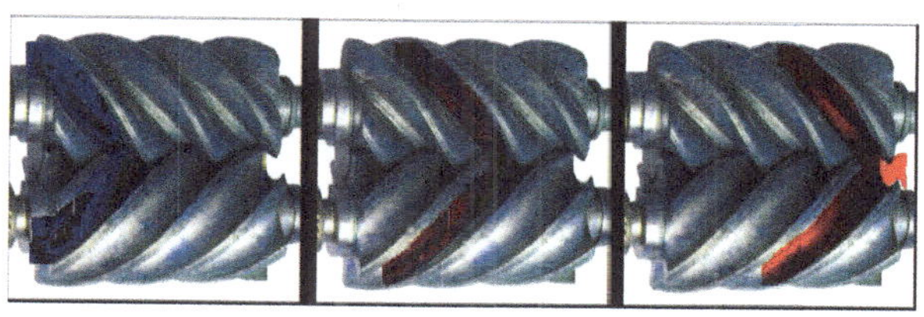

Figure 9-28 How a screw compressor works: Pockets created by the intermeshing of the rotor.

Performance: Actual Flow Rate, Power, and Efficiency: Volumetric, Mechanical, and Overall

Actual Flow Rate

Users of fluid power equipment in the pneumatic arena mostly use air in their systems. In terms of resources, air is relatively free, and any leak from the machinery and/or the equipment and piping is harmless to the environment. Nevertheless, there are some applications that demand an inert gas, most commonly nitrogen, as the fluid to transmit the power such as in hydraulic accumulators or shock absorbers and dampers. In both cases, designers and users have to take into account the fact that the fluid is readily compressible. In compressors, the fluid will reduce considerably in volume as it is compressed from atmospheric pressure. In motors, the fluid will expand significantly as it is exhausted back to the atmosphere.

Figure 9-29 shows International Organization for Standardization (ISO) diagrams of an isolated air compressor and another feeding a single-acting pneumatic cylinder directly. Pneumatic systems designers and users both need to know the volume of air needed, regarding the atmospheric air taken in and the volume of the compressed air being delivered to the system. The flow rate or volume taken in on the atmospheric side is normally referred to as free air. The flow rate or volume delivered compressed to the system is normally referred to as actual volume or actual flow rate.

Both free and actual volume or volumetric flow rates are calculated using the pressure and temperature conditions of the air on either side of the compression system. Free air and actual volumes can be related using the perfect gas laws. Pressure and temperature levels of typical compressed air systems are such that perfect gas laws can be used to calculate properties of pressurized air volumes and flow rates without incurring significant errors.

As discussed earlier in the book, by using perfect gas laws, we can calculate one property of the gas, either temperature or pressure or volume, when two of the three are known. Restating the perfect gas law:

$$\frac{P_1 \cdot V_1}{T_1} = \frac{P_2 \cdot V_2}{T_2}$$

If rearranged to calculate the intake of free air volume, V_1:

$$V_1 = \frac{P_2}{T_2} \times V_2 \times \frac{T_1}{P_1} \rightarrow V_1 = \frac{P_2}{T_2} \times \frac{T_1}{P_1} \times V_2$$

The volume at the intake of the compressor V_1 is related to the volume at the discharge V_2 by:

$$V_1 = \frac{P_2}{P_1} \times \frac{T_1}{T_2} \times V_2$$

To relate the volume at the discharge V_2 with the volume at the intake of the compressor V_1, the new formula can be written by simply flipping the subindexes:

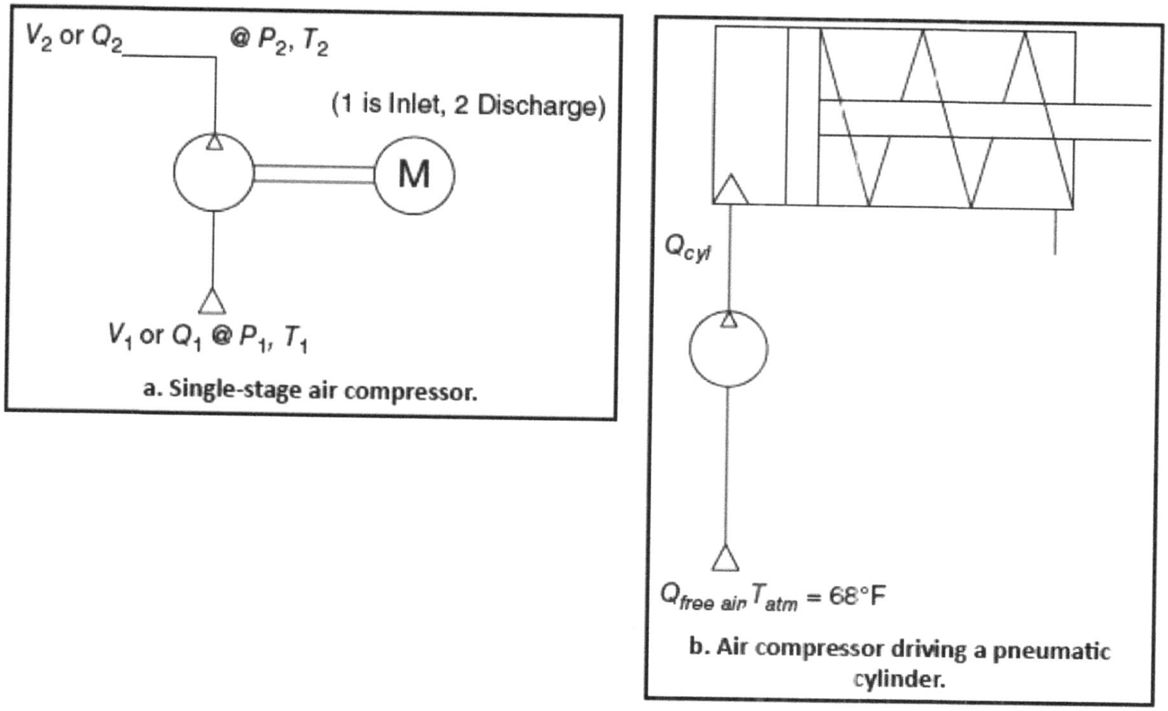

Figure 9-29 ISO diagrams of a single-stage compressor and feeding pneumatic actuator.

$$V_2 = \frac{P_1}{P_2} \times \frac{T_2}{T_1} \times V_1$$

To derive the volumetric flow rate, both formulas can be updated by recalling that volumetric flow rate is the volume of air that the compressor takes in or delivers per unit of time or:

$$Q = \frac{V}{t} \rightarrow Q \cdot t = V \rightarrow Q_1 \cdot t = V_1 \rightarrow Q_2 \cdot t = V_2$$

Therefore, the flow rate at the intake Q_1 is related to the flow rate at the discharge Q_2 by:

$$Q_1 \cdot t = \frac{P_2}{P_1} \times \frac{T_1}{T_2} \times (Q_2 \cdot t)$$

$$Q_1 = \frac{P_2}{P_1} \times \frac{T_1}{T_2} \times Q_2$$

Conversely, the flow rate at the discharge Q_2 is related to the flow rate at the intake Q_1 by:

$$Q_2 = \frac{P_1}{P_2} \times \frac{T_2}{T_1} \times Q_1$$

Power

The theoretical power required to drive a compressor or the one that can be extracted from a pressurized air stream in a pneumatic cylinder can be derived using the first law of thermodynamics (Turns & Pauley 2020). Such derivation is out of the book's scope (Hanlon 2020). Suffice it to say that either the compression or the expansion process can be assumed as ideal—i.e., isentropic and adiabatic—no heat transfer or cooling of the gas.

The theoretical formula to estimate the power required compressing a gas derived from the assumption of an isentropic and adiabatic process is:

$$P_{O_T} = \frac{k}{k-1} \times (P_1 \cdot Q_1) \times \left[\left(\frac{P_2}{P_1} \right)^{\frac{k-1}{k}} - 1 \right]$$

Recall that:

- $P_1 = P_{atm}$ and k = 1.4 (specific heat ratio for air)

The isentropic assumption idealizes the process, neglecting friction and other losses, thus underestimating the real power required to compress the gas. On the other hand, the adiabatic assumption neglects the fact that the gas is cooled down during the compression process in most applications, even in small single-stage reciprocating compressors, therefore over-predicting the real power required to compress the gas.

The theoretical formula to estimate the power that a pneumatic motor can produce, also derived from the assumption of an isentropic and adiabatic process, is:

$P_{O_T} = \frac{1}{k-1} \times (P_2 \cdot Q_2) \times \left[\left(\frac{P_1}{P_2} \right)^{\frac{k-1}{k}} - 1) \right]$

In this case, assuming that the motor exhaust is connected to the atmosphere:

- $P_2 = P_{atm}$ and k = 1.4 (specific heat ratio for air)

Volumetric, Mechanical, and Overall Efficiency

The compressibility nature of the fluid handled by pneumatic machines makes it more challenging to derive relationships for volumetric and mechanical efficiencies. As previously shown in Figure 9-15, a piston compressor—and, with few exceptions, all positive-displacement machines—have dead-end spaces. The size of these dead-end spaces or volumes determines the amount of fluid that is being delivered at any given pressure. Hence, both volumetric efficiency and mechanical efficiency depend on the actual operating pressure of the machine. The derivation of such relationships is beyond the scope of this book, see Hanlon if an estimation is required.

In pneumatics, it is more common to look at and evaluate the compressor's overall efficiency. For a compressor, the overall efficiency can be estimates as the ratio of the theoretical power required to compress the air flow and the actual power required to operate the compressor.

$\eta_O = \frac{P_{OUT}}{P_{IN}}$

- P_{OUT} is the theoretical power required to compress the air flow at the given conditions.
- P_{IN} is the actual power used to drive the compressor.

The theoretical power is estimated with the formula discussed in the previous section. The actual input power depends on the actual driver of the compressor. In small applications, the compressor is typically driven by an electric motor.

In a single-phase electric motor, the power delivered by the motor can be estimated by:

In a three-phase electric motor, the power delivered by the motor can be estimated by:

1. V_L = line voltage (zero-peak),
2. I = motor current, and
3. pf = motor power factor, typically between 0.8 and 0.9.

Like with hydraulic machines, the overall efficiency is a good indicator of the machine's integrity. Nevertheless, values have to be compared at similar inlet and outlet conditions and pressure ratios because the volumetric efficiency of pneumatic machines varies in a much larger magnitude due to the fluid compressibility.

Key Equations

Perfect Gas Laws: Air Volumetric Flow

Perfect gas law:

$$\frac{P_1 \cdot V_1}{T_1} = \frac{P_2 \cdot V_2}{T_2}$$

Compressor flow rate at the intake Q_1 related to the flow rate at the discharge Q_2:

$$Q_1 = \frac{P_2}{P_1} \times \frac{T_1}{T_2} \times Q_2$$

Compressor flow rate at the discharge Q_2 related to the flow rate at the intake Q_1:

$$Q_2 = \frac{P_1}{P_2} \times \frac{T_2}{T_1} \times Q_1$$

Theoretical Power: Pneumatic Machines

Power required to compress a gas (isentropic and adiabatic process):

$$P_{O_T} = \frac{k}{k-1} \times (P_1 \cdot Q_1) \times \left[\left(\frac{P_2}{P_1}\right)^{\frac{k-1}{k}} - 1\right]$$

For an air compressor:

1. $P_1 = P_{atm}$ and k = 1.4 (specific heat ratio for air)

Power produced by a pneumatic (isentropic and adiabatic process):

$$P_{O_T} = \frac{1}{k-1} \times (P_2 \cdot Q_2) \times \left[\left(\frac{P_1}{P_2} \right)^{\frac{k-1}{k}} - (k-1) \right]$$

For an air motor with atmospheric exhaust:

1. $P_2 = P_{atm}$ and k = 1.4 (specific heat ratio for air)

Overall Efficiency

$\eta_O = \frac{P_{OUT}}{P_{IN}}$

Power Input: Electric Motors

In a single-phase electric motor:
In a three-phase electric motor:

Exercises

1. Air is used at a rate of 0 cfm from a receiver at 110 degrees Fahrenheit (°F) and 140 psi. If the atmospheric pressure is 14.7 psi absolute (psia) and the atmospheric temperature is 75°F, how many cubic feet per minute of free air must be provided to the compressor (in cfm)?
2. Air is used at a rate of 30 cfm from a receiver at 100°F and 150 psi. If the atmospheric pressure is 14.7 psia and the atmospheric temperature is 80°F, how much free air must be provided to the compressor (in cfm)?
3. You are looking to purchase an 80-gallon (gal) air compressor. Assume that the compressor operates at 175 psi with air at 100°F. If the atmospheric pressure is 14.7 psia and the atmospheric temperature is 70°F, how big of a room would you need to place the compressor in if you wanted the room to be 10 times the size of the free air used by the compressor (in cubic feet)?
4. A typical portable garage compressor has a 30-gal tank and operates at 125 psi (gauge pressure). The delivery flow capacity of the compressor is approximately 5.5 standard cfm at 125 psi. Answer the following:
    a. What is the minimum (theoretical) horsepower required to drive the compressor before the compressor turns off (125 psi; in hp)?
    b. Home Depot website claims that an electric motor consumes 15 A @ 120 V. What is the efficiency? Hint: Electric AC power (single-phase) = voltage × current · $\sqrt{2}/2$ or P = V·I/$\sqrt{2}$; 1 hp = 746 watts.

IMG 9.1

5. If the overall efficiency of the previous compressor is 35%, what is the actual power required to drive the compressor (in hp)?

References

Hanlon, Robert. 2020. *Block by Block: The Historical and Theoretical Foundations of Thermodynamics*. Oxford University Press.
Turns, Stephen R. and Laura Pauley. 2020. *Thermodynamics: Concepts and Applications*.Cambridge University Press.

Figure Credits

Fig. 9-22a: Dresser-Rand, "Back-to-back multistage centrifugal compressor." Copyright © by Siemens.
Fig. 9-22b: Dresser-Rand, "Axial compressor." Copyright © by Siemens.
Fig. 9-22c: Atlas Copco, "Turbo Expander." Copyright © by Atlas Copco.
Fig 9-25: Copyright © by Fábio Teixeira (CC BY-SA 3.0) at http://commons.wikimedia.org/wiki/File:Compressor3.jpg.
Fig. 9-26a: Hydrovane, "Vane Compressor Rotor." Copyright © by Hydrovane.
Fig. 9-26b: MAPRO International, "Natural Gas Vane Compressor." Copyright © by MAPRO International.
Fig. 9-26c: Ingersoll Rand, "Compact Air Vane Motor," http://www.ingersollrandproducts.com/am-en/products/air-motors/direct-drive-multi-vane-air-motors/mva-and-movo-series. Copyright © by Ingersoll Rand.
Fig. 9-27a: Copyright © by Winfried Recker (CC BY-SA 3.0) at http://commons.wikimedia.org/wiki/File:Screw_Compressor.jpg.
Fig. 9-27b: Copyright © by Motorhead (CC BY-SA 2.5) at http://commons.wikimedia.org/wiki/File:Lysholm_screw_rotors.jpg.
Fig. 9-27c: Ingersoll Rand, "Oil Flooded Screw Compressor." Copyright © by Ingersoll Rand.
Fig. 9-27d: Ingersoll Rand, "Dry Screw Compressor." Copyright © by Ingersoll Rand.
Fig. 9-28: Copyright © by Motorhead; Adapted by Larry Villasmil and Robert Garrick (CC BY-SA 2.5) at http://commons.wikimedia.org/wiki/File:Lysholm_screw_rotors.jpg.
IMG 9-1: Grikalmis / German Wikipedia, https://commons.wikimedia.org/wiki/File:Kolbenkompressor.jpg, 2008.

CHAPTER 10

Fluid Power Auxiliary Devices

Overview and Learning Objectives

This chapter deals with the auxiliary devices and components required to operate a fluid power system. Although fluid power actuators and machinery are at the heart of any fluid power system, there are many other components required to smoothly operate this type of system, from the valves that direct the flow and protect the system from failure, storing and supplying a clean fluid, to those that provide auxiliary power in the case of main power outages or to minimize costs in variable=duty systems.

The chapter focuses first on the valves that protect the system from overpressurization and maintain a constant operating pressure in the system. Then we analyze the valves used to direct and control the flow supplied to the system.

The next theme is the storage of fluid, either oil or air, and how we provide clean fluid to all components in the system and maintain that cleanliness. The main conditioning measures are trapping solid particles and removing moisture from both oil and air.

The last section of this chapter focuses on hydraulic accumulators. Like an air receiver, the hydraulic accumulator allows storing fluid under pressure that can be supplied to the system when the demand exceeds the capacity of the pump. The accumulators also serve as a pressure pulsation dampener or as an emergency source of power.

After completing this chapter, students should be able to answer the following questions:

- What is a pressure relief valve (PRV)?
- What is the difference between a direct-acting and a compound PRV?
- What is cracking and full-flow pressure of a PRV?
- How does one calculate the cracking and full-flow pressure setting of a PRV?
- How does a pressure regulator work?
- What is a directional control valve?
- How does a directional control valve work?
- What is a flow control valve?
- How do flow control valves work?
- What is a hydraulic reservoir, and what is its function?
- What is a pneumatic receiver, and what is its function?
- How does one calculate the size of a pneumatic receiver?
- How does a filter work?
- How does an air dryer work?
- How does one estimate the moisture content in compressed air?
- What are the different types of accumulators?

- How does a hydraulic accumulator work?
- How does one size a hydraulic accumulator?

Valves

Pressure Relief

We discussed extensively in Chapter 9 that the great majority of machinery used in the fluid power industry is of the positive-displacement type. We also discussed the fact that a pump and sometimes a compressor will "push" fluid into the hydraulic or pneumatic system continuously. If any of the actuators in the systems does not demand the amount of fluid that the pump or compressor is pushing, the system pressure will increase very rapidly. With no demand at all, the pressure will rise almost instantly. To prevent failure of either the machinery, piping, or any other components in the fluid power circuit under pressure, a mechanism to release the excess pressure must be provided. This excess pressure is alleviated by a **pressure relief valve** (**PRV**). The PRV is the most widely used single valve or component in any fluid power circuit or system.

Figure 10-1 shows the schematic of a direct-acting or single PRV. The schematic sequence represents the basic three states in which the valve operates. In normal operation, the PRV is closed. The compressed spring (7) force holds the valve seat (2) against the inlet nozzle (1). If the operating conditions of the fluid power circuit cause the pressure to rise, the force (F_P) exerted by the pressure on the valve seat will increase until it matches the compressed spring force (F_K). In this condition, the valve is said to be cracked; the force due to pressure is equal to the force exerted by the spring. As the pressure continues to rise, the spring will be compressed further, and the valve will open, releasing fluid back to the reservoir, in the case of hydraulic circuits, or to the atmosphere, in the case of pneumatic systems.

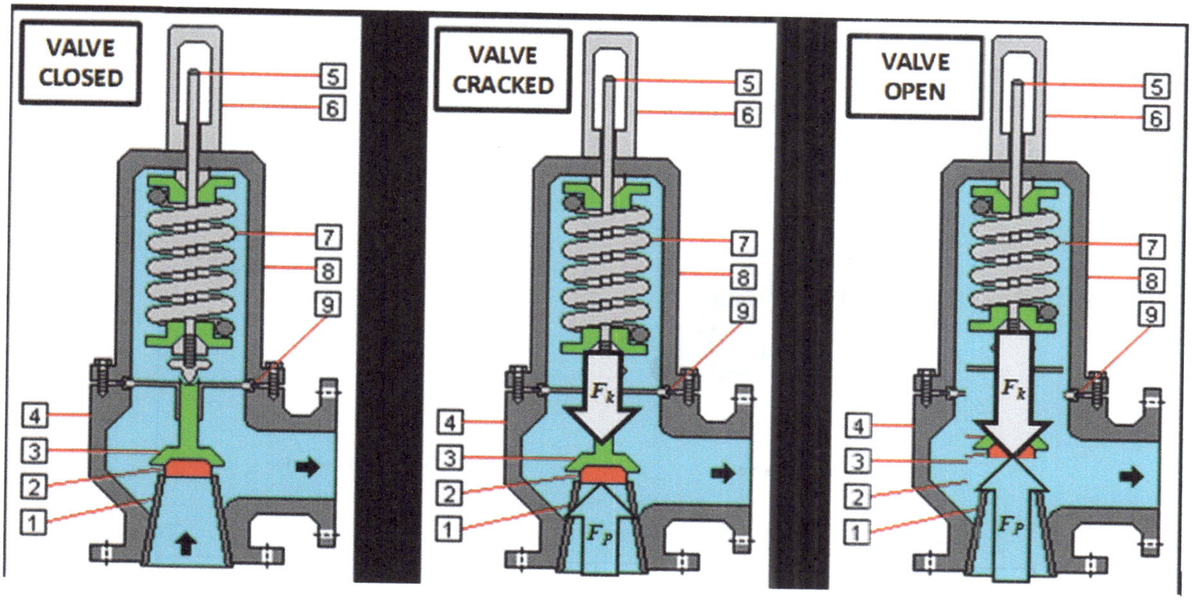

Figure 10-1 Schematic and operation of a direct-acting pressure relief valve.

Figures 10-2a and 2b show an actual direct-acting PRV and its schematic with the addition of a cylinder lock, which secures the cracking pressure set by an adjusting screw. As shown in the schematic, port P (circled in red) serves as the inlet nozzle, and if pressure within the hydraulic circuit rises above a preset pressure, port T (circled in green) is opened, allowing flow to return to the reservoir and thus limiting the pressure in the system. Figures 10-2c and 2d show the operating curves of a medium-pressure stage valve (up to 70 bars or 1,000 pounds per square inch [psi]) and a high-pressure stage valve (up to 320 bars or 4,600 psi). The operating characteristics of a direct-acting PRV are relatively simple to determine. With the principles learned in early chapters, we can estimate the cracking pressure of the valve if the spring constant and inlet nozzle area are known. Conversely, we can estimate the spring constant required to keep a valve of a particular size closed. The cracking pressure ***P***C of a valve with a nozzle diameter ***D***N and a spring with constant ***k*** is given by:

$$A_N = \frac{\pi}{4} D_N^2$$

$$P_C = \frac{F_k}{A_N}$$

$$F_K = k \cdot \delta_O$$

where:

- A_N = valve nozzle area,
- D_N = diameter of the valve nozzle,
- P_C = cracking pressure,

- F_k = spring compression force,
- k = spring constant, and
- δ_O = spring deflection (length as installed in the valve versus the no-load length).

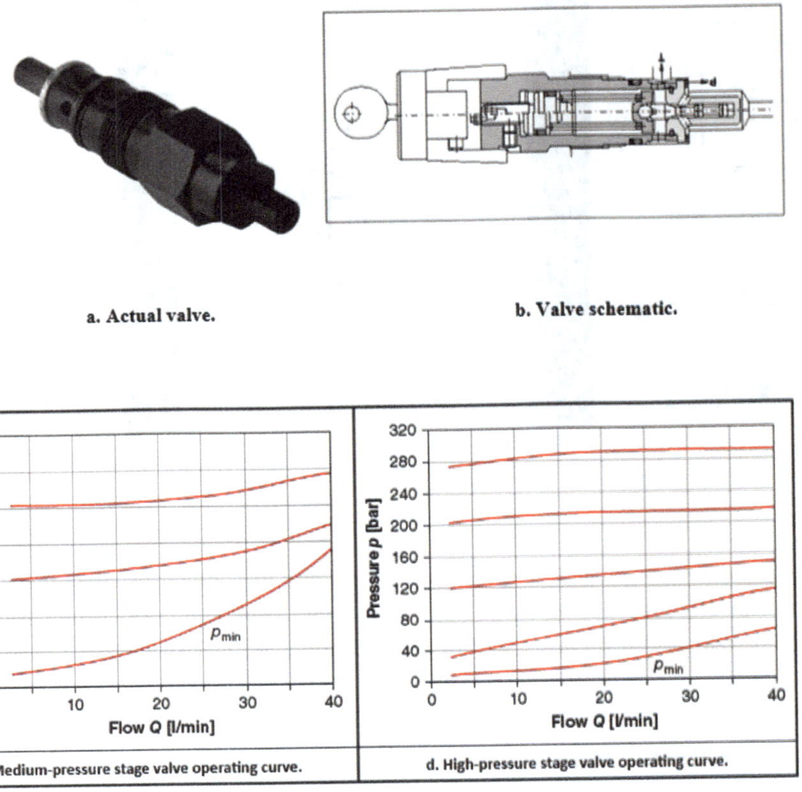

a. Actual valve.

b. Valve schematic.

c. Medium-pressure stage valve operating curve.

d. High-pressure stage valve operating curve.

Figure 10-2 NG06 direct-acting pressure relief valve, Series EVSA.

Similarly, the fully open or full-flow pressure **P**PRV of the valve is given by:

$$p_{PRV} = \frac{F_{K_{max}}}{A_N}$$

$$F_{k_{max}} = k \cdot \delta_{max}$$

$$\delta_{max} = \delta_O + \delta_{FF}$$

where:

- P_{PRV} = PRV setting,
- F_{kmax} = maximum spring compression force,
- δ_{max} = total spring deflection when the valve is fully open, and

- δ_{FF} = additional spring deflection required to fully open the valve from the closed state.

Direct-acting pressure relief valve construction, operation, and maintenance are relatively simple. They have big internal passages (no clogging) and very few moving parts (one spring and seal). The drawback of this type of PRV is the significant difference between the cracking pressure and the fully open pressure. Taking the middle curve in Figure 10-2c as an example, with the valve set to have a cracking pressure of 30 bars, the pressure will increase up to 45 bars for the valve to be fully open (Q = 40 liters per minute [lpm]). The difference in pressure represents a 50% increase. Although the relative difference between the cracking pressure and full-flow pressure reduces as the pressure rises, the difference remains fairly large. The middle curve in Figure 10-2d shows a cracking pressure setting of 120 bars. At full flow, the pressure in the high-pressure stage valve will increase up to 150 bars, or 25%.

Figure 10-3 shows the schematic of a pilot-operated or compound PRV. The schematic sequence represents the two basic states in which the valves operate. During normal operation, the PRV is closed (Figure 10-3a). The compressed pilot spring force maintains the pilot valve dart closed. If the operating conditions of the fluid power circuit cause the pressure to rise, the force (F_P) exerted by the pressure on the dart will increase until it matches the pilot spring compression force (F_K; Figure 10-3b). In this condition, the valve is said to be cracked; the force due to pressure is equal to the force exerted by the pilot spring. As the pressure continues to rise, the pilot spring will be compressed further, and the valve dart will open, allowing fluid back to the reservoir through the pilot drain, but the valve spool will momentarily remain in the same position, closing the valve. Indeed, the pilot spring only controls the pressure in the valve's upper chamber. The release of fluid from the upper chamber creates a pressure difference that drives fluid to flow to the upper chamber through a small orifice drilled in the spool skirt (Figure 10-3c). This pressure difference is determined by the main spool spring. Once the pressure difference that builds across the spool skirt generates enough force (ΔF) to compress the main spool spring, the valve spool is lifted, and the valve opens.

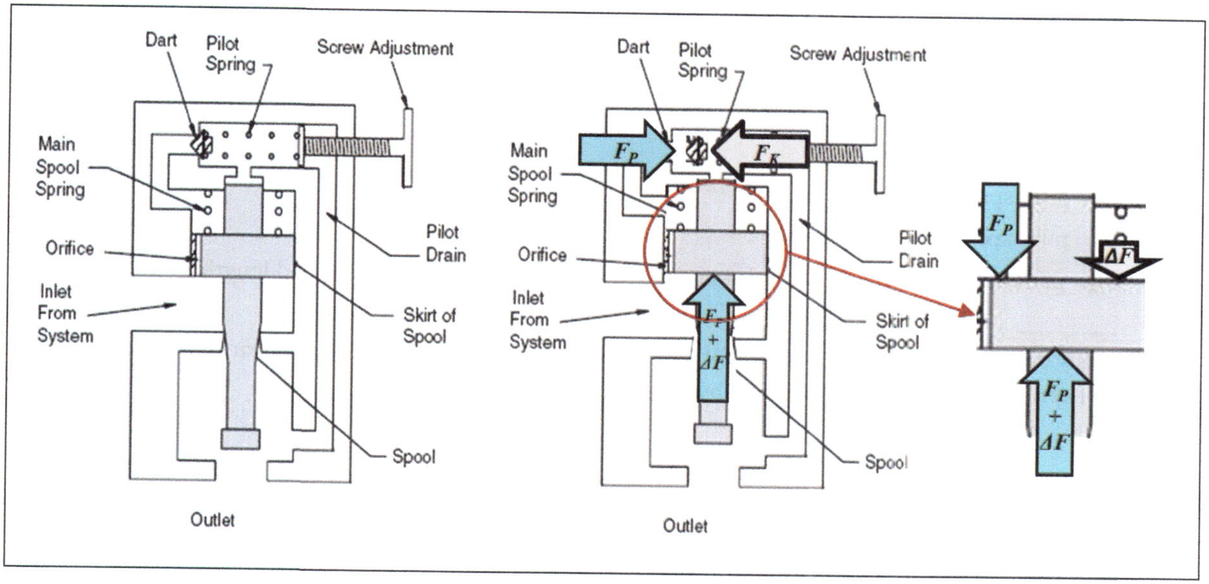

Figure 10-3 Schematic and operation of a pilot-operated pressure relief valve.

Figures 10-4a–c show an actual pilot-operated PRV and its schematic both with and without a directional control valve (DCV). The pilot-operated PRV depicted in Figure 10-4a operates in much the same manner as the simplified version shown in Figure 10-3 with the inlet of pressurized fluid at port P and the outlet at port T, as seen in figures 10-4b and 10-4c. If the pressure of the fluid entering at port P exceeds that set by the pilot spring (9), holding the dart (or, in this case, ball [8]) closed, the pilot will open, allowing the fluid to flow through to the pilot drain via the tank control line (13) and thus reducing the pressure behind the main spool and permitting it to open so that the fluid may flow directly from port P to port T, maintaining the set operating pressure. The DCV seen on the PRV in Figure 10-4a and c essentially allows for the pilot operation of the valve to be "turned off" by blocking the flow through the orifice bore (11) to the dart (8) and subsequently pilot spring (9) while still allowing it behind the main spool through the secondary control line (7) to hold it closed. Figure 10-4e shows the operating curves of a medium-pressure stage valve (up to 70 bars or 1,000 psi) and a high-pressure stage valve (up to 320 bars or 4,600 psi). The operating characteristics of a pilot-operated or compound PRV are also simple to determine. Following the analysis done for the direct-acting PRV, we can estimate the cracking pressure of the valve if the pilot spring constant and pilot dart opening area are known. Conversely, we can estimate the spring constant required to keep a valve of a particular size closed. The cracking pressure **P**C of a valve with a dart opening diameter **D**D and a pilot spring with constant ***k***D is given by:

$$p_C = \frac{F_K}{A_D}$$

$$F_K = k_D \cdot \delta_D$$

$$A_D = \frac{\pi}{4} D^2{}_D$$

where:

- A_D = pilot valve dart opening area,
- D_D = diameter of the *pilot* valve dart opening,
- P_C = cracking pressure,
- F_k = pilot spring compression force,
- k_D = pilot spring constant, and
- δ_D = pilot spring deflection (length as installed in the valve versus the no-load length).

The fully open or full-flow pressure **P**PRV of a compound relief valve with a spool skirt effective area **A**S and a main spool spring with constant ***k***S is given by:

$$P_{PRV} = \frac{F_k}{A_D} + \frac{\Delta F}{A_S}$$

$$\Delta F = k_S \cdot \delta_S$$

where:

- P_{PRV} = PRV setting,
- A_S = effective spool skirt area,
- ΔF = spool skirt spring compression force,
- k_S = spool skirt spring constant, and
- δ_S = spool skirt spring deflection (length as installed in the valve versus the no-load length).

It should be noted that in this derivation, we are neglecting the additional pressure required to fully open the pilot valve. The pilot valve needs to be only slightly open to allow the flow that trickles through the small orifice in the main valve spool skirt.

a. Actual valve.

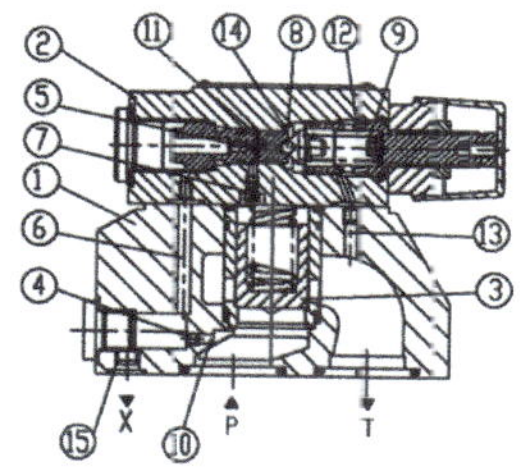

b. Valve schematic.

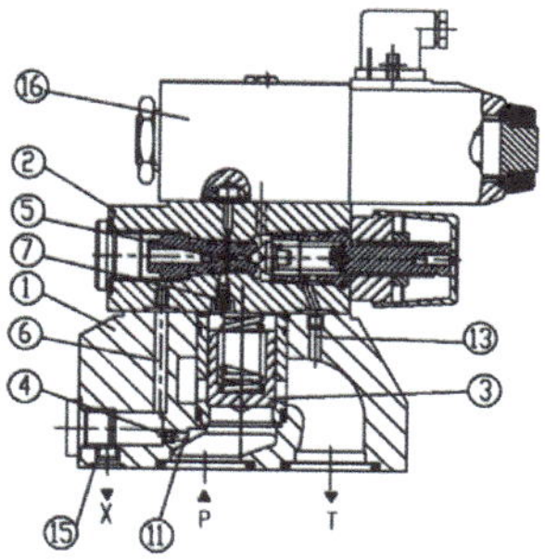

c. Valve schematic (with DCV).

1. Main Valve
2. Pilot Assembly
3. Spool Assembly
4. Main Orifice
5. Secondary Orifice
6. Main Control Line (passage)
7. Secondary Control Line (passage)
8. Ball
9. Spring
10. Direct Control Line (passage)
11. Orifice Bore
12. Spring Chamber
13. Tank Control Line (Passage)
14. External Control Line (Passage)
15. Unloading Control Line (Passage)
16. Built-in Directional Control Valve

d. Valve schematic nomenclature.

Figure 10-4a-d

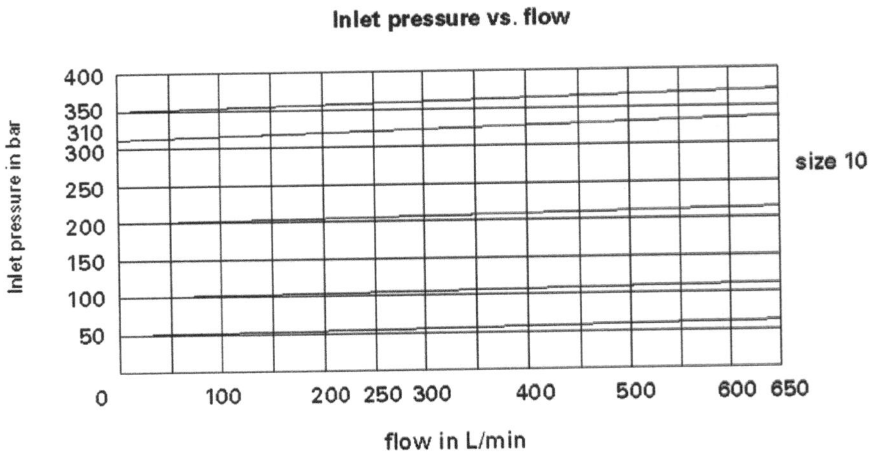

e. Valve operating curve.

Figure 10.4e Pilot-operated pressure relief valve, Series CG2V.

Pilot-operated PRV construction, operation, and maintenance are more complex than that of a direct-acting PRV. Although they also have big internal passages (no clogging), they have more moving parts (two springs and a spool skirt) and small passages and/or tight clearances (spool skirt orifice and the spool skirt itself). The main advantage of this type of PRV is the reduced difference between the cracking pressure and fully open pressure. Taking the second curve in Figure 10-4e as an example, with the valve set to have a cracking pressure of 100 bars, the pressure will increase up to 110 bars for the valve to be open (Q = 650 lpm). The difference in pressure represents only a 10% increase. An additional advantage of this compound PRV is that the pilot can be connected to a second pilot valve, either directly or through a DCV. This second pilot valve installed on the valve assembly or remotely located would allow overriding the main relief valve to set a lower-pressure-relief value. In this way, the compound PRV becomes in practice a dual-pressure-setting relief valve.

Pressure Regulator

A **pressure regulator**, or a pressure-reducing valve, is a device that maintains the fluid power system, hydraulic or pneumatic, pressurized up to a level previously set. As both names imply, this type of valve reduces the pressure of the fluid that the pump or compressor provides to the system and regulates its operating value regardless of pressure fluctuations or changes in flow demands. As discussed in previous chapters, the pressure in any fluid power system is determined mostly by the demands and loading of cylinders, motors, and, to an extent, fittings and accessories. A regulated constant pressure facilitates the design, the selection, and the operation of all components in the fluid power system. A typical fluid power system has multiple actuators operating separately and uncoordinated, so the flow demand is highly variable. Pressure regulators and pressure-reducing valves ensure the smooth operation of any fluid power system.

Figure 10-5a shows an actual direct-operating pressure regulator or pressure-reducing valve with manual screw set adjustment. The direct-operating pressure reducing valve has three connections. The fluid power system supply is connected to the pressurized port P, circled in red. The actuators and components of the system seen in the section view schematic presented in Figure 10-5b are connected to the regulated port A, circled in green. Port

T, circled in blue, is connected back to the tank or reservoir of the fluid power system. In terms of design and operation, the pressure-reducing valve is essentially a normally open direct-acting pressure relief valve, as can be seen in the International Organization for Standardization (ISO) symbol depicted in Figure 10-5c, where it is shown that under normal operation, fluid is allowed to flow freely from port P to port A. It is only when pressure rises above the set level established by the screw preloading the spring that the spool shifts positions to divert flow to port T until the pressure reduces back to operating levels. In Figure 10-5d, the supply pressure acts upstream on both ends of the valve spool, creating opposite forces (F_P) of equal magnitude that cancel each order. The regulated pressure acts downstream on one side of the valve spool, creating a force (F_R) that compresses the valve spring, moving the valve spool upward—indeed, opening port T of the valve—and releasing excess fluid back to the tank. The regulated pressure force (F_R) on the valve spool is balanced by the force exerted by the valve spring (F_K) and a residual force (F_T) generated by possible back pressures present in the lines on the tank side of the valve. The compression of the valve spring will determine the value of the regulated pressure.

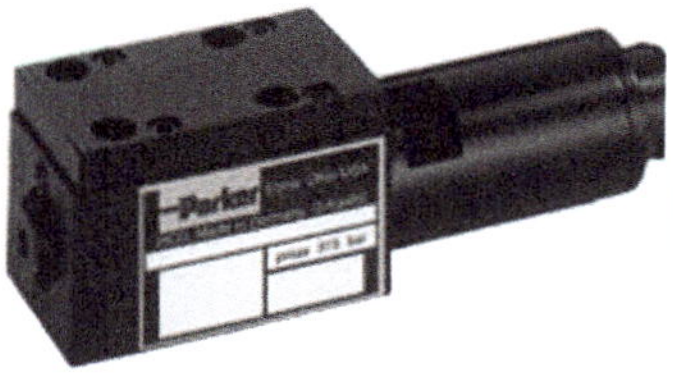

Figure 10-5a. NG06 actual reducing valve.

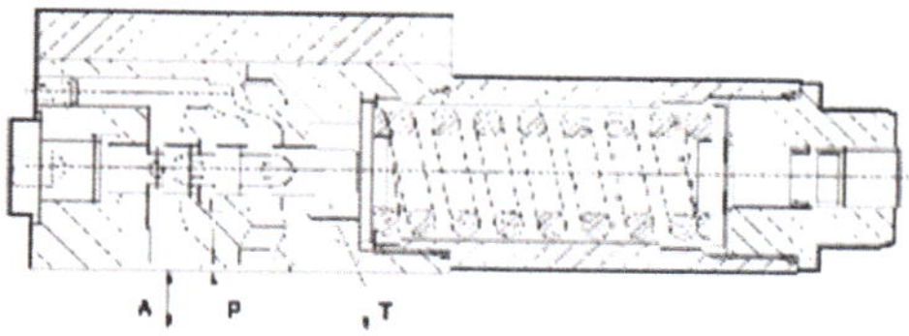

Figure 10-5b. NG06 reducing valve schematic.

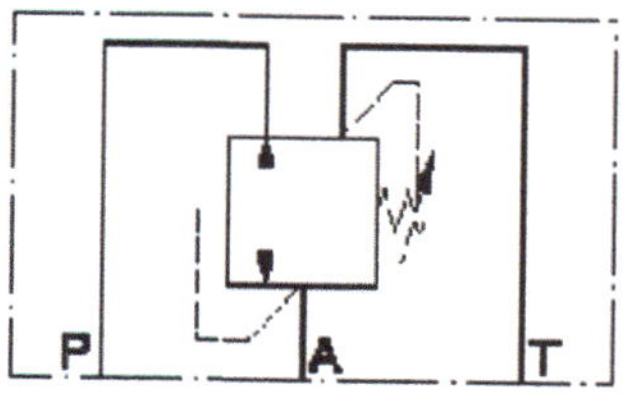

Figure 10-5c. NG06 regulator ISO symbol.

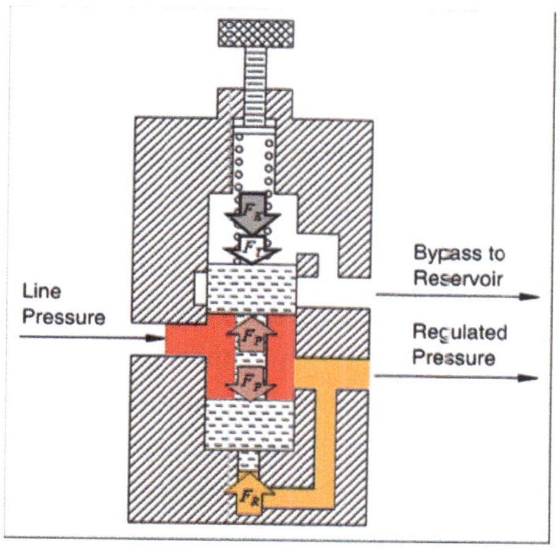

Figure 10-5d. Reducing valve operation. Fluid power circuits and controls. John S. Cundiff.

Figure 10-5 Direct-operating pressure regulator or reducing valve, Series VM.

Like direct-acting PRVs, the construction, operation, and maintenance of a direct operating pressure-reducing valve are relatively simple. They also have big internal passages (no clogging) with very few moving parts (the valve spool, one spring, and seals). Similarly, the drawback of this type of pressure-reducing valve is the significant variation in the regulated pressure as the flow demand of the system changes. With no flow demand or supply pressure, the force in the spring keeps the valve open. In normal operating conditions, the regulated pressure force is balanced by the spring force, and any reduction or increase in flow demand would either open or close the tank port. Hence, the tank port opening is determined by the compression of the spring. Fully opening the tank port requires a significantly larger force than keeping the port closed. The difference in the spring compression force will directly translate into an equivalent variation in the regulated pressure.

Figure 10-6 presents the operating curve of a typical pressure regulator or reducing valve. Taking as an example the upper curve in the figure, the pressure set to be regulated at 25 bars (Q = 0 lpm) changes to approximately 21 bars at maximum flow (Q = 25 lpm). This variation represents a 15% reduction from operating at no flow to providing the maximum flow. In this case, the fluid power system shall be designed to accommodate the inherent operating pressure variation.

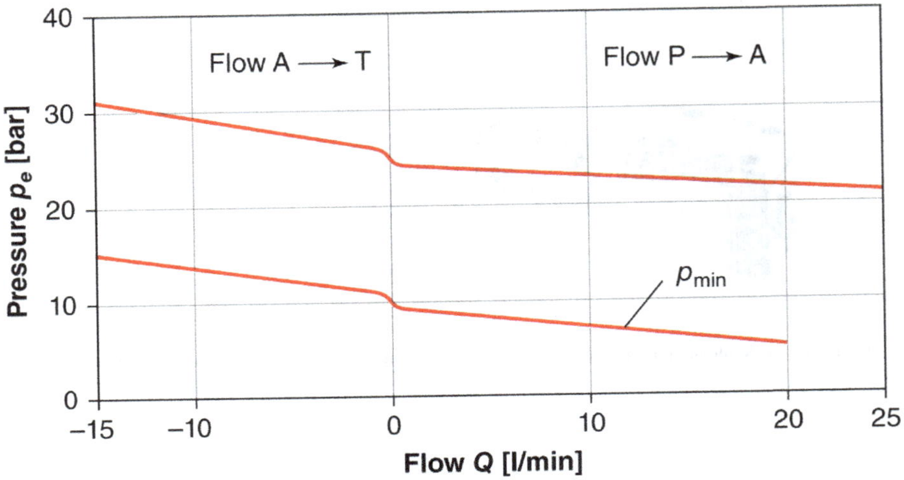

Figure 10-6 NG06 reducing valve operating curve.

Figures 10-7a and 7b show a pilot-operated pressure regulator or reducing valve and its schematic. Like in a pilot-operated PRV such as that depicted in figures 10-4a and 10-4b, the pressure to be regulated is set by the pilot spring compression (11) force. In normal operation, the pilot valve (2) seen in Figure 10-7b is closed, and the main spool insert (3) is open. If the operating conditions in the fluid power circuit cause the pressure to rise, the pilot valve opens, allowing fluid in the spring chamber (12) to flow back to the tank or reservoir via the control line (15). This secondary flow generates a pressure difference across the main spool insert orifice (4), creating a force that displaces the insert and regulating the main flow. As in the PRV, the pressure-reducing valve opening and closing is regulated by the spring in the main spool insert and not by the pilot valve spring. The only function of the pilot valve is to signal the main spool insert to start regulating the flow once the valve setting is reached.

Pilot-operated pressure regulator/reducing valves' construction, operation, and maintenance are more complex than that of a direct operating pressure regulator/reducing valve. Although they also have big internal passages (no clogging), they have more moving parts (e.g., two springs, spool insert, check valves) and small passages and/or tight clearances (spool insert orifice and the spool insert itself). The main advantage of this type of pressure regulator/reducing valve is the reduced difference between the zero-flow and the full-flow pressure. Figure 10-7c demonstrates this, presenting the operating curve of a typical pressure regulator or reducing valve. At full flow (Q = 400 lpm), the operating pressure is 315 bars. At zero flow, the operating pressure is 320 bars. The difference in pressure represents only a 2% increase, which is just a fraction of the operating pressure variation of a direct-operating pressure regulator valve. An additional advantage of this valve is that the pilot can be connected to an external valve, either directly or through a DCV to override the regulating function or to add a secondary pressure setting. Figure 10-7d presents the ISO symbol of a pilot-operated pressure-reducing valve with a check valve. It shows how the controlled pressure (A) is fed back to the regulator or could go back to the pressurized side (B) through the check valve in case of a system malfunction. The symbol also shows the internal port that communicates the supply pressure to the adjustable spring that sets the pressure value to be maintained.

a. Actual regulator.

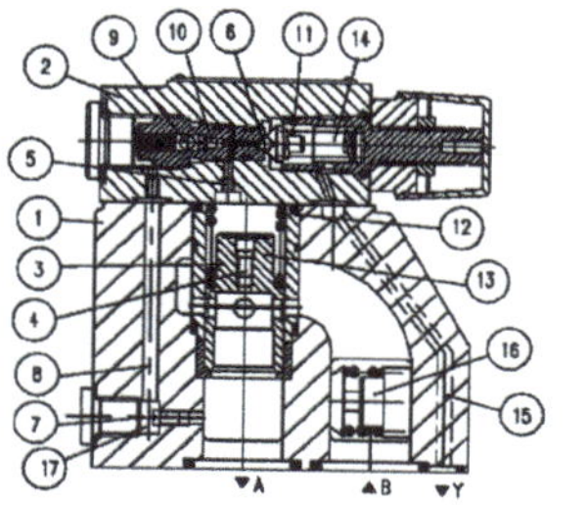

1. main valve
2. pilot valve
3. main spool insert
4. orifice
5. port
6. ball
7. orifice
8. control line
9. check valve
10. orifice
11. spring
12. spring chamber
13. control piston
14. spring chamber
15. control line
16. optional check valve
17. pressure gauge connection

b. Regulator schematic.

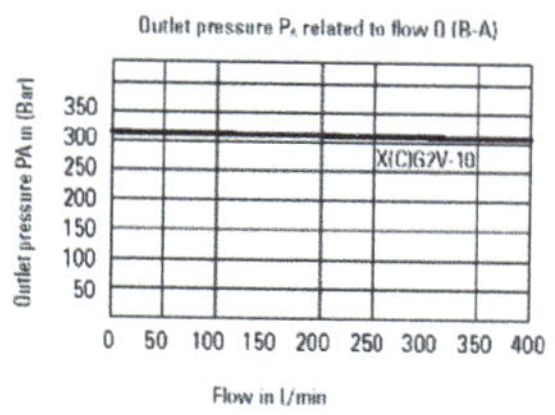

c. G2V-10 regulator operating curve.

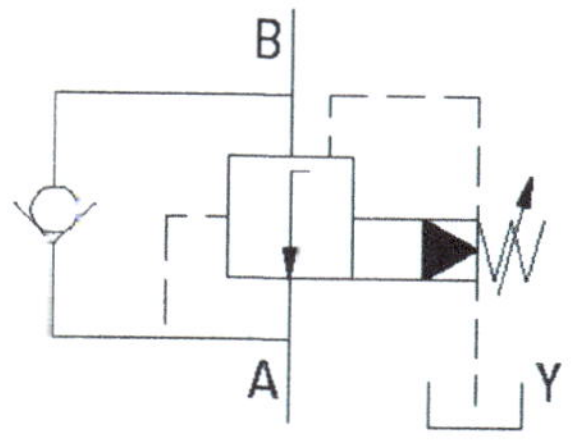

d. G2V-10 regulator ISO symbol.

Figure 10-7 Pilot-operated pressure regulator (reducing valve), Series XCG-2V-10.

Figure 10-8a shows a direct-operating air pressure regulator of the diaphragm type. The operation of an air or pneumatic pressure regulator is similar to the hydraulic pressure of reducing valves' regulators, discussed previously. Although the term pressure-reducing valve is applicable to the air pressure regulator, such terminology is seldom used. In the figure, the left port is connected to the air supply. The position of the valve disk is determined by the forces acting on the diaphragm. In normal operation, the air-regulated pressure acting on the bottom of the diaphragm compresses the spring on top of it until both forces equalize. When the pressure drops on the discharge or regulated side, the adjusting spring pushes down the diaphragm, increasing the valve opening by displacing the valve disk and thus allowing more air flow. Conversely, when the pressure increases on the discharge side, the large pressure on this side pushes up the diaphragm, compressing the adjusting spring, reducing the valve opening by displacing the valve disk, and restricting the air flow. Like hydraulic direct-acting PRVs and reducing valves, the setting of an air pressure regulator is set by adjusting the main valve spring. The typical operating curve of an air pressure regulator is shown in Figure 10-8b.

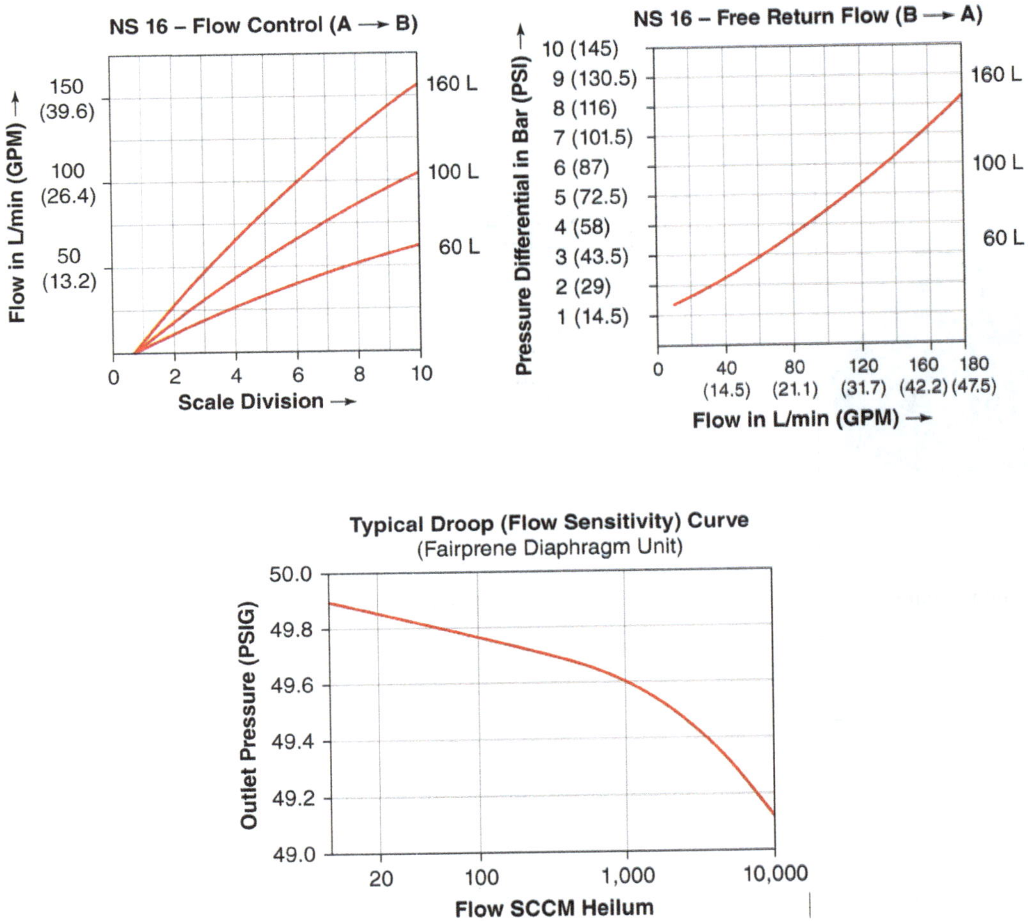

Figure 10-8 Direct-acting air pressure regulator, Model 8310.

Figures 10-9a and 10-9b show a pilot-operated air pressure regulator of the diaphragm type. The main difference between a direct regulator and a pilot regulator is the replacement of the main adjusting spring by a pilot signal. Rather than manually adjusting the spring to set the pressure at the required value, a pilot signal at this pressure is externally sent to the valve. Similarly, the position of the valve disk is determined by the forces acting on the diaphragm. In normal operation, the air-regulated pressure acting below the diaphragm matches the pilot signal pressure acting above it. When the pressure drops on the discharge side, the pilot signal pressure pushes down the diaphragm, opening the valve further until the discharge pressure acting below the diaphragm increases to match the pilot signal pressure. Conversely, a pressure increase on the discharge side pushes the diaphragm up, closing the valve restricting the air flow to regulate the discharge pressure and match the pilot signal pressure acting above the diaphragm. The main advantages of a pilot-operated air pressure regulator are the remote adjustment of the regulated pressure or valve setting and minimizing the pressure variation between the no-flow and full-flow operating conditions. Figure 10-9c shows the typical operating curve of a pilot-operated air pressure regulator.

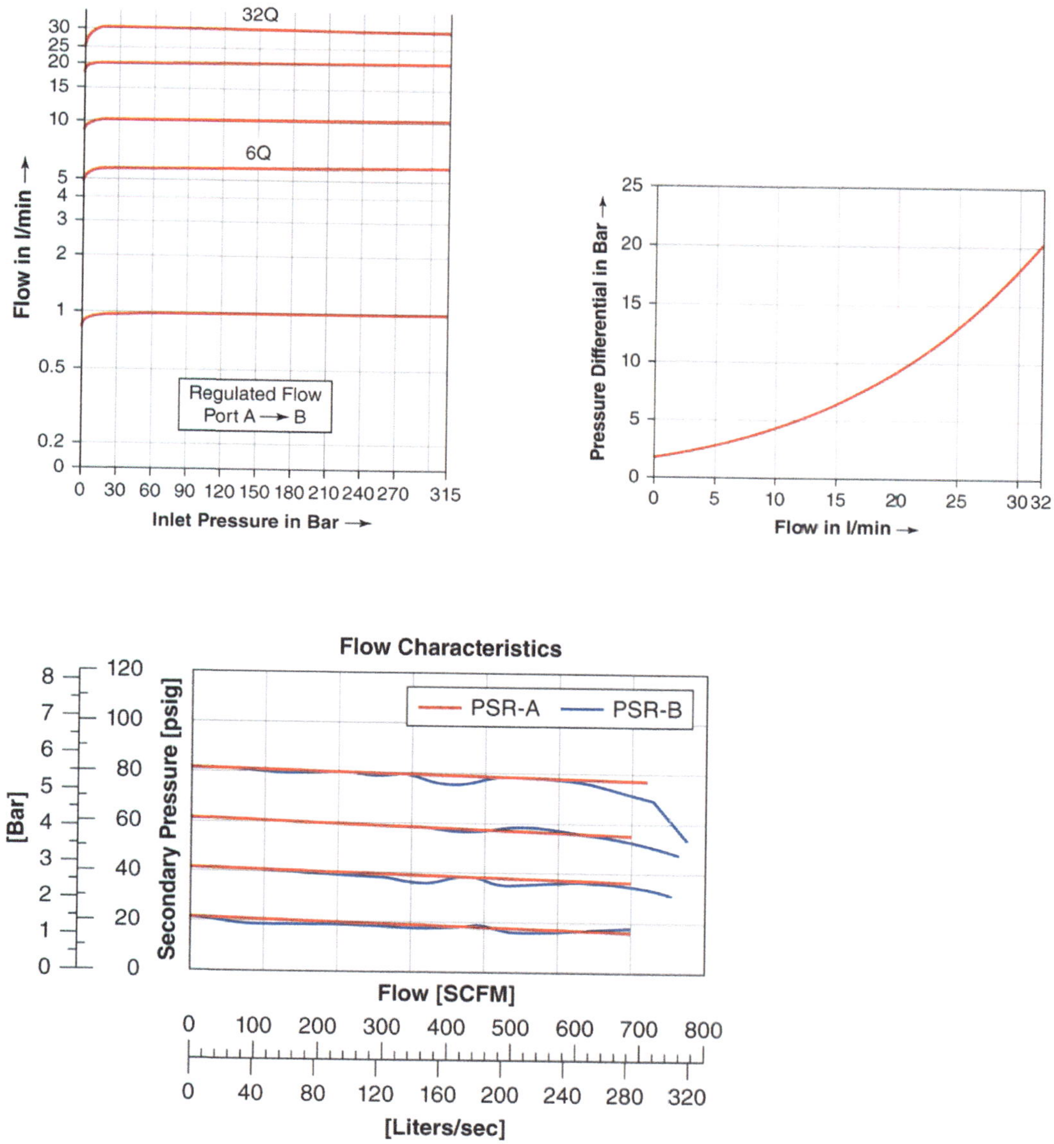

Figure 10-9 Pilot-operated air pressure regulator, Pro Air Series.

Directional Flow Control Valves

The DCV is one of the most important components of any fluid power system, second only to the pump or compressor supplying the power to the system. The function of a **DCV** is to direct the fluid—oil or air—to the various components in the system. The DCV directs the movement of the fluid so that it can do work where needed. The more common DCV designs are two-way, including two ports and two internal passages; three-way, with three ports and three internal passages; and four-way, including four ports and four internal passages. A DCV can be operated manually, mechanically, electrically, or remotely by a pilot arrangement.

Figure 10-10a shows a mechanically actuated directional spool valve. This valve controls the start, stop, and direction of the flow. The valve basically consists of a housing (1), the actuating element (2; in this case, a straight roller), a control piston (3), and one or two return springs (4; two or three position valves). In the de-energized (deactivated) condition, the return spring (4) maintains the control piston (3) in the stable or initial position of this two-position DCV. If the roller (2) is actuated, the control piston (3) is moved to the desired spool position.

In a three-way two-position (3/2) DCV like the one depicted by the ISO symbol presented in Figure 10-10b, the pressure port P connects directly to the straight supply port A in the stable or initial position, spool position a in Figure 10-10c, while the cross-supply port B is blocked. When the 3/2 DCV is actuated, spool position b in Figure 10-10c, the pressure port P directs the flow to the cross-supply port B, while the straight supply port A is blocked. In a four-way two-position (4/2) DCV, both pressure P and return T ports are connected directly to the straight-supply A and cross-supply B ports, respectively, when in the stable or initial position, spool position a in Figure 10-10c. When the 4/2 DCV is actuated, spool position b in Figure 10-10c, the pressure port P directs the flow to the cross-supply port B, while the straight-supply port A returns the flow to the tank T port.

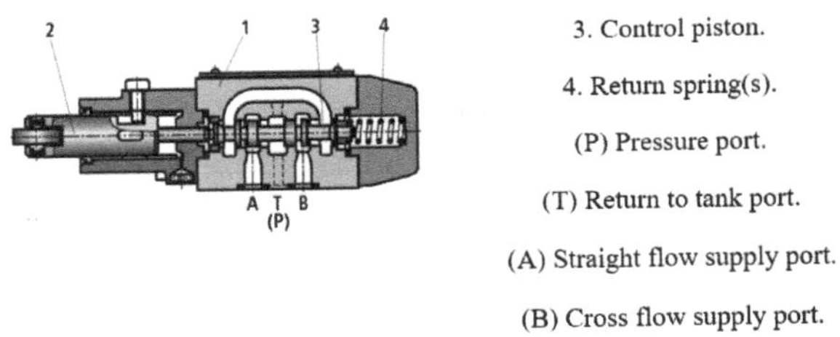

a. DCV schematic and main components.

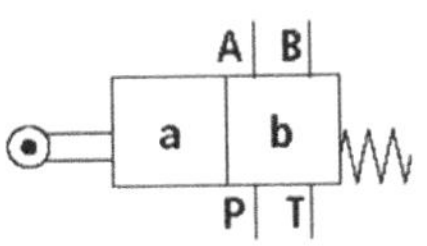

b. DCV ISO symbol, undefined control paths.

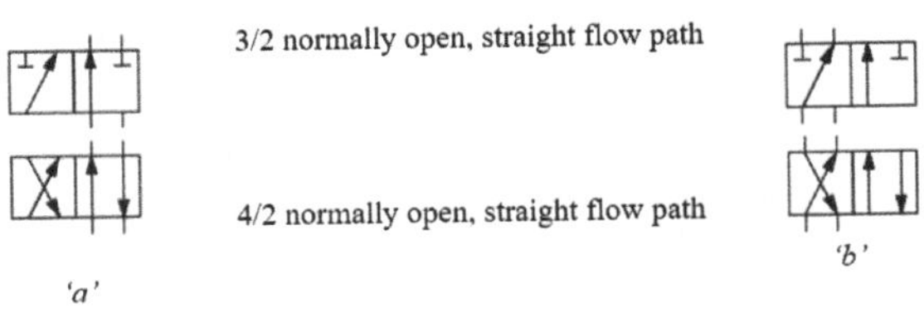

c. DCV control paths' ISO symbols, spool positions "a" and "b."

Figure 10-10 4/3 straight roller–actuated DCV.

Figure 10-11a shows a manually actuated directional spool valve. Similarly to the valve described previously, this valve controls the start, stop, and direction of the flow. This valve also consists of a housing (1), the actuating element (2; in this case a manual lever), a control piston (3), and two return springs (4; a three-position valve). In the de-energized (deactivated) condition, the return springs (4) maintain the control piston (3) in the stable or center position of the three-position DCV.

In a four-way three-position (4/3) DCV like the one depicted by the ISO symbol presented in Figure 10-11b, the design of the control piston defines the type of valve. In a normally closed center-spool design, all DCV ports are closed or blocked when the valve is in the stable position. When the DCV is activated by pulling the manual lever, the pressure port P connects directly to the straight-supply port A, spool position b in Figure 10-11c, while the cross-supply port B returns flow to the tank port T. When the DCV is activated by pushing the manual lever, the flow path is redirected, the pressure port P connects now to the cross-supply port B, spool position a in Figure 10-11c, while the straight supply port A is the one returning flow to the tank port T. Figure 10-11c presents the three most common 4/3 (ports/ways) spool valve configurations, normally closed center, both with tank return or without it, and normally open center, also with tank return.

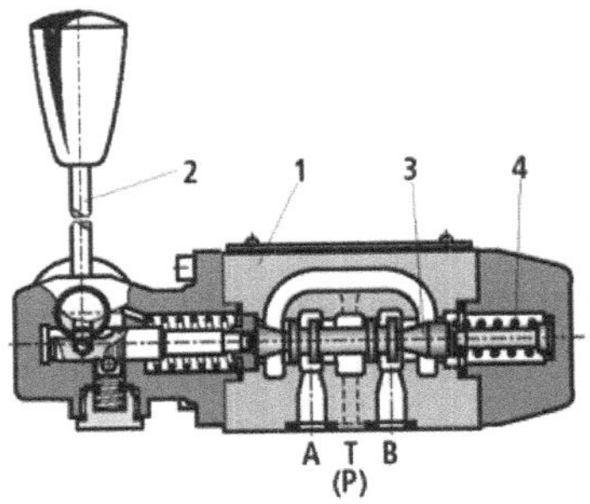

1. Housing.
2. Actuating element (lever).
3. Control piston.
4. Return spring/s.

(P) Pressure port.

(T) Return to tank port.

(A) Straight flow supply port.

(B) Cross flow supply port.

a. DCV schematic and main components.

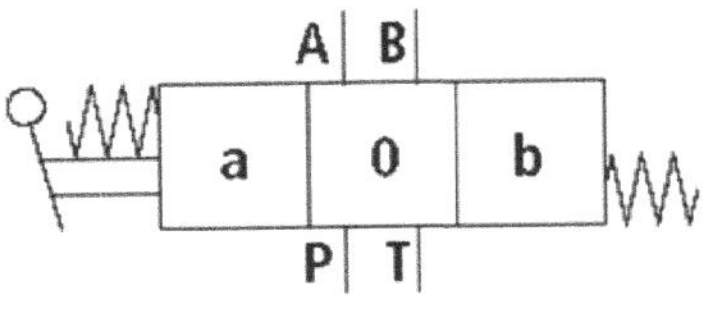

b. DCV ISO symbol, undefined control paths.

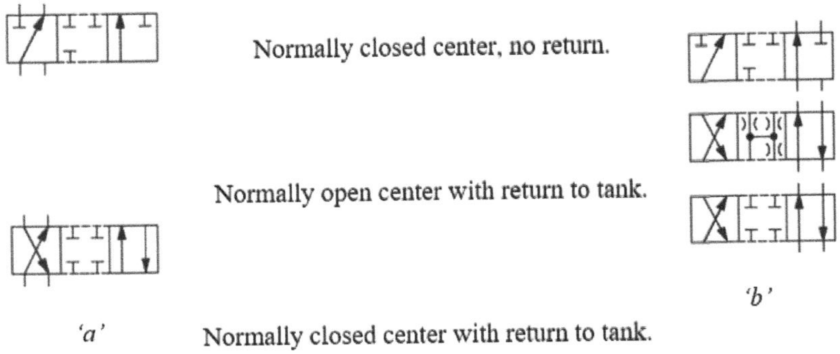

c. DCV control paths' ISO symbols, spool positions "a" and "b."

Figure 10-11 4/3 lever-actuated DCV.

Figure 10-12a displays a solenoid-operated directional spool valve of the "sandwich," or stackable type. Like the two previous valves, this stackable valve controls the start, stop, and direction of the flow and consists of a housing (1), a control spool (3), and two return springs (4) but, in this case, with either one or two solenoids (2) as the actuating elements. In the de-energized condition, the control spool (3) is held in the stable or initial position by the return springs (4). Energizing the solenoids (2) actuates the control spool (3) by their wet-pins. The force of the solenoid (2) acts via plunger (5) on the control spool (3) and pushes the latter from its rest position to the required end position. This opens up the required flow direction. When the solenoid (2) is de-energized, the return spring (4) pushes back the control spool (3) to its rest position. If required, a manual override (6) would allow the control spool (3) to be actuated without energizing the solenoid (2). In the configuration depicted by the ISO symbol shown in Figure 10-12b, the DCV valve acts as an isolator. When the solenoid (2) is de-energized, the straight-supply ports A, component and plate sides, are connected (②↔①) and the cross-supply ports B are also connected (②↔①). When the solenoid (2) is energized, both component side supply ports (①), straight A and cross B, are blocked; while the flow gets redirected between both plate side supply ports (②), straight A and cross B.

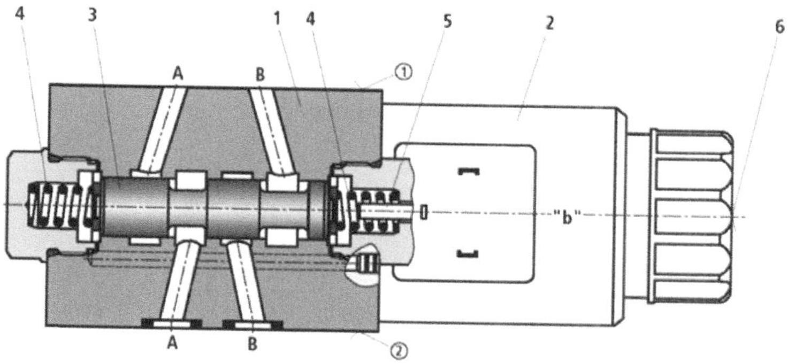

a. DCV Schematic (manual override).

1. Housing.
2. Solenoid(s).
3. Control spool.
4. Return springs.
5. Plunger.
6. Manual override (optional).

(A) Straight flow supply port.

(B) Cross flow supply port.

b. DCV main components.

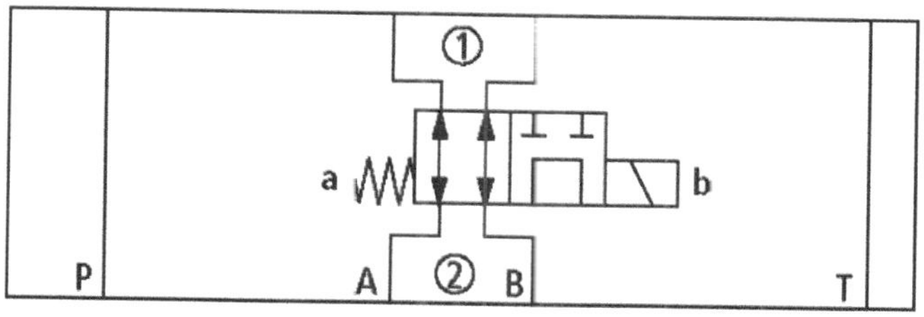

c. Isolator valve ISO drawing.

Figure 10-12 4/2 directional isolator valve.

Figure 10-13a shows an actual open-center sectional valve block system (often referred to as an open center mono-block valve system), which is widely applied in compact machinery (e.g., excavators, mini-excavators, back-hoe loaders, horizontal directional drilling, forklifts, and agricultural machinery). Figure 10-13b offers an exploded view of just one of the DCVs seen connected in series in the "mono-block" depicted in Figure 10-13a. Each of the DCVs operates similarly to the spool valve described above and seen in Figure 10-12a, with several of them being "stacked" having two common central passages running through them all to supply and return the hydraulic fluid. When all the valve spools are centered, the pump flow will simply run straight through and return to the tank at low pressure through the open center bypass. When a spool in one of the individual DCVs is actuated, the open center bypass is blocked and the inlet is connected to one of the work ports with the opposite work port being connected back to the tank, completing the circuit. Figure 10-13c shows a schematic for a potential use of the mono-block valve system in an excavator where it can be seen that in essence each of the separate valves within the system is responsible for controlling the flow to discrete actuators such as the Dozer, Arm, Boom, and Bucket. This allows for the control of the entire mechanism from a centralized location in a substantially more compact manner than would be possible if each DCV was a distinct unit, simplifying the overall plumbing of the machine by effectively reducing the number of inlet and return lines needed.

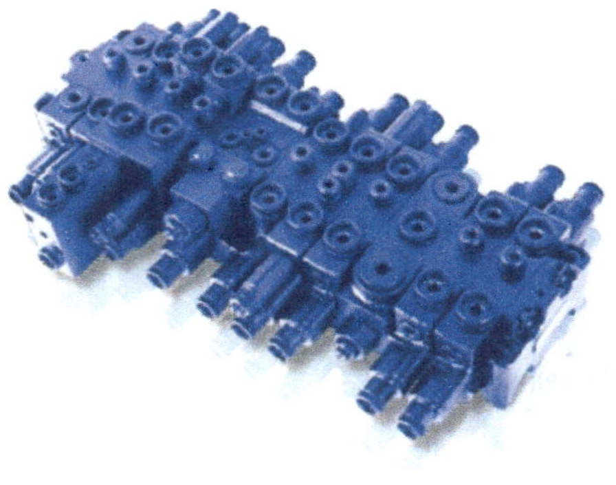

a. Open-center block system.

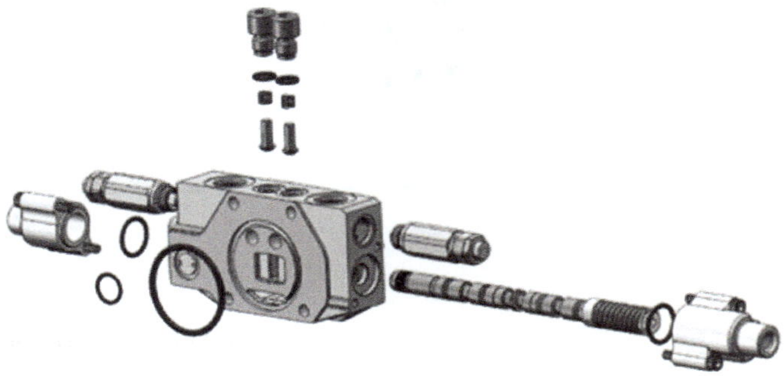

b. CMJ60 DCV, exploded view.

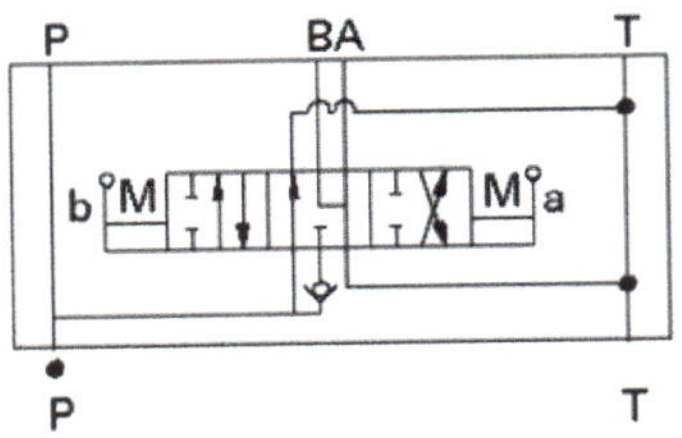

c. Typical schematic function. Manual control.

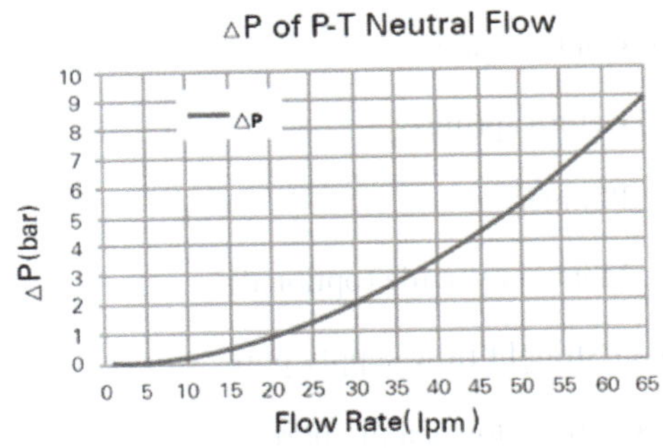

d. Typical pressure drop curve.

Figure 10-13 Schematic of typical mini excavator application. Mobile directional mono-block control valve.

Flow Regulators

As the name implies, the primary function of a **flow regulator** is to regulate the amount of flow delivered to a component or to the entire system itself. A flow regulator is essentially a valve that has a means of adjusting the free-flow opening, restricting or facilitating the fluid flow. Figure 10-14a shows an actual semi-rotary flow restrictor valve (screw-in cartridge type). This valve will increase or decrease flow based on the adjustment of the variable orifice via the rotation of an adjustment handle. Although the traditional port labeling convention suggests the flow should go from ports 1 to 2, the manufacturer (Eaton Hydraulics) recommends the flow path of 2 to 1 for this particular valve; as demonstrated in its ISO symbol, as shown in Figure 10-14c. The screw-in cartridge design offers maximum mounting flexibility and flow capabilities with this particular valve also offering added flexibility in mounting the adjustment handle and the ability to lock-in its position through the addition of a spring-loaded detent allowing for the regulator to be placed in tight spaces effectively. Figure 10-14b depicts a section view of the restrictor valve that offers insights into its operation as it can be seen that when the adjustment handle is rotated about the central axis, the orifice pin, through which the flow must pass to get from port 2 to port 1, is moved up or down, reducing or increasing the cross-section and thus modulating the flow allowed to pass through it.

Figure 10-14d shows typical performance curves for the MRV2-10 flow restrictor valve. Lines A, B, and C refer to different nominal orifice sizes (i.e., different maximum flow rates for a given pressure drop across the valve). As can be seen in the figure, the flow is directly proportional to the valve knob position (linear relationship)—that is, the actual orifice opening in the valve.

Estimating the theoretical performance or flow rate of any given flow restrictor valve is straightforward. In Chapter 5, we introduced the principle of conservation of energy and its most simple expression, **Bernoulli's equation**. Literally, a flow restrictor valve restricts the flow by having a variable opening that is significantly smaller compared with the diameter of the supply line (a pipe, conduit, or hose).

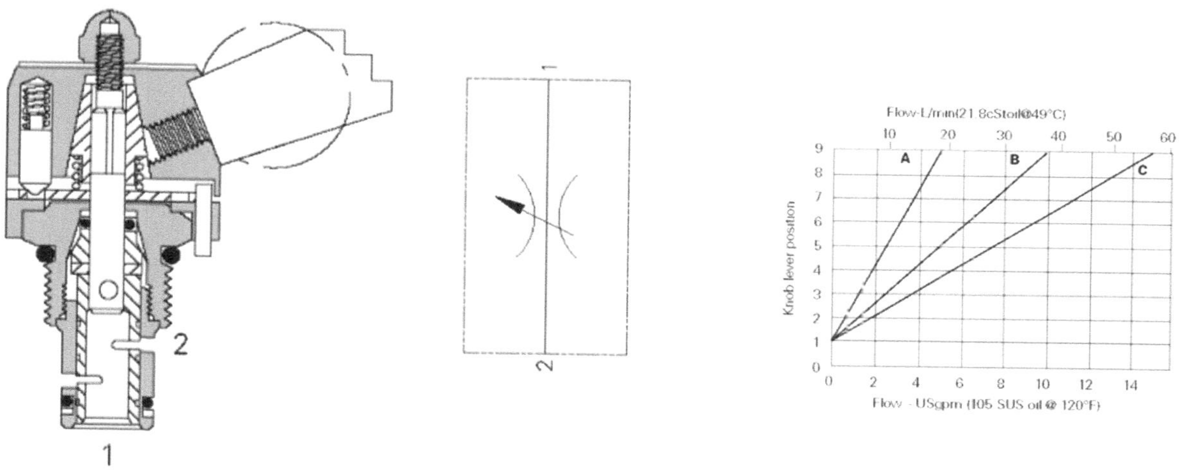

a. Sectional view. **b. ISO symbol.** **c. Typical performance curves.**

Figure 10-14 Semi-rotary flow restrictor valve, MRV2-10 screw-in cartridge type.

Restating Bernoulli's equation:

$$\frac{P_1}{Y} + \frac{V_1^2}{2_g} + z_1 = \frac{P_2}{Y} + \frac{V_2^2}{2_g} + z_2$$

with:

- P_1 = pressure at the valve inlet,
- P_2 = pressure at the orifice,

- V_1 = velocity at the valve inlet,
- V_2 = velocity at the orifice, and
- Q = volumetric flow rate.

Using the continuity equation:

$$Q = V \cdot A$$

$$Q_1 = V_1 \cdot A_1$$

$$Q_2 = V_2 \cdot A_2$$

$$Q_1 = Q_2 = Q$$

Hence:

$$V_2 = \frac{A_1}{A_2} \cdot V_1$$

We also need to recall the definition of specific weight and specific gravity:

$$\gamma = \rho \cdot g$$

and

$$SG = \rho / \rho_{ref}$$

Bernoulli's equation can be rearranged and used to estimate the regulator inlet flow velocity if the pressure drop across it and the size of the orifice and supply line are known:

$$V_1 = \sqrt{\frac{1}{\left[\left(\frac{A_1}{A_2}\right)^2 - 1\right]} \cdot \frac{2}{\rho_{ref}}} \times \sqrt{\frac{\Delta P}{SG}}$$

Hence, the theoretical flow rate is given by:

$$Q = Q_1 = V_1 \cdot A_1 = A_1 \cdot V_1 = A_1 \times \sqrt{\frac{1}{\left(\frac{A_1}{A_2}\right)^2 - 1} \cdot \frac{2}{\rho_{ref}}} \times \sqrt{\frac{\Delta P}{SG}}$$

In reality, the actual flow rate is somewhat less than the theoretical flow because of friction (due to the fluid viscosity) and turbulent losses in the flow restrictor valve. In practice, the fluid power industry "dumps" all variables in front of the pressure drop across the valve, including unit conversion factors, into what is referred to as the valve coefficient. The valve coefficient is determined by testing the flow restrictor valve and fitting the data to the following simplified equation:

$$Q = C_v \times \sqrt{\frac{\Delta P}{SG}}$$

Figure 10-15 shows a two-way flow control valve cartridge type. In this flow regulator design, the flow is regulated in one direction only, from right to left, as seen in *Figure 10-15a,* as there is free flow in the opposite direction. The flow is regulated by the spring rate of the spring pressing on the plunger and the size of the cross-holes. In operation and regulation mode, the actual compression in the spring determines the actual opening of the cross-holes, causing a restriction that regulates the predetermined flow rate of the flow control. Figure 10-15b depicts the ISO symbol for a two-way flow control valve and gives a simplified view of its functional properties. As the fluid flows from port B to A, the flow can be controlled because the check valve will close, forcing the flow into the restrictor direction. When the fluid flows from port A to B it will be allowed to pass freely through the check valve and thus will not see any restriction. The term check valve must be used in the abstract here as it is not a discrete component within the flow control valve; rather, the operation of the valve simply mimics the presence of a check valve functionally.

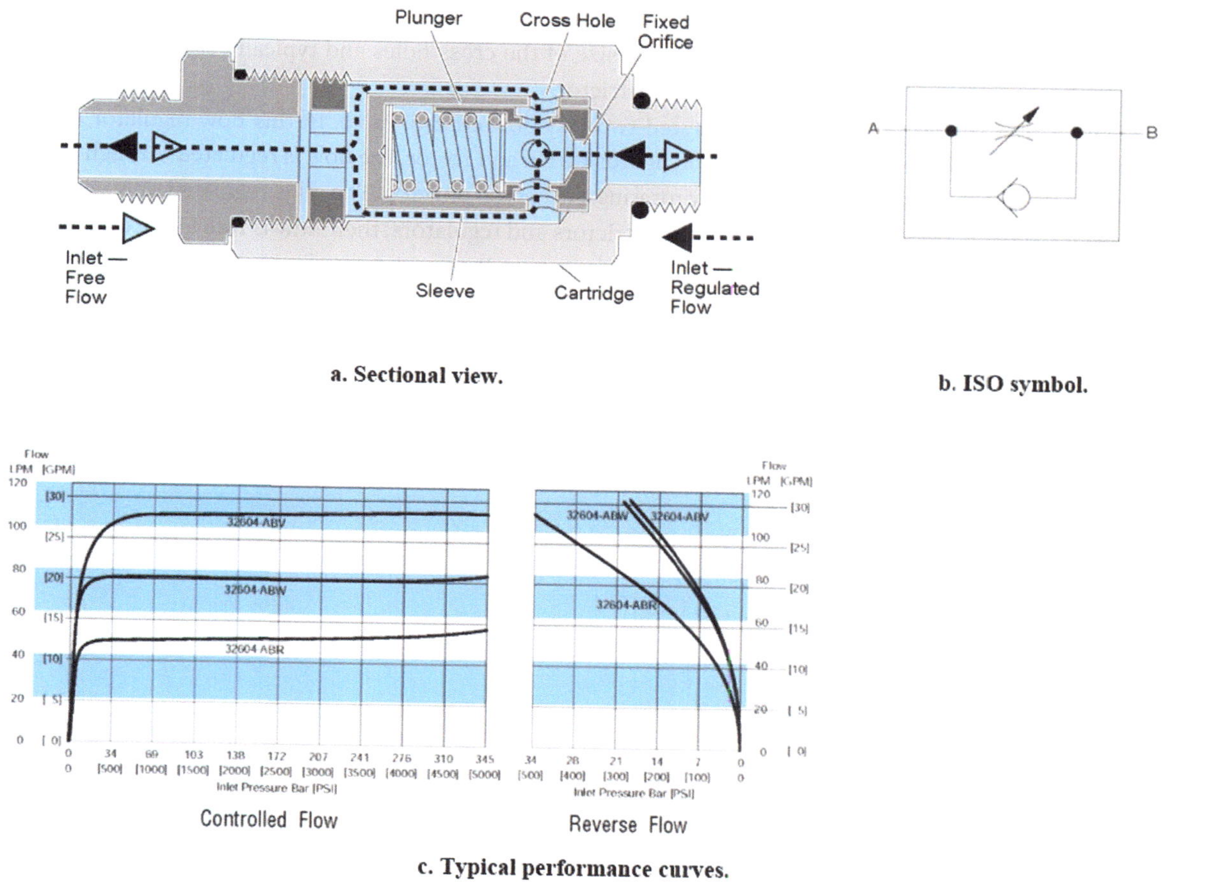

Figure 10-15 In-line flow control valve (Model 32604).

Following the oil flow path as shown in Figure 10-15a, the oil enters through the fixed orifice in the plunger, then gets out of the plunger and sleeve through the cross holes. The oil then flows around the sleeve and out of the flow regulator. As we just learned from the previous flow regulator example, the oil flowing through an orifice—fixed in size, in this case—creates a pressure differential. As the oil flow increases, the pressure differential increases, causing the plunger to move and compress the spring to gradually close off the cross-holes. This flow regulator is flow-sensitive and not pressure-sensitive. The flow rate is typically set by the manufacturer; it is not adjustable on site.

The most common application of inline flow regulators is to control the flow in and to a hydraulic circuit. These are also used as speed control for multiple motors connected in parallel, as motor over speed protection, and in hydrostatic limited slip differentials. Figure 10-15c shows typical performance curves for an inline flow control valve. Lines ABV, ABW, and ABR refer to different flow settings (different spring stiffness and different orifice sizes). In the controlled flow direction, after the minimum pressure differential required to initiate the spring compression is reached, the flow is effectively regulated at a fixed value in the entire operating pressure range. As the pressure is increased, the spring is compressed further, increasingly blocking the cross-holes and actually regulating the flow passage. In the reverse flow direction, the plunger is in the nonregulated flow position. The pressure drop across the valve is determined by the size of the cross-holes and typically varies with roughly the square of the flow, just as in a typical orifice flow restrictor.

Figure 10-16 shows a mechanically operated two-way flow control valve. In this flow regulator design, the section view of which is displayed and labeled in figures 10-16a and 10-16c, the flow is regulated through the adjustment of the throttling area in the orifice (9) by rotating the curved pin (10). The oil is regulated only when flowing from port A to port B, because, like in other flow restrictors and regulators, their flow is free in the direction opposite from the regulated path. This free return occurs from port B to port A via the check valve (4), which is also shown in both the simplified and detailed ISO symbols for the design seen in Figure 10-16b. In order to maintain a constant flow across the orifice (9), a pressure compensator (3) is installed upstream from the orifice (9). The pressure compensator maintains not only a constant flow virtually independent of the pressure but also the oil temperature.

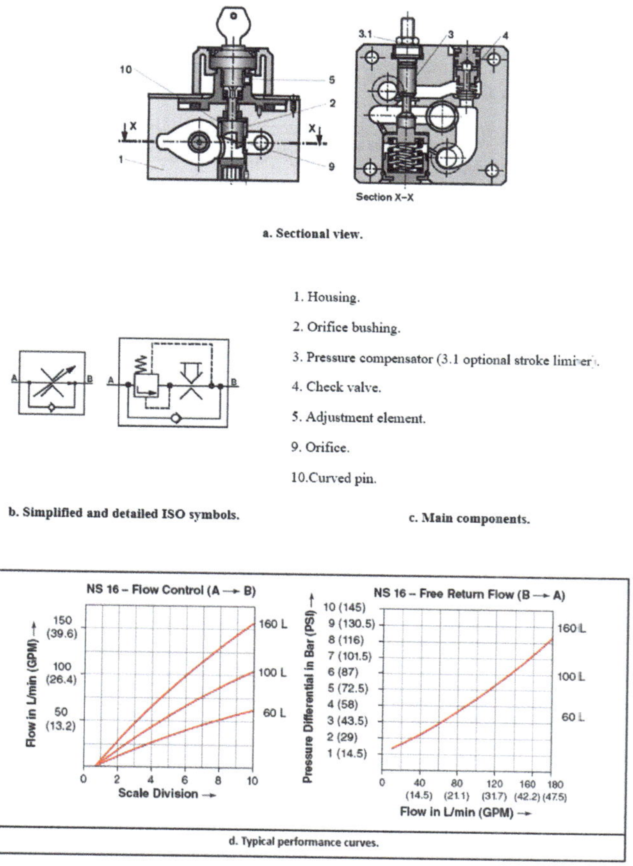

Figure 10-16 Two-way flow control valve, mechanical operation (Type 2FRM Series 3X).

Figure 10-17 displays a mechanically operated two-way flow control valve suitable for fitting into manifold systems. In this flow regulator design, the throttling of the flow occurs at the throttle area (3) shown in the section view offered by Figure 10-17a. The throttle cross-sectional area is changed by turning the adjustment element (2). This takes place between the throttle area (3) and the throttle bolt (4). In order to hold the flow constant, a pressure compensator (5) is also fitted to the regulator—in this case, downstream from the throttle area (3). Like the previous flow restrictors and regulators discussed earlier, the opposite direction flow from port B to port A is a free return as a result of the included check valve (6) seen clearly in both the simplified and detailed ISO symbols for the design shown in Figure 10-17b.

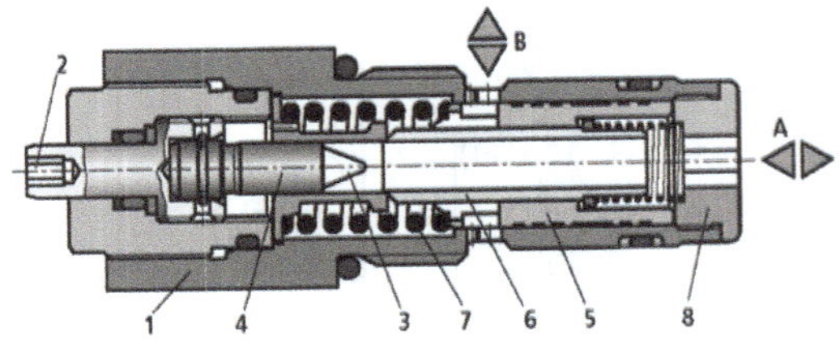

a. Sectional view.

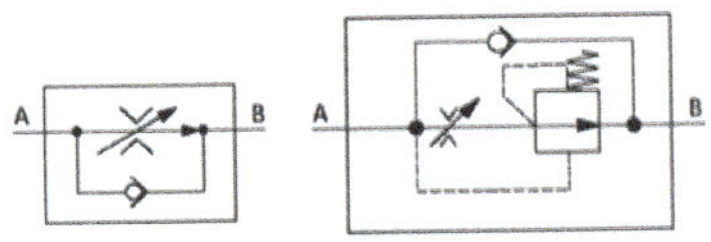

b. Simplified and detailed ISO symbols.

1. Housing.
2. Adjustment element.
3. Throttling area.
4. Throttle bolt.
5. Pressure compensator.
6. Check valve.
7. Compression spring.
8. Plug.

c. Main components.

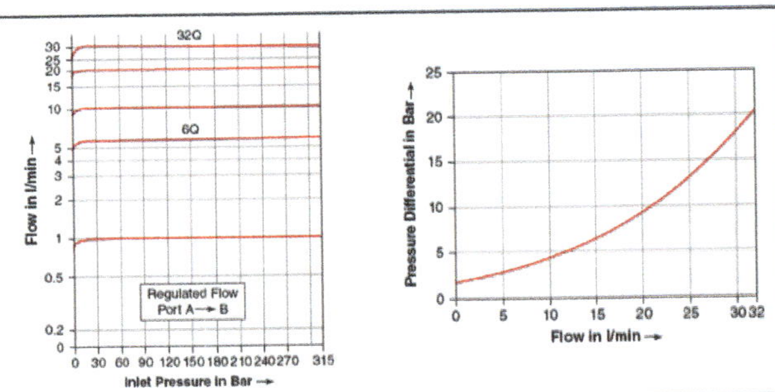

d. Typical performance curves.

Figure 10-17 Two-way flow control cartridge manifold valve (Type 2FRM Series 1X).

As most flow regulators currently used in the industry are pressure-compensated, it is worthwhile to briefly describe the operation of the pressure compensator. In this particular flow regulator (*Figure 10-17a*), the pressure compensator (5) is pressed against the plug (8) by the compression spring (7). When there is no flow through the valve, the pressure compensator stays in the open position. When there is oil flow through the valve, the pressure present in port A applies a force onto the pressure compensator (5). The pressure compensator moves into the compensating position until the forces, both upstream and downstream, are balanced. If the pressure increases in port A, the pressure compensator (5) moves toward its closed position until the forces are again balanced. Conversely, when the pressure decreases in port A, the pressure compensator (5) moves toward its open position until the forces are balanced again. This continuous compensating action maintains a constant flow through the flow regulator, which is set by the adjustment element (2).

Figure 10-17d shows typical performance curves for a two-way flow control cartridge manifold valve. Lines 6Q and 32Q refer to the flow limit of a particular valve size. Like other pressure-compensated regulators, after a minimum pressure differential is reached, the flow is effectively regulated at a fixed value in the entire operating pressure range. In the nonregulated direction, the pressure drop across the flow regulator also varies with the square of the flow.

Fluid Storage

Hydraulic Reservoirs

The amount of fluid required to power a hydraulic system varies significantly, from portable and mobile systems holding a few gallons to large stationary equipment that uses thousands of cubic feet. Whatever the amounts of fluid, all hydraulic systems have a tank or reservoir that holds the fluid needed to ensure all lines and components in the system are completely filled with fluid at all times. In addition to holding the required amount of fluid to operate the system, the reservoir has the following functions:

1. Separates entrained air: The tank must be sized so that enough dwell time is provided for air bubbles to rise to the top and burst on the surface.
2. Traps contaminants: The tank is the safety net to catch solid particles by settling when the filters in the system are bypassed.
3. Provides cooling: Acting as the heat sink of the system, the hydraulic fluid returns hot to the reservoir, where it can cool down when properly sized.
4. Dissipates pressure pulses: Possible back-pressure pulsations on return lines end in the reservoir.

Pneumatic Receivers

There is no need of a tank or reservoir that would store the supply air of a pneumatic system. The atmosphere is a continuous source of air supply. On the other hand, the compressibility of gases makes it relatively easy to store large amounts of air on the pressure side of a pneumatic system (i.e., on the compressor's discharge.) Large industrial pneumatic systems typically supply air to multiple applications at once. For example, in a food processing plant, the pneumatic system provides the fluid that powers the actuators driving conveyors, food dispensers, and container-sorting machines. However, such a system would also provide the air required by the plant instrumentation and the service air required to handle maintenance and service tools. In such cases, the air demand for compressed air varies very widely, from minimum or no demand to very high demand at times. An air tank or receiver can store air under pressure to complement the air supply when the demand exceeds the capacity of the compressors. In addition to providing a capacity of reserve air for variable demand operations, the receiver has the following functions:

1. Pulsation dampener: The receiver suppresses the pressure pulses coming from the compressors. Pressure pulsations may damage the compressed air system and it may be noisy enough to disturb

nearby workers. Pressure pulsations are significantly larger when reciprocating compressors are used.

2. Provides cooling: In large systems, the compressed air will remain in the receiver long enough to dissipate the heat, gained by the air during the compression process, to the surroundings.
3. Water trap: As the compressed air cools down to the temperature of the surroundings, moisture in the compressed air will condense. The receiver must be installed near the compressors, not only to dampen flow pulsations but also to trap the condensed moisture carried by the air. Automatic drains at the bottom of the receiver are frequently installed to take out the water trapped in the receiver.

Figure 10-18 shows a typical pneumatic receiver. Pneumatic receivers are considered by manufacturers and engineering standards as pressure vessels. Large receivers like the one shown in Figure 10-18 pose a safety problem. The amount of energy stored in compressed air is extremely large. Failure of a storage vessel under pressure might lead to an explosion that would cause significant damage to facilities and bystanders. To prevent failure due to overpressure, receivers have safety valves installed.

As already indicated, most compressed air installations are concerned with variable air demand. The size of the receiver determines the peak demand that can be met. An undersized receiver limits the air supply. On the contrary, there is no risk associated with an oversized receiver. In general, large receivers improve the performance and operation of the entire pneumatic system. The sizing of the receiver must take into account all the system requirements such as the operating pressure, the flow demand, the compressor capacity, and the duty operation.

When the demand exceeds the air supplied by the compressors, the size of the receiver can be estimated by the following relationship:

$$V_r = \frac{P_{atm}}{P_{max} - p_{min}} \times t \times (Q_d - Q_c)$$

a. Horizontal tank with access door.

Figure 10-18 Tank and air receiver.

- V_r = volume of the receiver,
- P_{atm} = atmospheric pressure,
- P_{max} = maximum operating pressure of the receiver,
- P_{min} = minimum operating pressure of the receiver,

- t = time in operation of the receiver; the demand exceeds the supply, and the compressor is on,
- Q_d = air demand (in standard units), and
- Q_c = compressor air capacity (in standard units).

In some applications, the demand is somewhat predictable, but it needs to be satisfied in a continuous manner. In such cases, the capacity of the compressors must exceed the demand, and the receiver is designed and sized to provide not only the demand but also standby time to the air compressors. In such events, the sizing of the receiver is coupled to the duty cycle of the compressor and can be estimated as follows:

$$V_r = \frac{P_{atm}}{P_{max} - p_{min}} \times t_{on} \times (Q_c - Q_d); \text{ compressor is on.}$$

$$V_r = \frac{P_{atm}}{P_{max} - P_{min}} \times t_{off} \times Q_d; \text{ compressor is off.}$$

- t_{on} = period of time when the compressor is on; the compressors fulfill both the system demand and the flow required to fill the receiver, and
- t_{off} = period of time when the compressor is off; the demand is only supplied by the receiver.

Fluid Conditioning

Oil Filtration

All hydraulic pumps, motors, and actuators have moving parts. Moving parts such as pump and motor rotors and/or valves and cylinder pistons have very tight clearances, and in some instances, the metal-to-metal contact is inevitable. Having very clean oil in the fluid power system ensures a long life for all these components. Oil cleanliness in a system is normally addressed by adding filters. Filters are essentially a screen with openings of a specific size able to retain solid particles that exceed the size of the openings. Filters and filter elements are manufactured in a great variety of sizes and materials, depending on the application.

Most fluid power systems have a reservoir where the majority of the oil is stored. Reservoirs are typically open, as they are normally used to treat the oil. Therefore, the first line of defense on maintaining oil cleanliness is to add breathers to the reservoir in order to prevent or minimize the amount of contaminants reaching the oil. Figure 10-19 presents a typical flange-on breather and filling filter, which can be easily mounted to a reservoir via the bolts (4) running through the flange (2) and will serve to both filter the air entering the reservoir or venting the oil vapors as it heats in operation as well as screen any oil being added to the system. The path of the air entering and exiting the breather is represented by the red arrows seen entering through holes under the cap (1) before going through the filter element within it and then down past the filling screen (3) into the reservoir itself.

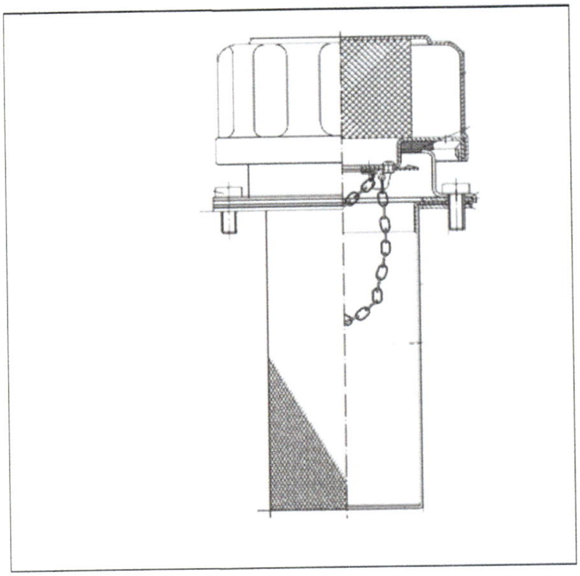

Figure 10-19 Combination of flange-on filling filter and removable cap.

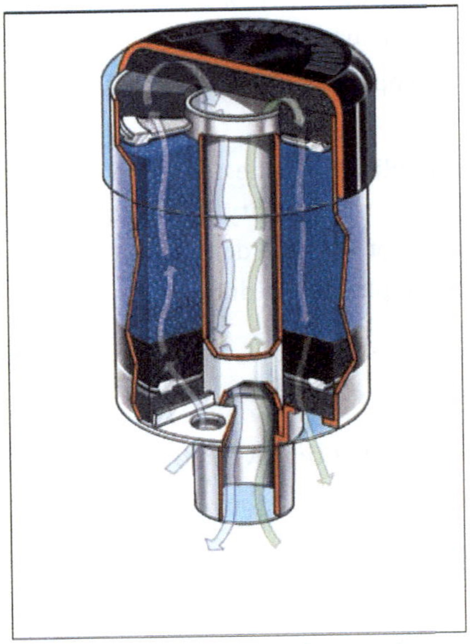

Figure 10-20 Breather filter with screw-off exchange cartridge and internal element. Air path.

Figure 10-20 also presents a breather-filter, but of the cartridge type. This breather-filter, as with the previous example, is designed to both filtrate and vent the reservoir drawdown and air exchange. On the other hand, the filter has a desiccant section designed to absorb and prevent moisture to get into the tank during the air/vent exchange. The filter section (boxed in red) removes solid particles on the upstream side (atmosphere) and then gets regenerated when the air flow reverses and releases those particles. The moisture removal section is typically a desiccant (boxed in blue), a material that removes water by adsorption. The primary difference between the design shown below in Figure 10-20 and that seen in Figure 10-19 is the manner by which they mount to the reservoir tank. The flange-on filter mounts via the relatively permanent bolts with the cap containing the filter being the removable element when the filter must be changed or oil must be added. With the cartridge type, on the other hand, the entire unit is simply screwed onto the reservoir and replaced wholesale when the filter is spent.

Figure 10-21 presents a duplex tank-mounted return line filter. Return line filters are designed for installation on fluid tanks. Their function is to separate solid materials from the hydraulic fluid. They are typically made of highly efficient filtering material like nonwoven glass fibers and are designed to filtrate very fine particles down to 3 µm. When they are in use, the hydraulic fluid entering at the inlet flows through one of the filtering elements (5) contained with the filter bowls (3); it is here that it is cleaned as it travels from the outside of the filter towards the inside with even very fine contaminants being captured and prevented from entering the tank at the outlet. The advantage of a duplex system such as the one displayed in Figure 10-21 is that when a filter becomes saturated with contaminants and must be changed, the system can simply be switched to the other filter by means of the switchover fitting (2), ensuring that continuous flow is preserved even when the system is undergoing maintenance.

Figure 10-22 offers a section view of a typical inline filter. The inline filter design is similar to that of a return line filter, except that inline filters are intended for attachment in pipelines. Like return filters, inline filters also remove very fine particles. Because they are installed on the supply line, which is typically significantly pressurized, the filter elements must have a high collapse resistance, meaning that it must be able to withstand the forces associated with the substantial pressure differentials generated as a result of the filtering process. During operation, hydraulic fluid enters at the inlet (1) before passing through the filter element (3), where contaminants are either captured by the filter itself or are taken out of the fluid stream and forced to settle in the filter bowl (2). Now clean, the hydraulic fluid continues to the outlet (O) and into the hydraulic circuit.

Compressed Air Treatment

Preparation

Any pneumatic system will need an air filtration system; atmospheric air is inherently dirty. Solid particles such as airborne contaminants, pipe scale, rust, and pipe dope must be removed from the airstream before the air gets stored and distributed to the pneumatic system. It is critical to clean the air, as these contaminants may plug small orifices or cause excessive wear and premature failure of pneumatic components.

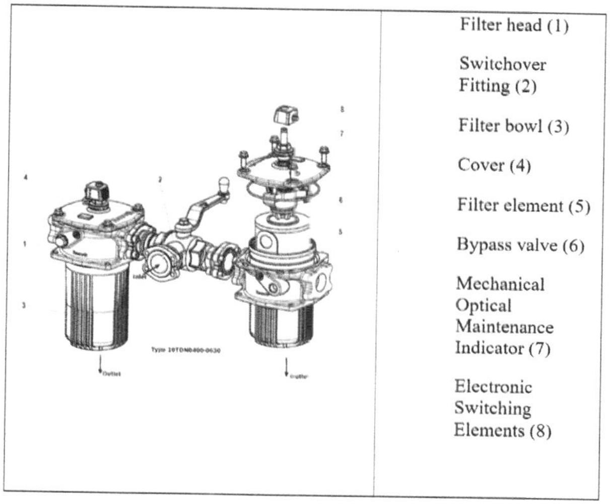

Figure 10-21 Duplex tank–mounted return line filter with filter element.

Figure 10-23 displays a typical medium pressure inline air filter and its labelled internal components. This type of filter is typically installed in facilities that require pressurized air for instrumentation and small- to medium-sized service tools and actuators. This filter also has excellent water removal capabilities, a metal bowl with sight gauge for strength and visual access, a large filter element to ensure low-pressure drops and increased lifetime, and a drain that can be automatic. In terms of the air path, this filter is considered a two-stage filter. The air enters at the inlet port, as seen in Figure 10.23, and flows through the deflector plate (A), which causes the air to swirl. By the centrifugal action of the swirling air, liquids and coarse particles are forced to the bowl interior wall (B), where they go down by the force of gravity and settle at the bowl bottom. The shroud (C) ensures that the proper swirling action prevents the air from passing directly through the filter element (D). The baffle (E) makes the lower portion of the

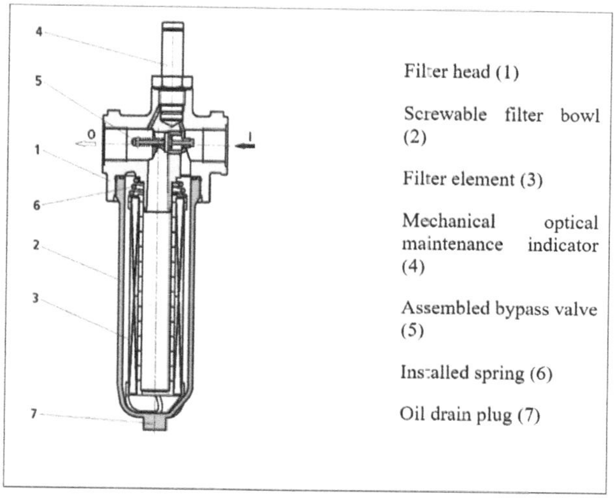

Figure 10-22 Inline filter with filter element.

bowl a stagnant flow region, where the removed liquids and particles collect and remain unaffected by the swirling air.

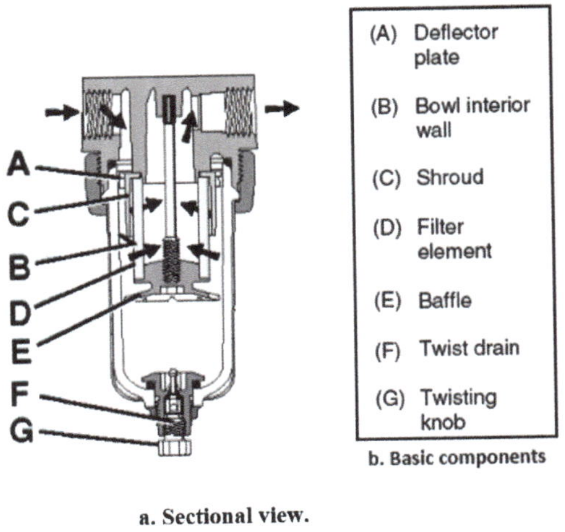

Figure 10-23 Air-line filter and water removal unit.

In general, the air filter type described above is part of a compact air preparation unit. The air preparation unit consists of an air filter (F), a pressure regulator (R), and a lubricator (L), the last of which provides an oil mist to protect from corrosion and lubricates all components in the pneumatic system. The combination of these discrete components into a single unit is shown in Figure 10-24. In the FRL shown, the air flows from left to right. First, the air enters the filter unit and goes through a 40-μm filter element. It then passes to a labyrinthine passage inside the metal bowl to promote water separation and collection, water that will be drained manually or semi-automatically drained from the bowl periodically. From the filter unit, the air goes through the pressure regulator, where, as previously discussed, the downstream pressure is controlled, provided that the pressure setting is lower than the supply pressure. Lastly, the regulated pressure flow goes into a lubricator unit. The metal bowl of this unit is filled with oil, and with the help of internal passages and nozzles, the pressurized air leaving the regulator generates and carries an oil mist. The oil mist is delivered to all components in the pneumatic system that use air from this FRL for the purpose of lubrication. In an FRL unit, the functionality that the individual components offer can be captured in a relatively small form factor that is typically easier to mount and maintain as all of the air treatment processes are carried out in a centralized location within the pneumatic system.

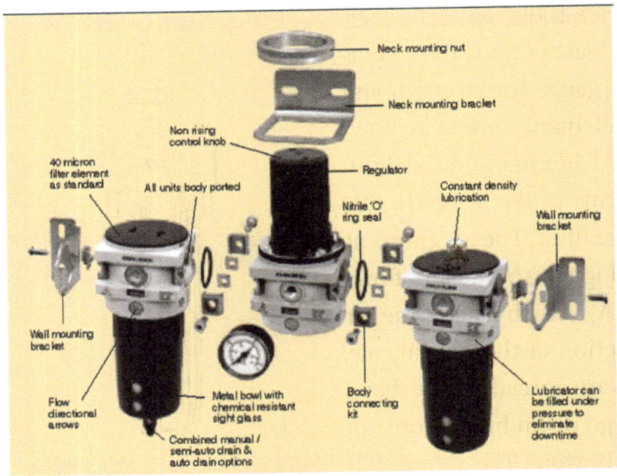

Figure 10-24 FRL air preparation unit.

Moisture Removal

Water and moisture are always present in compressed air. Atmospheric air always contains some water vapor, which will begin to condense into liquid water in any compressed air system when the air cools down below the dew point. The presence of water and moisture is damaging for any compressed air system. Moisture causes rust and increased wear of moving parts as it washes away lubrication. Moisture also adversely affects the color, adherence, and finish of paint applied by compressed air.

Moisture can also cause malfunctioning of pneumatic controls and instrumentation. Additionally, in cold weather, water and moisture could freeze in control lines, also causing faulty operations.

Large compressed-air systems like the one shown in Figure 10-25 require special equipment to remove water and moisture. Tanks, coalescent filters, and dryers are combined to treat the incoming air and deliver a continuous supply of technically clean and dry air. Different methods can be used to reduce the moisture content in compressed air. Current drying technologies include the following:

1. After-coolers: These cool back down the warm air, leaving the compressors at ambient temperature. Water condenses because of the lower dew point due to the increase in pressure.
2. Water traps: These remove bulk water condensed by the after-coolers.
3. Coalescing filters: These remove aerosol water and other liquids, which bypass the water traps.
4. Refrigeration: This cools down the compressed air well below ambient temperature. Water again condenses because of the lower dew point of cold air. A typical dew point is 37°F (3 degrees Celsius [°C]).
5. Deliquescent chemical dryers: The deliquescent dryer uses a hygroscopic desiccant material that absorbs the water vapor and is dissolved in the liquid formed, reducing the air dew point to about 50°F (10°C).
6. Regenerative desiccant dryers, heated and heatless: These dryers use a desiccant, which absorbs the water vapor in the airstream. Contrary to absorption dryers, where the desiccant is dissolved and needs to be replenished periodically, during absorption, the moisture adheres only to the desiccant and gets collected in the thousands of small pores within each desiccant bead. The composition of the desiccant is not changed, and the moisture can be driven off in a regeneration process by applying dry purge air or heat. Dew points in a range of –40°F to –100°F (–40°C to –73°C) can be attained.
7. Membrane dryers: A specially designed membrane allows water vapor (a gas) to pass through the membrane pores faster than the other gases (air), reducing the amount of water vapor in the airstream at the outlet of the membrane dryer and suppressing the dew point. The typical dew point achieved is 40°F (4°C), but lower dew points down to –40°F (–40°C) can be attained.

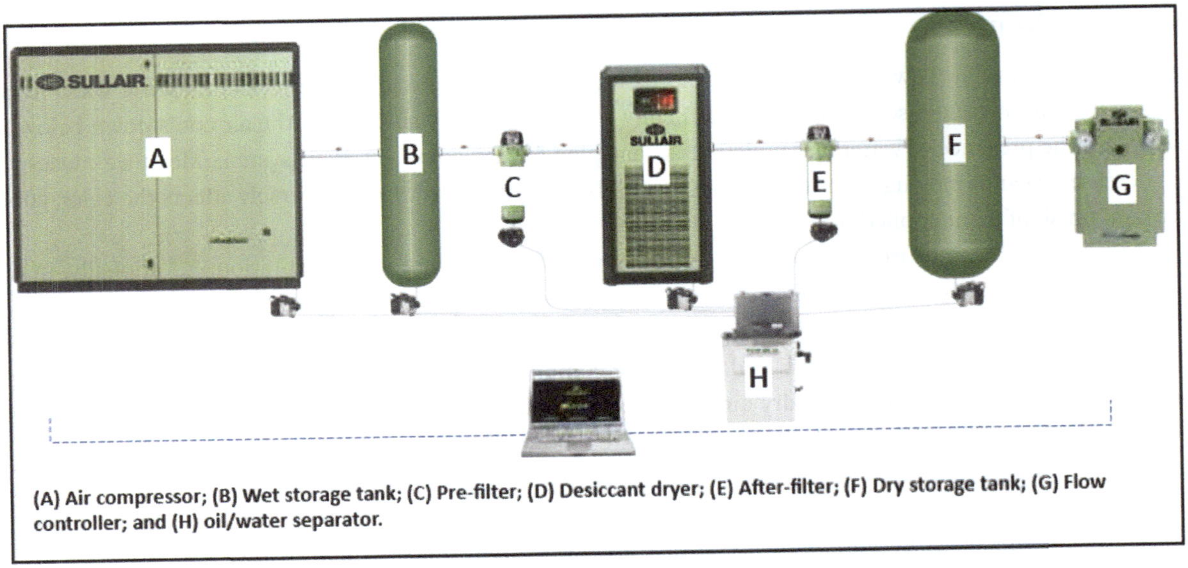

Figure 10-25 Compressed-air system.

According to figures by the Compressed Air and Gas Institute (CAGI), the desiccant regenerative dryer is the most efficient in removing water and moisture from compressed air, as shown in Figure 10-26a. Figure 10-26b presents a graph to estimate the moisture content in saturated compressed air at given temperature and pressure conditions. As the temperature increases, the ability of air to hold moisture also increases. On the contrary, as the pressure increases, the ability of air to hold moisture diminishes.

To estimate the moisture content in the air, we would enter Figure 10-26b with the gauge pressure of the air and the temperature. For example, at typical room conditions, 14.7 psi absolute (psia; 0 psi gauge [psig]) and 70°F, the maximum amount of moisture in 1,000 cubic feet (ft^3) of air is 1.00 lbm. The amount of moisture the air could hold more than doubles when warmed to 90°F, up to 2.13 lbm.

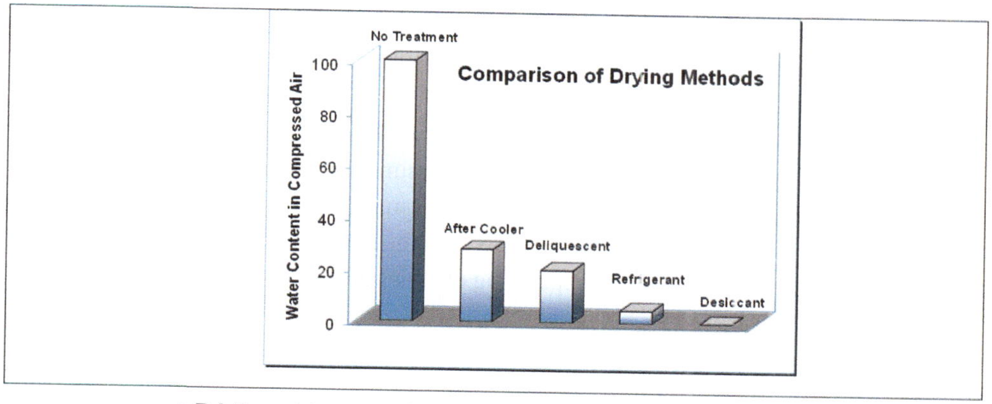

a. Relative moisture removal performance of several drying methods.

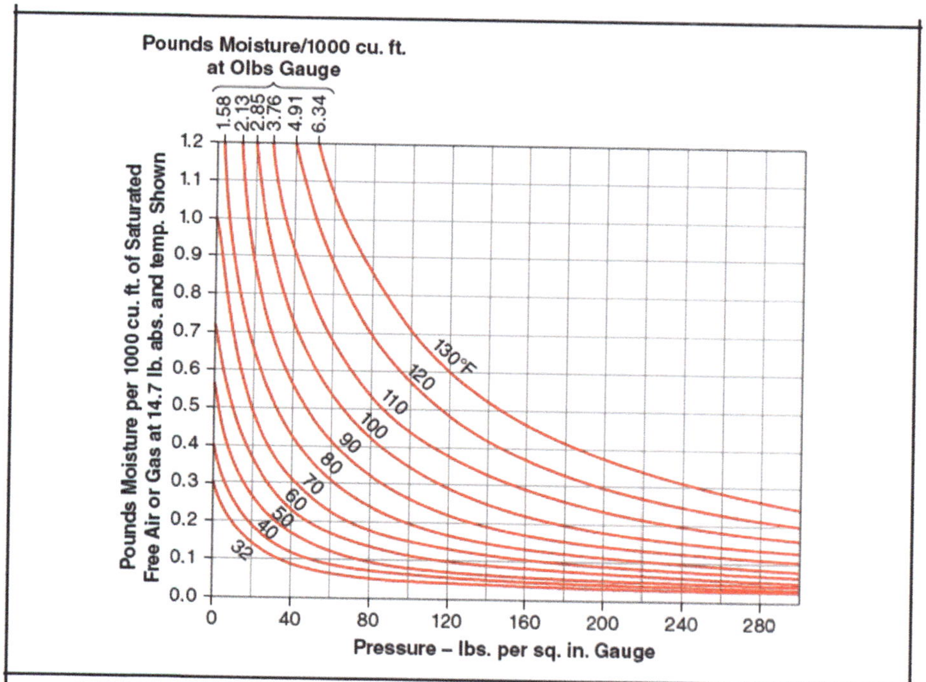

b. Moisture remaining in saturated air when compressed isothermally to the pressure shown.

Figure 10-26 Drying methods' efficiency and water content in compressed air.

Hydraulic Accumulators

Accumulators are very common in modern hydraulic and industrial lubrication systems. As the name suggests, **accumulators** are a storage device. They carry out multiple functions, among them energy storage, reserve, and recovery; emergency power and lubrication systems, delivering steady pressure, shock absorption, and pulsation attenuation; leakage; and thermal compensation. In simple terms, the main task of a hydraulic accumulator is absorbing and storing under pressure a certain amount of fluid taken from the fluid power system and returning it to the system when required. Depending on the application, the accumulator would need to store and release small amounts of fluid extremely fast or a large volume of fluid in a controlled manner.

Hydraulic accumulators basically consist of a gas-tight section and a fluid section. The gas-tight and fluid sections are separated by either a flexible or a rigid element. The **fluid section** is connected to the hydraulic circuit. The gas-tight section is typically precharged at a high pressure with an inert gas, such as nitrogen. Air is not normally used due to risk of fire and explosion. When the pressure in the hydraulic system rises, the gas in the accumulator is compressed and fluid from the system gets into the fluid section. When the pressure in the system falls, the compressed gas expands, pushing the separating element and displacing the accumulated fluid into the hydraulic circuit.

The energy storage method and the built-in element separating the hydraulic fluid from the gas essentially define the accumulator type. They are either a gas type or spring loaded and weight-loaded design. In **gas-tight** accumulators, a bag or bladder, diaphragm, or piston is used. Although not very common, some applications do not use or require a separating element. Spring-loaded and weight-loaded accumulators are almost exclusively of the piston type.

The most common of the gas-tight accumulators is the bladder type. Bladder-type accumulators, such as that seen in the section view given in Figure 10-27, consist of a seamless cylindrical or spherical pressure vessel (1) made of high-tensile steel and an elastic bladder (2) mounted in the interior of the vessel. They also have a fluid port assembly, a poppet (3), and a gas valve (4) used to charge the accumulator to the specified gas precharge pressure. Bladder accumulators feature fast response, a medium gas compression ratio of about four to one and moderate flow rates.

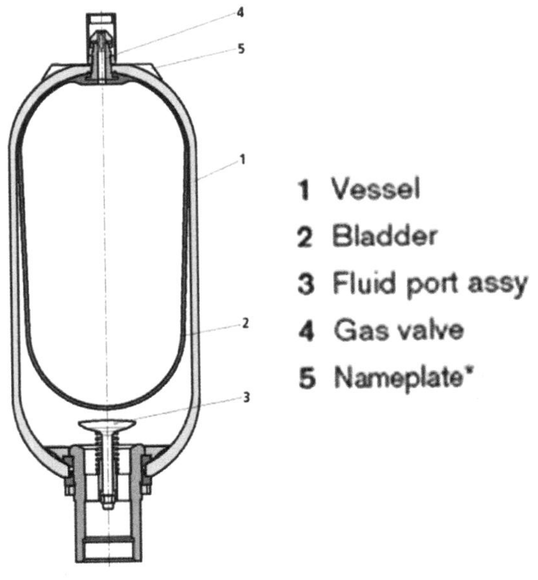

Figure 10-27 Reparable hydraulic accumulator, bladder type.

The design of a diaphragm-type accumulator is similar to that of a bladder-type accumulator. The only significant difference between the two designs is the elastic element that separates the gas from the fluid in the system (see Figure 10-28). A diaphragm-type accumulator consists of two hemispheres made from steel forgings that form the cylindrical or hemispherical pressure-tight vessel (1); a flexing convoluted and elastic diaphragm clamped to the vessel periphery (2); a closing poppet normally affixed to the diaphragm (2) and plug screw or gas valve (4); and the fluid port (5). Diaphragm accumulators are able to sustain larger gas compression ratios than bladder-type accumulators, but they are inherently limited to smaller volumes, as the elastic element, the diaphragm, has significantly less stretching surface area than the bladder. In addition, diaphragm accumulators can sometimes be affected by gas permeation across the diaphragm (Brendan Casey, H&P).

In contrast with elastic separating element accumulators, bladder or diaphragm, the inherent rigidity of a piston allows accumulators of this type to handle much higher gas compression ratios and very large flow rates. Another advantage of piston accumulators is that they can be mounted in any position. Nevertheless, the major drawback of piston accumulators is the need of sealing elements in the piston. These piston seal elements create friction against the vessel walls, which has to be machined with precision. The piston-wall friction translates into variations in the operating pressure (hysteresis) and slower response times caused by accelerating and decelerating (during both charging and discharging) the heavy piston. The constant friction also demands higher levels of fluid cleanliness.

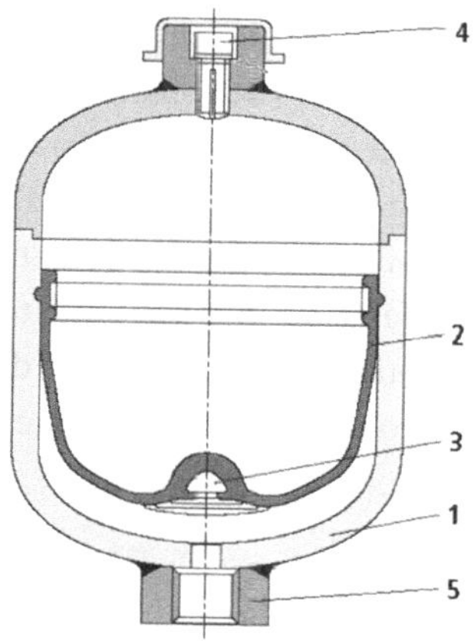

Figure 10-28 Reparable hydraulic accumulator, diaphragm type.

Hydraulic Accumulator Operation

Hydraulic fluids, mostly oils, are nearly incompressible. They compress a tiny fraction, only about 2% under very high pressures, at 5,000 psi. This incompressibility translates into small variations in flow rate or leakage, causing large fluctuations in the operating pressure of the hydraulic system. In contrast, gases are readily compressible; an initial large volume of gas can be reduced significantly by increasing the gas pressure. Potential energy is stored in the compressed gas, energy that can be released upon demand to the hydraulic system.

In addition to energy storage, accumulators are also designed to absorb flow pulsations. Both reciprocating and rotary pumps providing the power to the system generate pulsating flow. Particularly piston pumps, commonly used in high-pressure applications, produce flow pulsations detrimental to high-pressure systems. When properly selected, the addition of an accumulator substantially attenuates the flow of pulsations in the system.

Similarly, most mobile fluid power applications, such as excavators, front-end loaders, lifts, and refuse trucks, have motors and cylinders driving heavy components that start and stop suddenly—starts and stops that create pressure waves that travel back through the system. These pressure waves, or shocks (pressure waves in a fluid are transmitted at the speed of sound), can develop peak pressures several times greater than normal operating pressures. In these cases, the accumulator acts as a shock-cushioning device while also reducing noise and preventing system failure.

Accumulators are also used to supplement the flow coming from the pump in systems that operate intermittently. The accumulator stores potential energy when the work cycle demands no flow and the system pump is idling, filling the accumulator. Then, the stored energy can be used to provide additional power to the system, complementing the amount that the hydraulic pump can deliver. This enables utilizing a smaller pump, resulting in savings in cost and power.

One of the nontraditional applications of fluid power currently under research, development, and testing is the propulsion of heavy trucks and buses. Private industry, federal and state funding agencies, and local governments have teamed up to evaluate the feasibility and benefits of using hydraulic hybrids in applications that require frequent stop-and-goes, such as delivery trucks and public transportation. Figure 10-29 shows the schematic of a hydraulic hybrid truck with two hydraulic accumulators of the gas type (precharged). In normal operation, both accumulators are pressurized by the main hydraulic pump coupled to the internal combustion engine (ICE). The ICE operates at constant speed near or at the point of maximum efficiency. The vehicle is propelled by a hydraulic differential (bidirectional pump/motor) that receives power from the main pump. During stop-and-go periods, the hydraulic differential plays the most important role. During stops, the hydraulic differential becomes a hydraulic pump, converting the braking power into fluid power to be stored in the accumulators. During periods of acceleration, the accumulators deliver high-pressure fluid to the system, complementing the flow sent by the main pump to the hydraulic differential, which acts regularly as a motor. It is well documented that in these stop-and-go transportation applications, hydraulic hybrids are more efficient than electric hybrids.

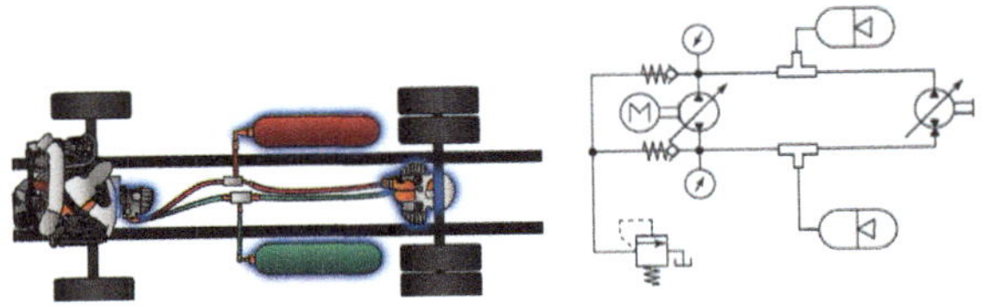

Figure 10-29 Hydraulic hybrid vehicle with gas-tight–type serviceable hydraulic accumulator.

Sizing of a Hydraulic Accumulator

When sized and precharged properly, hydraulic accumulators normally cycle between stages d and f (see Figure 10-30). The bladder will not expand completely so as to contact the poppet or be totally compressed so that it becomes destructively folded into the top of the vessel. In the case of piston accumulators, the piston will not contact the cap, gas, or fluid end. As the accumulator operates on the base of expanding and compressing an inert gas, manufacturers design and size the accumulator analyzing the gas side. During the precharge, pressurized gas fills the entire accumulator, expanding completely the bladder, the diaphragm, or pushing down the piston. The volume at stage b in Figure 10-30 is essentially the maximum volume of the accumulator, denoted by V_1. Once installed, the accumulator is pressurized further by the hydraulic fluid in the fluid power system. The bladder or diaphragm is compressed or the piston is pushed up by storing hydraulic fluid until the maximum operating pressure is reached, stage d in Figure 10-30 or volume V_2, which represents the minimum volume occupied by the precharged gas. Finally, when required, the accumulator will provide fluid to the system by allowing the precharged gas to expand up to the minimum operating pressure required by the system, but not further to avoid damages to the flexible elements, bladder or diaphragm, or the piston itself, stage f in Figure 10-30 or volume V_3. In contrast, the fluid power system engineer sizes the accumulator analyzing the fluid side, essentially calculating the amount of fluid that the accumulator will need to store and release when required.

The volume of fluid required by the fluid power system determines the volume occupied by the precharge gas in stages d and f (Figure 10-30). The volume required by the system as V_{SYS} can be related to the volumes on the gas side as:

$$V_{SYS} = V_3 - V_2$$

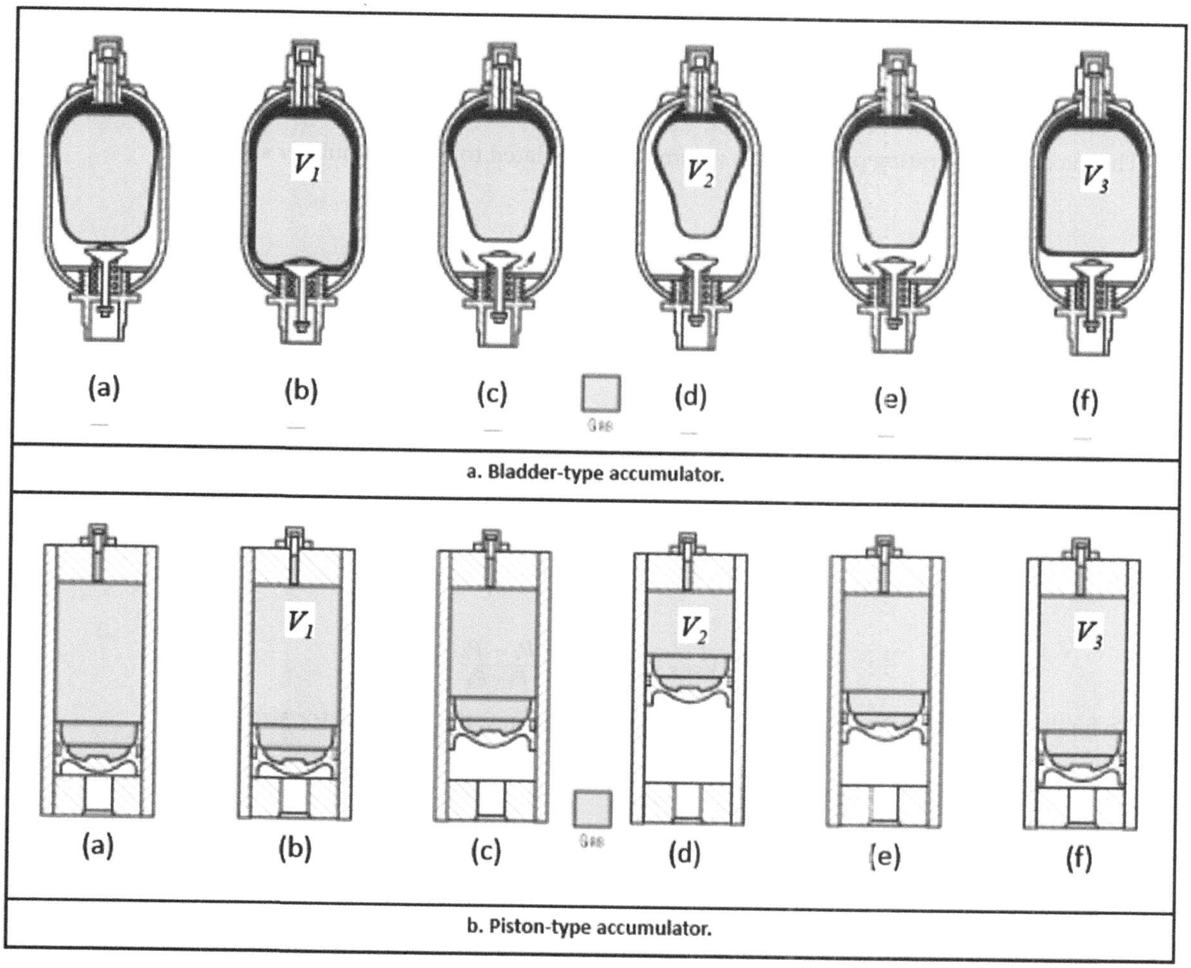

a. Bladder-type accumulator.

b. Piston-type accumulator.

Figure 10-30 Schematic operation of a bladder-type and a piston-type accumulator. Hydraulics & Pneumatics.

The volumes on the gas side can be related using Boyle's law for gases if the accumulator precharge and operating pressures of the fluid power system are known.

$$P_1 \cdot V_1 = P_2 \cdot V_2 = P_3 \cdot V_3$$

- P_1 = precharge pressure,
- V_1 = precharge gas volume and equivalent to the physical volume/size of the accumulator,
- P_2 = maximum operating pressure in the system,
- V_2 = pre-charge gas volume at the maximum operating pressure (P_2),
- P_3 = minimum operating pressure in the system, and
- V_3 = precharge gas volume at the minimum operating pressure (P_3).

The minimum operating pressure gas volume can be related to the accumulator size as:

$$V_3 = \frac{P_1}{P_3} \times V_1$$

The maximum operating pressure gas volume can be related to the accumulator size as:

$$V_2 = \frac{P_1}{P_2} \times V_1$$

Upon substitution of both operating pressure volumes into the amount of fluid required by the system:

$$V_{SYS} = \frac{P_1}{P_3} \times V_1 - \frac{P_1}{P_2} \times V_1$$

$$V_{SYS} = P_1 \cdot V_1 \times (\frac{1}{P_3} - \frac{1}{P_2})$$

$$V_{SYS} = P_1 \cdot V_1 \times (\frac{P_2 - P_3}{P_2 \cdot P_3})$$

To finally get the volume occupied by the precharge gas (i.e., the accumulator size):

$$V_1 = \frac{P_2}{P_1} \times \frac{P_3}{P_2 - P_3} \times V_{SYS}$$

Manufacturers typically recommend precharging the flexible-element accumulators, bladder, and diaphragm to 80% of the minimum operating pressure in the hydraulic system pressure. In contrast, piston accumulator manufacturers typically specify a fixed margin (e.g., 100 psi) below the minimum system pressure. The accumulator precharge pressure determines the amount of fluid that remains in the accumulator at the minimum pressure in the system. Denoting as ξ the ratio between the accumulator precharge pressure and the minimum operating pressure, P_3:

$$\xi = \frac{P_1}{P_3}$$

$$V_1 = \frac{P_2}{\xi \cdot P_3} \times \frac{P_3}{P_2 - P_3} \times V_{SYS} = \frac{1}{\xi} \times \frac{P_2}{P_2 - P_3} \times V_{SYS}$$

Finally, denoting by r_P the pressure ratio between the maximum and minimum operating pressures in the system:

$$r_P = \frac{P_2}{P_3}$$

$$V_1 = \frac{1}{\xi} \times \frac{P_2}{P_2 - P_3} \times V_{SYS} = \frac{1}{\xi} \times \frac{r_p \cdot P_3}{P_3 - P_3} \times V_{SYS}$$

$$V_1 = \frac{1}{\xi} \times \frac{r_P}{r_P - 1} \times V_{SYS} = \frac{1}{\xi} \times \frac{1}{1 - 1/r_P} \times V_{SYS}; \xi < 1.0 \text{and} r_P > 1.0$$

A correct precharge pressure is the most important factor in prolonging the accumulator life. The way precharging is accomplished and maintained must be considered when choosing the type of accumulator for the application. Piston accumulators are more forgiving in the amount of care and maintenance required but are designed to operate with very clean fluid, whereas flexible elements can tolerate some dirt in the hydraulic oil.

Hydraulic accumulators operate at extremely high pressure. For design purposes, they are considered pressure vessels, and as such, they must be manufactured, tested, and certified according to statutory standards. In the United States, for example, the relevant standard is the American Society of Mechanical Engineers (*ASME*) *Boiler & Pressure Vessel Code VIII, Division 1.*

Summary

Pressure relief valves are the most widely used single valve or component in any fluid power circuit or system. The main function of a pressure relief valve is to release any overpressure in the system caused by a sudden change in the flow demand. If a pressure rise above a specific setting is detected, the PRV opens, alleviating the pressure in the system. For hydraulic systems, the hydraulic fluid returns to the reservoir. In the case of pneumatic systems, the excess air is released to the atmosphere. There are basically two types of PRVs: the direct-acting PRV and the compound or pilot-activated PRV.

Pressure regulators or pressure-reducing valves are devices that maintain the fluid power system, hydraulic or pneumatic, pressurized up to a level previously set. A pressure regulator is essentially a normally open PRV. While the PRV senses only the pressure in the system, a pressure regulator senses both the pressure provided by the pump or compressor and the pressure in the system. The pressure regulator reduces the pressure in the pump or compressor side to the level required in the system side. Like PRVs, there are two types of pressure regulators: the direct-acting pressure-reducing valve and the compound or pilot-activated pressure-reducing valve.

The DCV is one of the most important components of any fluid power system. The primary function of a DCV is to direct the fluid—oil or air—to the various components in the system. A flow regulator controls the amount of flow delivered to a component or to the entire fluid power system itself. A flow regulator is essentially a variable opening valve, restricting or facilitating the fluid flow.

A hydraulic reservoir is a tank or vessel that holds the fluid needed to ensure all lines and components in the fluid power system are completely filled with fluid at all times. In addition to holding the required amount of fluid to operate the system, the reservoir separates entrained air, traps contaminants, provides cooling, and dissipates pressure pulsations. Similarly, an air tank or receiver stores air under pressure to complement the air supplied by the compressor when the demand exceeds the capacity. In addition to providing a reserve air for variable demand operations, the receiver also acts as a pulsation dampener, provides cooling, and acts as a water trap.

Oil and air filters are designed to retain solid particles. Depending on the location in the fluid power system, they must be designed to sustain low or high pressures. The particle retention capacity of the filter is determined by the size of the opening in the filtering element. Both oil and air can carry water in liquid or vapor form. In the case of oil, the fact that water and oil do not mix makes it relatively easy to separate the water from the oil. Some additives in the oil and enough dwelling time in the reservoir allow fluid stratification in the reservoir, with the water going to the bottom, where it can be easily drained. In the case of air, the challenge is to remove the moisture in the gaseous form. Water vapor cannot easily be mechanically separated from the air. Several methods, mostly based on temperature change or chemistry, are available to remove moisture from the compressed air. On the thermal side, after-coolers and refrigeration dryers are used. On the chemical side, deliquescent dryers and regenerative desiccant dryers are used. Mechanical methods such as water traps and coalescing filters only work on removing water droplets. Only recently have membrane dryers been developed to mechanically separate water vapor from the air by controlling the permeation rate.

A hydraulic accumulator is a device that stores potential energy and hydraulic fluid under pressure. The potential energy is stored in a chamber that contains a compressed gas. The fluid is stored under pressure when the pump in the system pushes the fluid into the accumulator, compressing the gas in the chamber. The energy storage method and the built-in element separating the hydraulic fluid from the gas define the accumulator type. They are either gas-type or spring-loaded and weight-loaded design. In gas-tight accumulators, a bladder, diaphragm, or piston is used. Spring-loaded and weight-loaded accumulators are almost exclusively of the piston type. Hydraulic accumulators operate at extremely high pressure. For design purposes, they are considered pressure vessels, and as such, they must be manufactured, tested, and certified according to statutory standards.

Exercises

1. What inlet (feeding the pump) and outlet (coming from the pump) of a nominal pipe size line size would you select for a pump system of 26 gallons per minute (gpm) using the recommendations (flow velocities) from the textbook and video using a Schedule 80 pipe in the figure titled "Common Pipe Sizes"?

Table 10.1

Nominal Pipe Size	Pipe Outside Diameter	Pipe Inside Diameter		
		Schedule 40	Schedule 80	Schedule 160
1/8	0.405	0.269	0.215	-
1/4	0.540	0.364	0.302	-
3/8	0.675	0.493	0.423	-
1/2	0.840	0.622	0.546	0.466
3/4	1.050	0.824	0.742	0.614
1	1.315	1.049	0.957	0.815
1-1/4	1.660	1.380	1.278	1.160
1-1/2	1.900	1.610	1.500	1.338
2	2.375	2.067	1.939	1.689

2. A steel tubing has a 1.50-inch (in) outside diameter and a 1.30-in inside diameter. It is made of American Iron and Steel Institute (AISI) 4130 steel, which has a rated tensile strength of 74,000 psi. What would be the safe working pressure for this tube, assuming a factor of safety of 8?
3. What minimal nominal pipe size and minimal schedule steel pipe using the figure titled "Common Pipe Sizes" is required for a pump discharge (outlet) line for a 120-lpm pump operating at 10 megapascals (MPa) with AISI 4130 steel, which has a rated tensile strength of 74,000 psi, using a safety factor of 6?
4. You need to determine the minimal nominal pipe size and minimal schedule using Table 10.1 that will work for the following application.
    a. Minimum inside diameter: 0.750 in; high-strength alloy ASTM International A514 steel, which has a rated tensile strength of 100,100 psi
    b. Operating pressure of 4,500 psi with a safety factor of 6
5. A simple pressure relief valve contains a poppet (piston) with an area of 1.000 square inches (in^2) on which the system acts. During assembly, a spring with a constant of 2,500 pound-forces (lbf)/in is installed in the valve to hold the poppet against its seat. The adjustment mechanism is then set so that the spring is compressed 0.75 in from its free-length condition. In order to pass full flow through the valve, the poppet must move 0.25 in from its fully closed condition. Determine the full flow pressure of the valve (equivalent to the setting of the valve). Answer in psi.
6. A pressure relief valve contains a poppet that is 4.20 square centimeters (cm^2). During assembly, a spring that is 3,200 newtons (N) per centimeter (cm) is installed in the valve to hold the poppet against the seat (see Image 10.1). The adjustment mechanism is then set so that the spring is compressed 0.50 cm from the free length condition (Δl_1). In order to pass full flow through the valve, the poppet must move 0.30 cm from its closed position (Δl_2). Find the following:

a. The cracking pressure (in MPa)
b. The full pump flow pressure or PRV setting (in MPa)

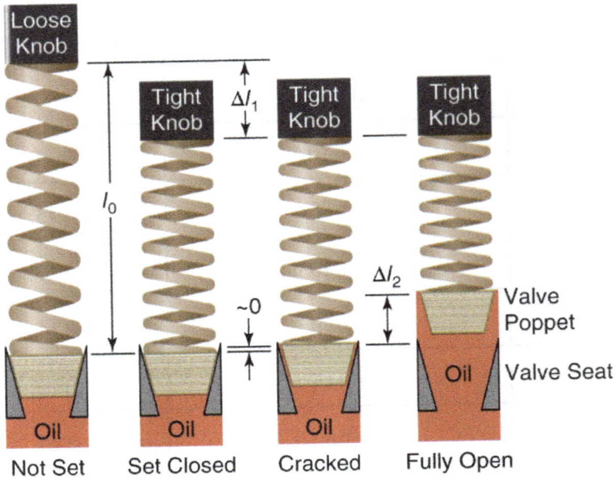

IMG 10.1

7. You are designing a hydraulic circuit with a maximum allowable working pressure of 1,000 psi. To optimize the operation of the circuit, the cracking pressure of the valve has to be set as close as possible to the full pump flow pressure. The poppet you chose has a diameter of 1.25 in and must move only 0.25 in from its closed position ($\boldsymbol{\Delta l_2}$). If the cracking pressure must be 950 psi, determine the following:
   a. The spring stiffness, or $\boldsymbol{k}$ (in pound-forces per inch)
   b. Then initial valve setting, or $\boldsymbol{\Delta l_1}$ (in inches)
8. Figures 10-15a–10-15c show the flow versus pressure drop curves, controlled flow and reverse flow, for three inline flow control valves when the fluid is hydraulic oil with a specific gravity of 0.90. Regarding the inline flow control valve identified as 32604-ABW, answer the following:
   a. The valve coefficient when controlling the flow at 2,000 psi
   b. The valve coefficient when the reverse flow is 25 gpm
9. Determine the flow rate through a flow control valve that has a valve coefficient of 2.5 lpm/√kPa. The pressure drop across the valve cannot exceed 1.0 MPa. The fluid is hydraulic oil with a specific gravity of 0.85.
10. The extension speed of a hydraulic cylinder is controlled with a meter-in flow restrictor arrangement, as shown in Image 10.2. The hydraulic cylinder has a diameter of 6 in and a rod of 2 in. The setting of the relief valve is 250 psi. Determine the extension speed of the cylinder when pushing a 2,000-lbf load if the valve coefficient is 0.25 gpm/√psi. Provide the answer in feet (ft) per minute (min).

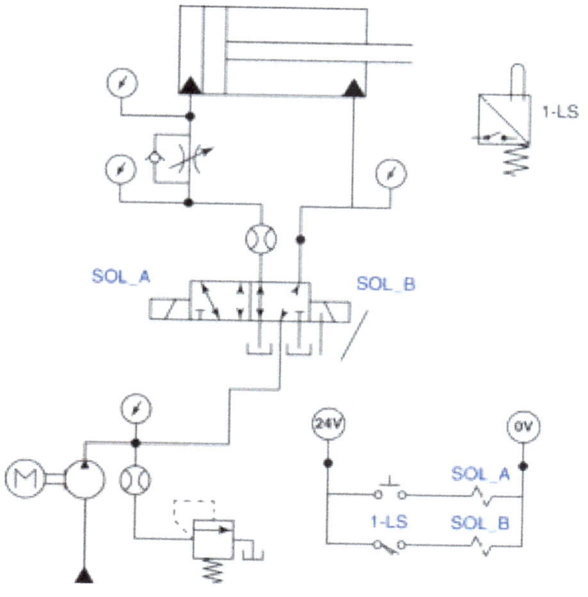

IMG 10.2

   a. Assume that the specific gravity of the hydraulic fluid is 0.95.

11. A compressor delivers 75 standard cubic feet per minute (scfm) of air at 160 psig to a pneumatic system. The compressor is installed in a facility near the lake shore with very high humidity, and the air temperature is 90°F. If the compressor operates for eight hours per day, determine the following:
    a. The amount of moisture delivered by the compressor to the pneumatic system in gallons per day
12. If an aftercooler is installed in the compressor discharge to bring the air back to 90°F, how much moisture is now delivered to the pneumatic system in gallons per day?
13. If an air dryer is also installed to cool the compressed air down to 50°F, how much moisture is now delivered to the pneumatic system in gallons per day?
14. Repeat problem 5 but answer the three questions a, b, and c from the perspective of the water being removed from the pneumatic system in each case. The water that condenses in lines and the air receiver must be removed with the installation of water traps. Knowing how much is removed from the compressed air allows the designer to select the size and number of these traps.
15. An accumulator must supply 425 cubic inches (in^3) of hydraulic oil to a system with a maximum pressure of 3,750 psig and a minimum pressure of 2,000 psig. If the nitrogen precharge pressure is 1,200 psig, find the minimum size of the accumulator.
16. An accumulator is originally filled up to a pressure of 12 MPa with 50 liters (L) of hydraulic fluid at room temperature, 30°C. In operation, the accumulator typically supplies up to 25 L to the hydraulic circuit. If the operating temperature of the hydraulic oil is 70°C, determine the pressure that the accumulator can supply when filled with 25 L. Assume that the precharge gas temperature is similar to that of the hydraulic fluid in the accumulator.
17. The direct drive hydraulic motor of a go-kart is powered by a hydraulic accumulator. Based on

equipment catalogs, you preselected pressure values (in absolute pressure) as indicated in Image 10.3 (10 MPa absolute, 30 MPa absolute, and 10 MPa absolute). If the size of the motor size you picked is 55 cubic centimeters (cm^3) per revolution, with a run volume (V_{run}) of 17.507 L, what is the minimum size of the accumulator required (in liters)? To maximize the use of the accumulator, minimum and precharge pressures are the same. (P_{atm} = 0.1 MPa; 1 MPa = 106 Pascals [Pa]; 1 meter [m] = 100 centimeters [cm]; 1 cubic meter [m^3] = 1,000 L.)

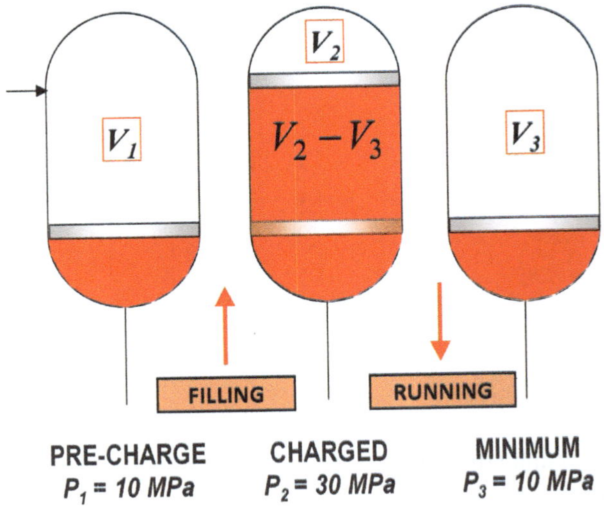

IMG 10.3

The circuit in Image 10.4 is used to crush old cars into metal bales in a nearby a junkyard. The diameter of the hydraulic cylinder is 4 in, and it has a stroke of 10 ft. Each car has to be crushed in about 15 seconds (s). Removing the metal bale and placing another car in the crusher takes about 5 min. The design characteristics of the accumulator are the following: precharge: 1,000 psia; maximum pressure (relief valve setting): 2,500 psia; and minimum pressure required to crush a car: 1,500 psia. Determine the following:

a. The minimum size of the accumulator required to operate. Assume that only the accumulator provides flow to the cylinder during the extension stroke. Provide the answer in gallons.
b. The pump hydraulic horsepower (hp) and flow requirements with and without the accumulator. Provide the answer in hp and gpm. Neglect the volume occupied by the cylinder rod. Comment on the benefits and/or drawbacks of adding the accumulator.

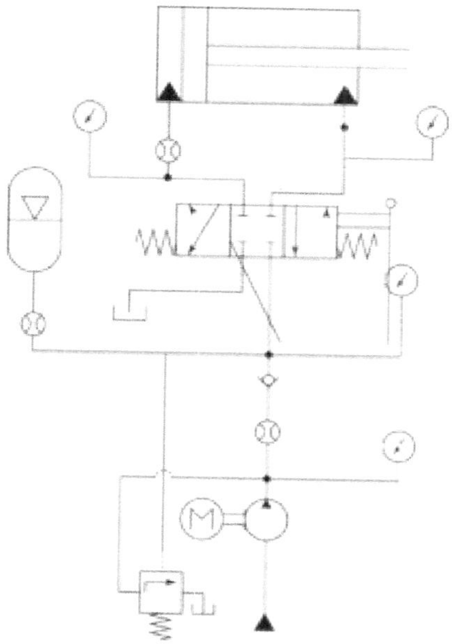

IMG 10.4

Figure Credits

Fig. 10-1: Copyright © by Milton Beychok; Adapted by Larry Villasmil and Robert Garrick (CC BY-SA 3.0) at http://commons.wikimedia.org/wiki/File:Relief_Valve.png.
Fig. 10-2a: Copyright © by Parker Hannifin Corporation.
Fig. 10-2b: Copyright © by Parker Hannifin Corporation.
Fig. 10-2c: Copyright © by Parker Hannifin Corporation.
Fig. 10-2d: Copyright © by Parker Hannifin Corporation.
Fig. 10-3: Copyright © by Parker Hannifin Corporation.
Fig. 10-4a: Copyright © by Eaton.
Fig. 10-4b: Copyright © by Eaton.
Fig. 10-4c: Copyright © by Eaton.
Fig. 10-4d: Copyright © by Eaton.
Fig. 10-4e: Copyright © by Eaton.
Fig. 10-5a: Copyright © by Parker Hannifin Corporation.
Fig. 10-5b: Copyright © by Parker Hannifin Corporation.
Fig. 10-5c: Copyright © by Parker Hannifin Corporation.
Fig. 10-5d: John S. Cundiff, "Reducing Valve Operation. Fluid Power Circuits and Controls," Fluid Power Circuits and Controls: Fundamentals and Applications. Copyright © 2001 by Taylor & Francis Group.
Fig. 10-6: Copyright © by Parker Hannifin Corporation.
Fig. 10-7a: Copyright © by Eaton.
Fig. 10-7b: Copyright © by Eaton.
Fig. 10-7c: Copyright © by Eaton.
Fig. 10-7d: Copyright © by Eaton.
Fig. 10-8a: Harry L. Stewart, "Typical Pressure Regulator Components," Pneumatics and Hydraulics. Copyright © 1984 by Audel.
Fig. 10-8b: Copyright © by Parker Hannifin Corporation.
Fig. 10-9a: Copyright © by Proportion-Air, Inc.
Fig. 10-9b: Copyright © by Proportion-Air, Inc.
Fig. 10-9c: Copyright © by Proportion-Air, Inc.
Fig. 10-10a: Copyright © by Bosch Rexroth Corporation.

Fig. 10-10b: Copyright © by Bosch Rexroth Corporation.
Fig. 10-10c: Copyright © by Bosch Rexroth Corporation.
Fig. 10-11a: Copyright © by Bosch Rexroth Corporation.
Fig. 10-11b: Copyright © by Bosch Rexroth Corporation.
Fig. 10-11c: Copyright © by Bosch Rexroth Corporation.
Fig. 10-12a: Copyright © by Bosch Rexroth Corporation.
Fig. 10-12c: Copyright © by Bosch Rexroth Corporation.
Fig. 10-13a: Copyright © by Eaton.
Fig. 10-13b: Copyright © by Eaton.
Fig. 10-13c: Copyright © by Eaton.
Fig. 10-14a: Copyright © by Eaton.
Fig. 10-14b: Copyright © by Bosch Rexroth Corporation.
Fig. 10-14c: Copyright © by Eaton.
Fig. 10-14d: Copyright © by Eaton.
Fig. 10-15a: Copyright © by Eaton.
Fig. 10-15b: Copyright © by Eaton.
Fig. 10-15c: Copyright © by Eaton.
Fig. 10-16a: Copyright © by Bosch Rexroth Corporation.
Fig. 10-16b: Copyright © by Bosch Rexroth Corporation.
Fig. 10-16c: Copyright © by Bosch Rexroth Corporation.
Fig. 10-16d: Copyright © by Bosch Rexroth Corporation.
Fig. 10-17a: Copyright © by Bosch Rexroth Corporation.
Fig. 10-17b: Copyright © by Bosch Rexroth Corporation.
Fig. 10.17c: Copyright © by Bosch Rexroth Corporation.
Fig. 10-17d: Copyright © by Bosch Rexroth Corporation.
Fig. 10-18: Copyright © by Milton Beychok (CC BY-SA 3.0) at https://commons.wikimedia.org/wiki/File:Modified_Hanson_steelwatertank.jpg.
Fig. 10-19: Copyright © by Bosch Rexroth Corporation.
Fig. 10-20: Copyright © by Parker Hannifin Corporation.
Fig. 10-21: Copyright © by Bosch Rexroth Corporation.
Fig. 10-22: Copyright © by Bosch Rexroth Corporation.
Fig. 10-23a: Copyright © by Parker Hannifin Corporation.
Fig. 10-23b: Copyright © by Parker Hannifin Corporation.
Fig. 10-24: Copyright © by Parker Hannifin Corporation.
Fig. 10-25: Sullair, LLC, "Compressed Air System." Copyright © by Hitachi Global Air Power US, LLC.
Fig. 10-26a: Copyright © by Compressed Air & Gas Institute.
Fig. 10-26b: Copyright © by Compressed Air & Gas Institute.
Fig. 10-27: Copyright © by Bosch Rexroth Corporation.
Fig. 10-28: Copyright © by Bosch Rexroth Corporation.
Fig. 10-29a: U.S. Environmental Protection Agency, http://www.epa.gov/otaq/technology/research/research-hhvs.htm.
Fig. 10-29b: Copyright © by Larry Villasmil and Robert Garrick; Automation Studio. Reprinted with permission.
Fig. 10-30a: Hydraulics & Pneumatics, "Bladder-type Accumulator," http://www.hydraulicspneumatics.com/200/TechZone/Accumulators/Article/False/6446/TechZone-Accumulators. Copyright © by Endeavor Business Media, LLC.
Fig. 10-30b: Hydraulics & Pneumatics, "Piston-type Accumulator," http://www.hydraulicspneumatics.com/200/TechZone/Accumulators/Article/False/6446/TechZone-Accumulators. Copyright © by Endeavor Business Media, LLC.
Table 10-1: Copyright © by Anthony Esposito.
IMG 10-2: Copyright © by Larry Villasmil and Robert Garrick; Automation Studio. Reprinted with permission.
IMG 10-4: Copyright © by Larry Villasmil and Robert Garrick; Automation Studio. Reprinted with permission.

CHAPTER 11

Internal Flow, Frictional Losses, Conductors, and Pipelines

Overview and Learning Objectives

In this chapter, we will estimate the energy lost caused by wall friction in fluids that flow inside a pipe/conductor. The friction in the fluid occurs within the fluid itself as it moves through the pipe/conduit. This friction is caused by the interaction of the moving fluid with the stationary internal walls of the pipe/conduit.

Upon completing this chapter, students should be able to do the following:

- Determine whether the flow is laminar or turbulent.
- Calculate the Reynolds number at any location in a pipe/conduit.
- Understand the significance of the Reynolds number.
- Define in their own words the term friction factor.
- Determine the friction factor when the Reynolds number and characteristics of the pipe/conduit are known.
- Determine the head loss or pressure loss in a pipe/conduit when the flow is either laminar or turbulent.
- Determine the head loss or pressure loss in valves and fittings.
- Define in their own words the term equivalent length and its use.
- Determine the equivalent length for a piping/conduit system.

Based on our previous experiences, we would all expect that when two surfaces are in contact and moving relative to each other, friction between the two surfaces would inhibit their movement and generate heat. The heat energy is a result of the energy losses caused by friction. From your experiences, you may have seen that rubbing two pieces of wood together can produce enough heat energy to cause a fire to start. In our fluid power systems, we want to minimize these energy losses/heat gains caused by the kinetic energy of the fluid being converted into heat. We therefore need to understand fluid friction and the subsequent energy losses in fluid power circuits caused by it.

Laminar and Turbulent Flow

Fluids, both gases and liquids, are composed of molecules that interact and collide with each other and any solid surfaces that they touch. We can visualize these individual molecules of fluid flowing in a conduit (e.g., a pipe, tubing) in individual layers or streams as the flow moves through the conduit. We will assume that the flow is a steady-state flow in which the fluid properties at any point in the system do not change over time. In other words, the flow has been established, and a steady-state condition is assumed. In this condition, we can have two types of flow in the conduit. The first type of flow is referred to as **laminar flow**. **Laminar flow** is characterized by having chunks of fluid molecules flowing mostly in smooth layers that can be traced and form streamlines that would not cross. As shown in Figure 11-1, the layers, or *laminae*, of fluid are smooth and parallel to each other.

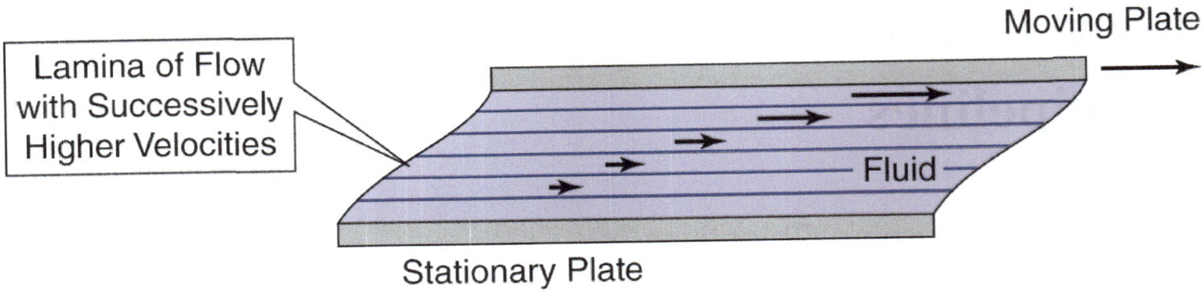

Figure 11-1 Fluid between two plates.

We can visualize this phenomenon as cars moving smoothly on an expressway on which the cars do not change lanes, moving only in parallel paths. In laminar flow, we have minimal collisions among the chunks of fluid molecules moving between contiguous layers or streamlines. In a pipe/conduit, the streamlines would resemble a deck of cards in which the adjacent layers slide past one another with the highest-velocity magnitude in the center of the pipe/conduit and the lowest-velocity (or zero-velocity) magnitude at the walls of the pipe/conduit, as shown in Figure 11-2.

Therefore, in laminar flow, the major source of fluid friction is the sliding of one fluid layer or streamline over another. When a fluid is flowing inside a pipe/conduit, laminar flow typically occurs while the fluid moves at low velocities.

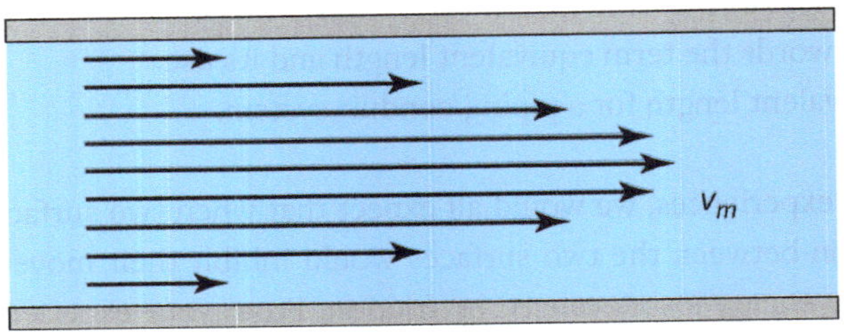

Figure 11-2 Laminar flow velocity profile in conduit.

As the velocity of the flow increases, the flow transitions from laminar (smooth) flow to turbulent (chaotic) flow. In Figure 11-3, the sketched streamlines illustrate the difference between the laminar and turbulent flows. While in laminar flow, the streamlines remain parallel, turbulent flow streamlines will intertwine, mixed, and eventually disappear.

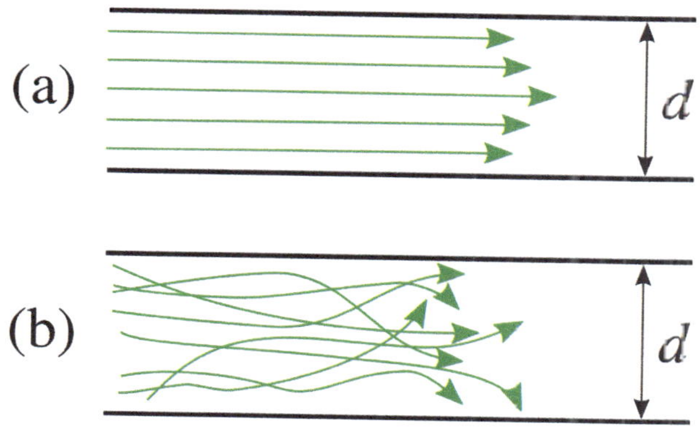

Figure 11-3 Laminar and turbulent flows.

An increased chaotic movement of the individual chunks of fluid molecules characterizes turbulent flow. Turbulent flows have higher irregularity and diffusion or mixing of the fluid molecules and layers in the flow. This increased irregularity and mixing in the fluid flow means that the individual fluid molecules have additional collisions as compared to the laminar flow. These additional collisions of the particles cause increased resistance in the flow and thus greater energy losses as compared to laminar flow. Therefore, the source of fluid friction in turbulent flow is caused by both the sliding of one layer over another and the individual collisions of the fluid molecules with each other as they chaotically mix within the flow. It is interesting to think of these laminar and turbulent/chaotic flows occurring within our body in our bloodstream, as shown in Figure 11-4. The figure illustrates the beginning of turbulent flow within an aorta.

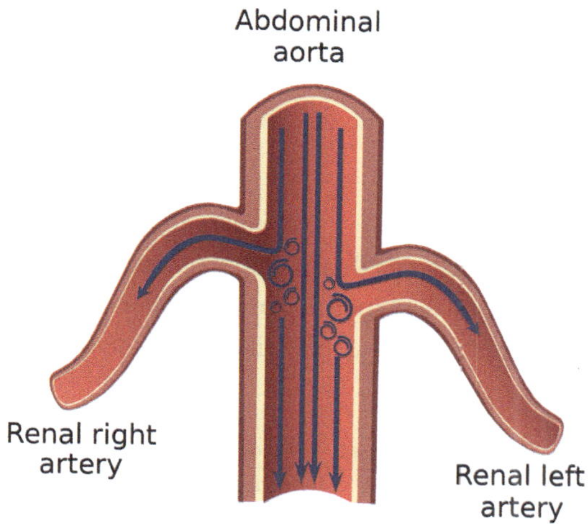

Figure 11-4 Schematic illustration of turbulent flow genesis inside the abdominal aorta.

Reynolds Number

In designing a fluid power system, it is important for us to know the flow pattern inside of our system. Based upon the work of George Stokes and Osborne Reynolds, a dimensionless number was developed that quantifies the conditions governing the transition between laminar and turbulent flows. Reynolds studied the flow condition of fluids in pipes and, as shown in Figure 11-5, the factors that affected the fluid flow patterns inside pipes using dyes. If the flow in the pipe were laminar, the dye would flow in a straight line, in layers without mixing. When turbulent flow occurred in the pipe, the dye dispersed chaotically, mixed, and disappeared downstream from the pipe.

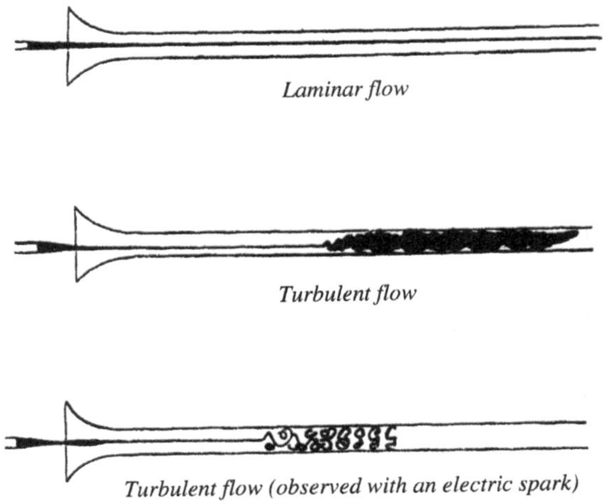

Figure 11-5 Water flow observed in a pipe, as drawn by Reynolds in his best-known experiment on fluid dynamics in pipes.

A conclusion of Reynolds' experiments was that the flow pattern that was exhibited (laminar versus turbulent) depended on the fluid velocity, size of the pipe (inside diameter), and fluid properties (density and dynamic viscosity). Reynolds related these parameters in the following relationship:

$$Reynolds\#\,(Re) = \frac{Density * Velocity * Diameter}{Dynamic_{Viscosity}} = \frac{\rho * \nu * D}{\mu}$$

where:

1. v = mean velocity of the object relative to the fluid (International System of Units [SI] units: meter per second [m/s]),
2. D = diameter for round pipes (SI units: m),
    a. = dynamic viscosity of the fluid (SI units: pascal-second [Pa·s] or newton-second per square meter [N·s/m^2]), and
        i. = density of the fluid (kilograms per cubic meter [kg/m^3]).

It is important to note that the **Reynolds number** is a dimensionless parameter; it has no units. If the Reynolds number is less than 2,300, the flow is laminar. If the Reynolds number is greater than 4,000, then the flow is turbulent. In the interval between 2,300 and 4,000, the flow transitions back and forth between laminar and turbulent with no single flow pattern. In designing a fluid power system, we would like to minimize losses as well as heat generated in our system, and we prefer to design systems with laminar flow.

Darcy-Weisbach Equation

As mentioned, the friction in the fluid and its interaction with a pipe/conduit is the main cause of energy losses/heat generated in a fluid power system. This energy loss can be seen as a loss of pressure as we move through our fluid power system. As reviewed in Chapter 3, we can express pressure in terms of pressure head. The Darcy-Weisbach equation relates this pressure head loss due to the fluid friction along a length of pipe to the average velocity of the fluid flow. The following equation expresses this pressure head loss:

$$PressureHeadLoss\,(h_f) = f_d * \frac{L}{D} * \frac{v^2}{2 * g}$$

where:

- h_f = head loss due to friction (SI units: m)
- f_d = dimensionless coefficient called the Darcy friction factor,
- L = length of the pipe (SI units: m),
- D = hydraulic diameter of the pipe (for a pipe of circular section, this equals the internal diameter of the pipe; SI units: m),
- v = mean velocity of the object relative to the fluid (SI units: m/s), and
- g = local acceleration due to gravity (SI units: meters per second squared [m/s^2]).

Recall from our previous work that the average velocity of the fluid flow equals the volumetric flow rate (Q) divided by the cross-sectional area (A) of the flow.

$$\text{or } v = \frac{Q}{A}$$

We should note in the Darcy-Weisbach equation that the pressure head loss increases to the square of the water's velocity. Thus, a doubling of the velocity will cause a fourfold increase in pressure loss. The Darcy-Weisbach equation can be used to determine the pressure head loss in a pipe/conduit for both laminar and turbulent flows. The key difference in determining pressure head losses in laminar and turbulent flows is the Darcy friction factor (f_d) in the above equation. The Darcy friction factor is also known as the Moody friction factor.

Friction Factor With Laminar Flow

The Darcy friction factor in pipe/conduit laminar flow (Reynolds number less than 2,300) is given by the following equation in which the friction factor equals 64 divided by the Reynolds number:

$$f_d = \frac{64}{Re}$$

Substituting this friction factor into our Darcy-Weisbach equation yields the following equation:

$$PressureHeadLoss(h_f) = \frac{64}{Re} * \frac{L}{D} * \frac{v^2}{2 * g}$$

This equation, known as the Hagen-Poiseuille equation, is named after the two scientists who experimentally and independently derived this relationship, Gotthilf Heinrich Ludwig Hagen and Jean Léonard Marie Poiseuille (1840). The equation gives us the pressure drop of a laminar flowing fluid in a long cylindrical pipe, as shown in Figure 11-6.

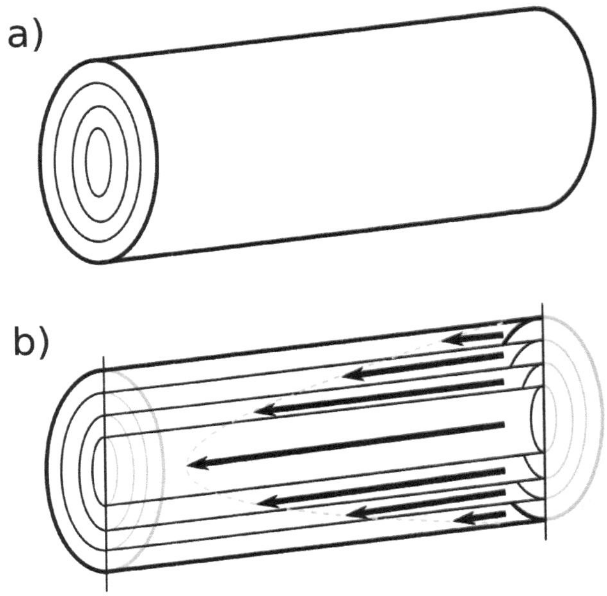

Figure 11-6 Hagen-Poiseuille laminar pipe flow.

Friction Factor With Turbulent Flow

The Darcy-Weisbach equation can also be used to calculate the pressure head loss in turbulent fluid flow. However, the Darcy friction factor cannot be represented in a simple equation as with laminar flow, since turbulent flow has chaotic, random, mixing, and fluctuating movement of the fluid molecules.

For turbulent flows, the Darcy friction factor has been experimentally shown to be a function of both the Reynolds number and the relative roughness of the pipe. The relative roughness of the pipe is defined as the pipe inside surface roughness (absolute roughness) divided by the pipe inside diameter. This relative roughness is expressed in the following equation:

$$Relativeroughness = \frac{PipeInsideDiameterRoughness}{PipeInsideDiameter} = \frac{\varepsilon}{D}$$

As noted above, the relative roughness depends on the pipe's inside diameter. As this diameter increases, the relative roughness decreases. Table 11-1 contains typical values of the pipe's inside diameter roughness (ε), which is called the pipe absolute roughness.

Table 11-1. Absolute Roughness Table (Pipe Inside Diameter Surface Roughness)

Type of Pipe	U.S. Customary Units (Feet)	SI Units (Millimeters)
Drawn tubing, brass/copper	8.2 E-06	0.0025
Glass, plastic, Perspex (plexiglass)	8.2 E-06	0.0025
Steel, structural or forged	8.2 E-05	0.025
Steel, rusted	0.0016	0.5
Steel, mortar lined	0.00033	0.1
Cast iron, asphalt dipped	0.00039	0.12
Cast iron, uncoated	0.00085	0.26
Concrete, new smooth	8.2 E-05	0.025
Concrete, coarse	0.00082	0.25
Riveted steel	0.0164	5.0
Sewers, old	0.00984	3.0

It is very important to note that the absolute pipe roughness data are for new pipes. As pipes age, they typically increase their apparent roughness due to deposits, erosion, and pitting, among other factors. This rise due to aging can be very significant; consequently, engineers typically use a safety factor in the design process to account for aging.

Based on experimental results in studying turbulent flow with pipes of different roughness, C.F. Colebrook (1939) developed the following equation to allow us to solve for the Darcy friction factor for turbulent flow. Note that this equation is valid only for turbulent-flow conditions.

$$\frac{1}{\sqrt{f}} = -2\log_{10}\left(\frac{\varepsilon}{3.7 * D} + \frac{2.51}{Re * \sqrt{f}}\right)$$

where:

1. f = Darcy friction factor,
2. ε = absolute roughness (SI units: m),
3. D = hydraulic diameter of the pipe (for a pipe of circular section, this equals the internal diameter of the pipe; SI units: m), and
4. Re = Reynolds number (dimensionless).

The above Colebrook equation is not factorable in terms of the friction factor (f), as the friction coefficient is involved in both sides of the equation and therefore is awkward and difficult to solve. This equation is generic and used to calculate the friction coefficient for various fluid flows, such as heating, ventilation, and air-conditioning ducts, pipes, and tubes with various liquids and gases. We can solve this equation using computer numerical iteration or by using an online calculator (e.g., The Engineering Toolbox, 2008) or smartphone app (e.g., PipeLife Design Portal [Colebrook-White PF], Fluid Mechanics Calculator, and many others).

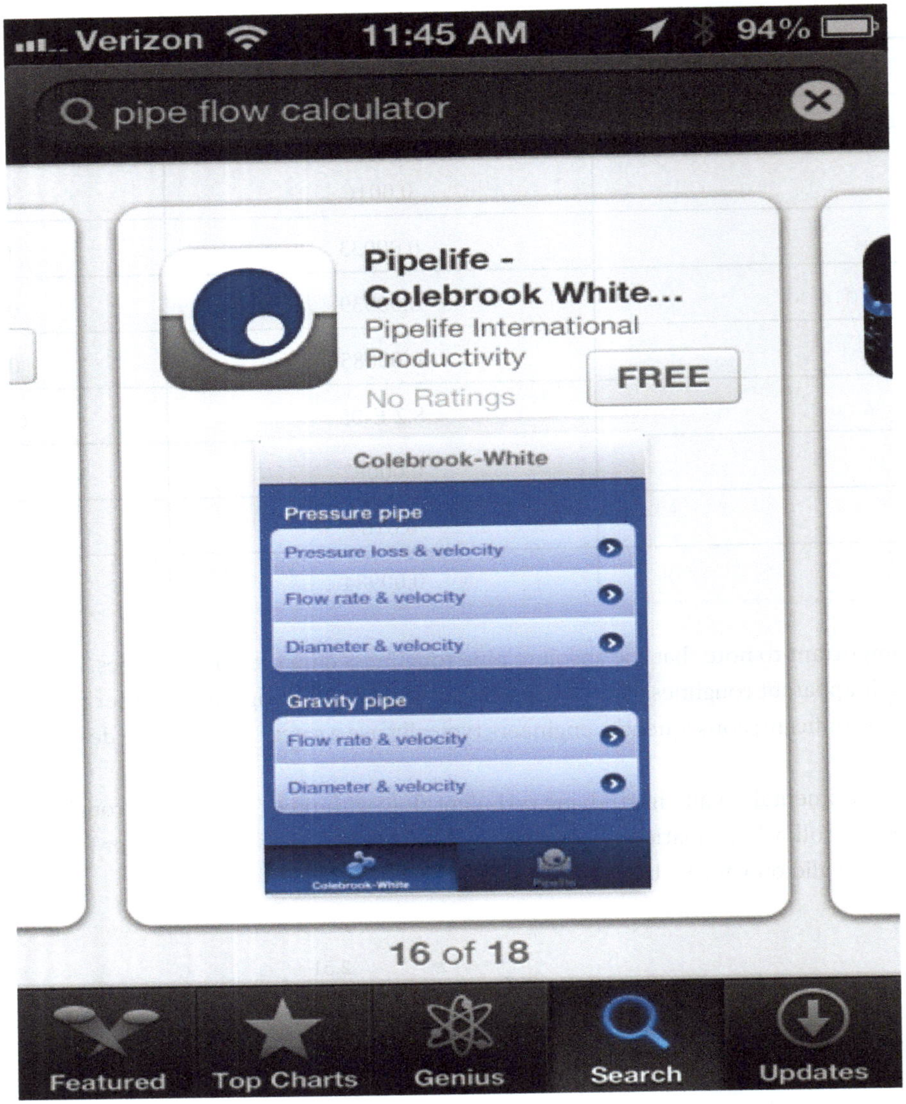

Figure 11-7 Typical app for calculating friction factor and pipe flow losses.

If a computer or smartphone is not available, we can determine the Darcy friction factor by using the Moody chart/ diagram. The Moody diagram is a graphical representation in nondimensional units that relate the Darcy friction factor, Reynolds number, and relative pipe roughness for *steady-state flow in circular pipes*. This diagram was developed by L.F. Moody (1944) based on Colebrook's work. While most students and practicing engineers would use computer online calculators or smartphone apps rather than the Moody diagram to calculate the Darcy friction factor, it is important for students to begin with the Moody diagram to establish a visual, graphical feel for the relationships of relative roughness, the Reynolds number, and the friction factor.

It is important to note the following:

- The Moody diagram is plotted on logarithmic scales due to the large ranges of relative roughness, Reynolds number, and friction factors encountered.
- We can also see on the diagram that for laminar flows, the friction factor simplifies to the single linear relationship, 64/Re, as previously discussed.
- The diagram shows the transition region between laminar and turbulent flows. In this region, engineers typically assume turbulent flow as the friction factor rises significantly from the laminar flow point. In assuming turbulent flow in this transition region, we are using the worst-case condition.
- For Reynolds numbers beyond the transition region (greater than 4,000), each of the curves plotted represents a specific relative pipe/conduit roughness value. For relative pipe/conduit roughness between plotted lines, interpolation is required.
- Also on the Moody diagram is a dashed line labeled complete turbulence. This line indicates that for values of Reynolds numbers beyond this dashed line, the change in friction factor is minimal. In other words, the friction factor is essentially constant depending only on the relative roughness and not on the Reynolds number; the plotted lines have relatively zero slopes once complete turbulence in the pipe is reached.

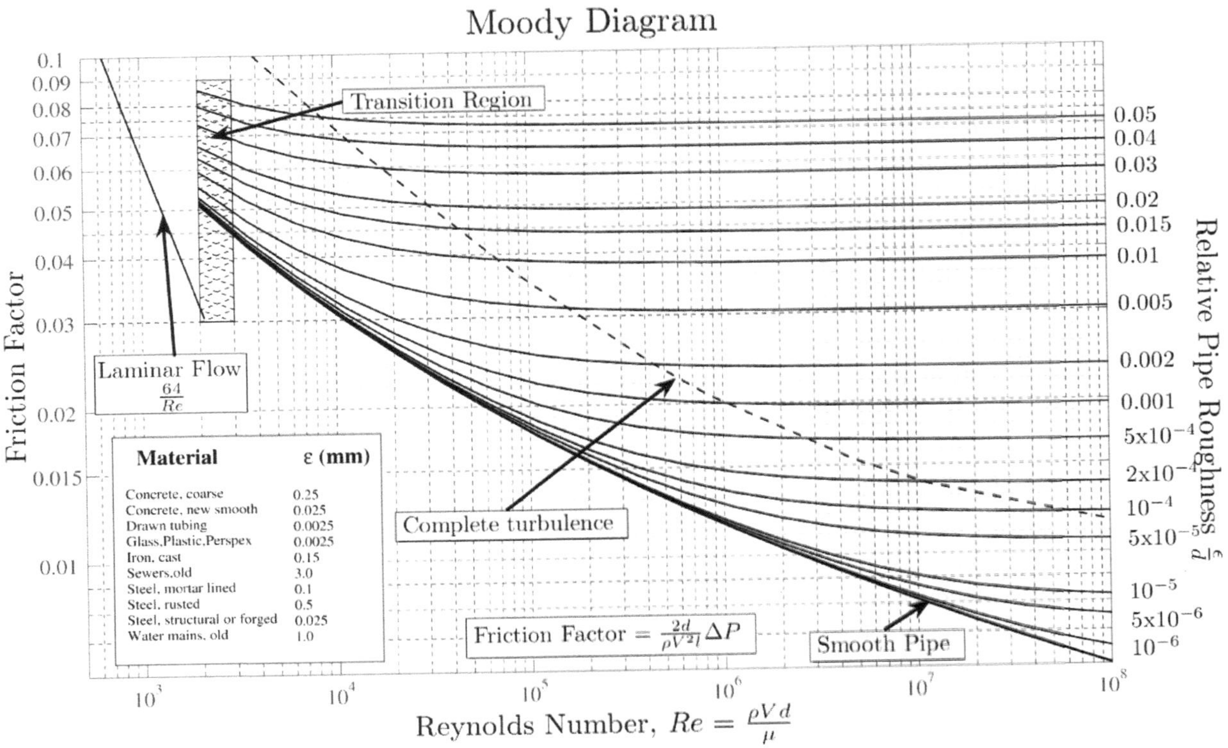

Figure 11-8 Moody diagram. Lines created using Swami and Jain formula. Plot created on MATLAB.

Losses With Fittings, Valves, and Connectors

In our fluid power circuits, we also have energy losses in fittings, bends, and valves. These losses can be very significant in fluid power systems. We also can experience energy losses due to sudden changes in pipe diameter. Experimental results in testing various fittings, valves, and connectors have shown that head loss is proportional to the square of the fluid velocity multiplied by a constant of proportionality (K), as shown in the following equation:

$$PressureHeadLoss(H_L) = K * (\frac{v^2}{2 * g})$$

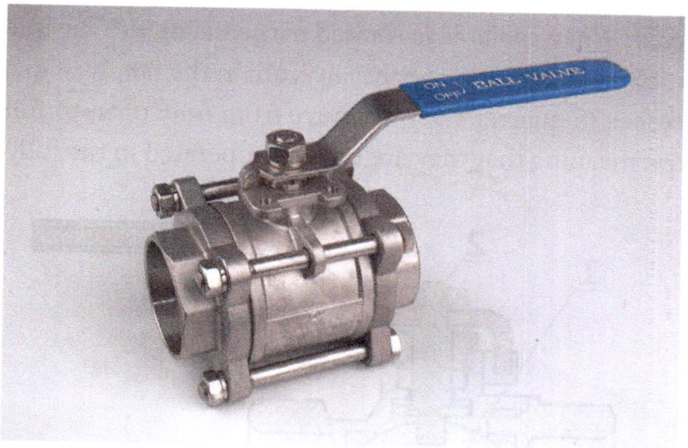

Figure 11-9 Standard hydraulic valve.

This constant of proportionality (K) is called the loss coefficient, or K-factor. The K-factor is typically provided by the device manufacturer and has no units (dimensionless).

Table 11-2. K-Factor for Pipe Fittings (Fully Open Unless Stated)

FITTING	Pipe Size 0.5 Inches (In)	Pipe Size 1 In	Pipe Size 2 In	Pipe Size 6 In
Ball valve	0.08	0.07	0.06	0.05
Ball valve one-third closed	6.2	6	5.7	5.5
Butterfly valve	NA	1	0.86	0.68
Gate valve	0.22	0.18	0.15	0.12
Gate valve half-closed	3.5	3	2.5	2.1
Globe valve	9.2	7.8	6.5	5.1
Standard elbow 9^{0} degrees	0.81	0.69	0.57	0.45
Standard elbow 45 degrees	0.43	0.37	0.30	0.24
Standard tee (through-flow)	0.54	0.46	0.38	0.30
Standard tee 90-degree flow path	1.62	1.38	1.14	0.90

Ball Valve

Figure 11.10 presents a ball valve. A ball valve has a spherical ball inside the valve, shown in stainless steel. The sphere is rotated by the red handle to control the flow in the valve. The sphere has a hole through the middle of the sphere that is only slightly smaller than the inside diameter of the pipe where the valve would be installed. When this hole is aligned with the inlet and outlet, as shown in the figure where the handle is aligned with the body of the valve, flow will occur. When the hole is rotated perpendicular to the inlet and outlet using the handle, no flow occurs. The ball valve, along with the butterfly valve, are in the family of quarter-turn valves. These valves only need to be turned 90 degrees (a quarter-turn) to move from fully open to fully closed. Most ball valves are used to provide or shut off flow, meaning that they are typically operated in the fully open or fully closed position.

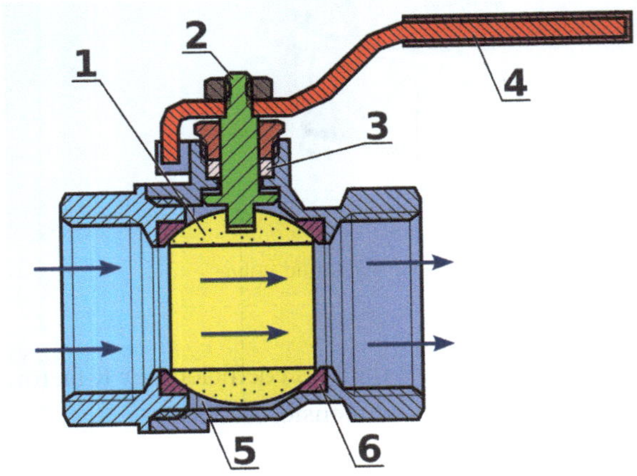

Figure 11-10 Sectioned ball valve.

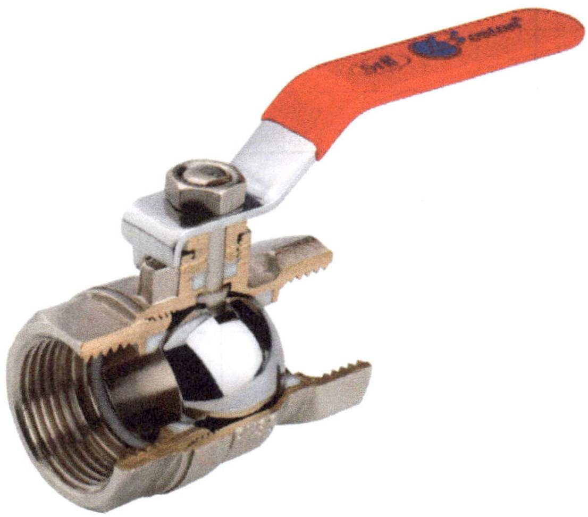

Figure 11-11 Sectional diagram of a ball valve: 1) ball; 6) seat; 2) stem; 3) compression seal fitting; 4) lever; 5) dead space.

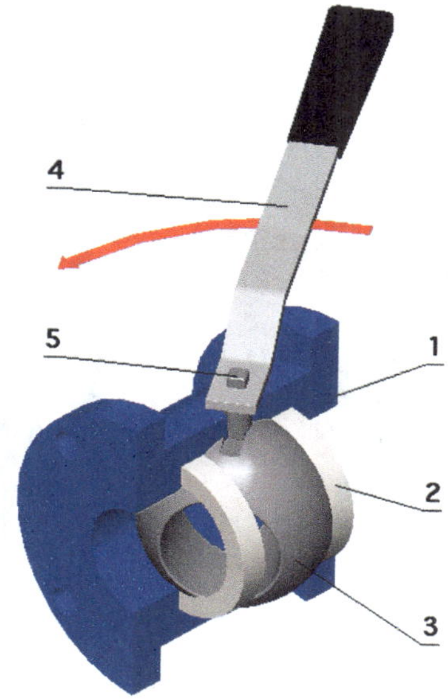

Figure 11-12 Cutaway of a simple manual ball valve: 1) body; 2) head; 3) ball; 4) lever handle; 5) stem.

Butterfly Valve

A butterfly valve is also a quarter-turn valve, as a disk is rotated around its center axis to regulate flow. The operation of a butterfly valve is similar to the ball valve, but a butterfly valve is typically lower in cost and lighter. The controlling disc is positioned in the center of the housing with a rod passing through or under the disc. Different from the ball valve, the disc when fully open is still part of the fluid flow path and therefore has a higher K-factor than the ball valve.

Figure 11-13 Butterfly valve, Yagisawa Dam, Japan.

Figure 11-14 Butterfly valve. Valves are generally made of plastic or steel.

Gate Valve

As shown in Figure 11-15, fluids flow in a straight line through the gate valve; thus, gate valves have little resistance to flow when fully wide open. The gate valve has a round disk, or gate, that moves 90 degrees to the flow to regulate fluid flow. The gate valve can also be called a sluice valve.

Figure 11-15 Molybdenum material gate valves, usually found in highly corrosive petrochemical processes.

Figure 11-16 Five- and six-foot gate valves shown partially open (gate moves vertically up and down).

Globe Valve

In the globe valve, the incoming fluid must change flow directions by approximately 90 degrees to flow through the globe and seat area. The flow exiting the globe and seat area must then change flow directions again by approximately 90 degrees to exit the valve. These rapid changes in direction cause energy losses in the fluid. The globe valve has a much higher K-factor than the ball, butterfly, or gate valve.

Figure 11-17 Globe valve (4 in).

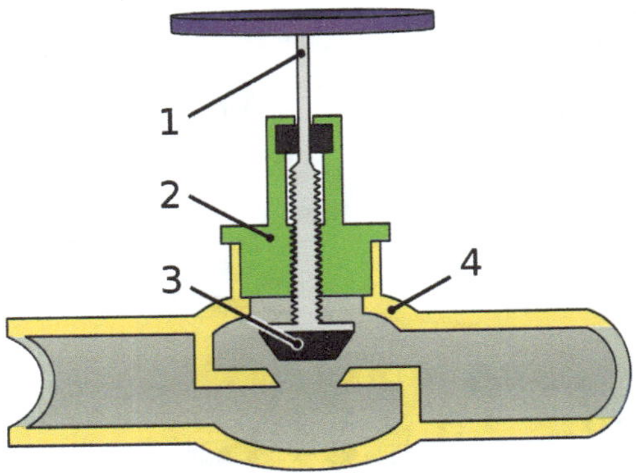

Figure 11-18 Drawing of the parts inside a globe valve with numbered parts: 1) stem; 2) bonnet; 3) globe; 4) body.

Elbows and Tees

An elbow is a pipe fitting that is installed to turn or redirect the fluid flow. A tee is used to split or combine fluid flow sources. Please note from the K-factor table (Table 11-2) that with a tee, the K-factor depends on the flow direction.

Figure 11-19 Ordinary thread pipe and elbow fitting.

Figure 11-20 Elbow fittings in use.

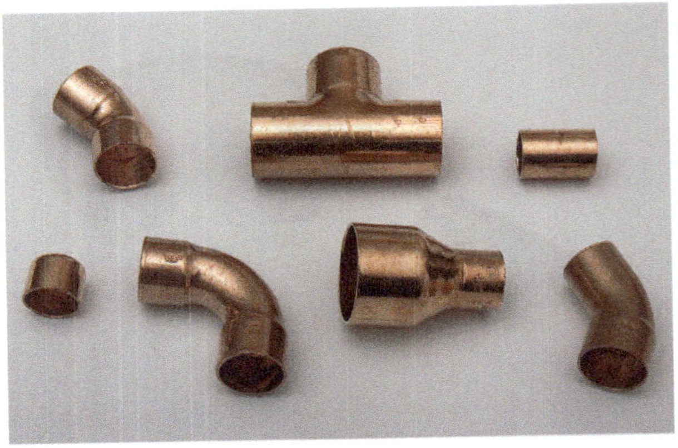

Figure 11-21 Copper fittings. Upper level: 45-degree elbow (female-female); tee; and coupler. Lower level: endcap; 90-degree sweep elbow (female-female); reducer; and 45-degree elbow (male-female).

Equivalent-Length Technique

Using the Darcy-Weisbach equation, we can calculate the pressure head loss in any section of pipe, and with the K-factor, we can calculate the pressure head loss in any valve/fitting. An effective and simple way to analyze the pressure loss in a fluid power system is to use the equivalent-length technique. With the equivalent-length technique, we find a length of pipe that is equivalent to the same head loss as the valve or fitting being analyzed. This equivalent length of pipe is found by equating head losses of the valve/fitting being analyzed and the pipe length. We can equate our Darcy-Weisbach head loss equation and our energy lost in fittings in the following equations:

$$Pressure\ Head\ Loss\ (h_f) = f_d * \frac{L_{eq}}{D} * \frac{v^2}{2*g} = K * \left(\frac{v^2}{2*g}\right)$$

$$f_d * \frac{L_{eq}}{D} * \frac{v^2}{2*g} = K * \left(\frac{v^2}{2*g}\right)$$

$$f_d * \frac{L_{eq}}{D} = K$$

$$L_{eq} = \frac{K*D}{f_d}$$

L_{eq} is the equivalent length of a valve/fitting for a given diameter pipe. With these techniques and the equivalent-length process, we can perform a complete analysis of a fluid power system to analyze the energy losses in the system due to friction.

Key Equations

$$Reynolds\#\,(Re) = \frac{Density * Velocity * Diameter}{Dynamic_{Viscosity}} = \frac{\rho * \nu * D}{\mu}$$

where:

- v = mean velocity of the object relative to the fluid (SI units: m/s),
- D = diameter for round pipes (SI units: m),
- μ = dynamic viscosity of the fluid (SI units: Pa·s or N·s/m^2), and
- ρ = density of the fluid (kg/m^3).

$$PressureHeadLoss(h_f) = f_d * \frac{L}{D} * \frac{v^2}{2 * g}$$

where:

- h_f = head loss due to friction (SI units: m),
- f_d = Darcy friction factor,
- L = length of the pipe (SI units: m),
- D = hydraulic diameter of the pipe (for a pipe of circular section, this equals the internal diameter of the pipe; SI units: m),
- v = mean velocity of the object relative to the fluid (SI units: m/s), and
- g = local acceleration due to gravity (SI units: m/s^2).

Recall from our previous work that the average velocity of the fluid flow equals the volumetric flow rate (Q) divided by the cross-sectional area (A) of the flow.

or $v = \frac{Q}{A}$

$$f_d = \frac{64}{Re}$$

Substituting this friction factor into our Darcy-Weisbach equation yields the following equation:

$$PressureHeadLoss(h_f) = \frac{64}{Re} * \frac{L}{D} * \frac{v^2}{2 * g}$$

$$Relativeroughness = \frac{PipeInsideDiameterRoughness}{PipeInsideDiameter} = \frac{\varepsilon}{D}$$

C. F. Colebrook (1939) equation, Darcy friction factor for turbulent flow. Note that this equation is valid only for turbulent-flow conditions.

$$\frac{1}{\sqrt{f}} = -2\log_{10}(\frac{\varepsilon}{3.7 * D} + \frac{2.51}{Re * \sqrt{f}})$$

where:

- f = Darcy friction factor,
- ε = absolute roughness (SI units: m),
- D = hydraulic diameter of the pipe (for a pipe of circular section, this equals the internal diameter of the pipe; SI units: m), and
- Re = Reynolds number (dimensionless).

$$PressureHeadLoss(H_L) = K * (\frac{v^2}{2 * g})$$

This constant of proportionality (K) is called the loss coefficient, or K-factor. The K-factor is typically provided by the device manufacturer and has no units (dimensionless).

Darcy-Weisbach head loss equation and our energy lost in fittings equations as shown below:

$$Pressure\ Head\ Loss\,(h_f) = f_d * \frac{L_{eq}}{D} * \frac{v^2}{2*g} = K * \left(\frac{v^2}{2*g}\right)$$

$$f_d * \frac{L_{eq}}{D} * \frac{v^2}{2*g} = K * \left(\frac{v^2}{2*g}\right)$$

$$f_d * \frac{L_{eq}}{D} = K$$

$$L_{eq} = \frac{K*D}{f_d}$$

L_{eq} is the equivalent length of a valve/fitting for a given-diameter pipe.

Exercises

1. You want to define what is the maximum velocity for a hydraulic system using Reynolds number (laminar flow) as a design guideline for your Go-kart design. Establish this velocity guideline using the following: Reynolds number below 2,300 (laminar flow), pipe diameter of 0.75 in, kinematic viscosity =

5.382 x 10^{-4} square feet per second (ft^2/s). Find the velocity in feet per second (ft/s). How does this guideline compare to the video's or book's guideline of less than 20 ft/s?

2. The kinematic viscosity of a hydraulic oil is 80 centistokes (cSt). If the oil is flowing in a 1.5-in-diameter pipe at a velocity of 30 ft/s, what is the Reynolds number? How is the flow classified (laminar, transition, or turbulent)?
3. The kinematic viscosity of a hydraulic oil is 0.0001 m^2/s. If the oil is flowing in a 30-mm-diameter pipe at a velocity of 9 m/s, what is the Reynolds number, and how is the flow classified?
4. A hydraulic system is operating at a Reynolds number of 1,500 (laminar) at 40 degrees Celsius (°C) with International Organization for Standardization (ISO) grade 32 hydraulic oil. If the temperature of the oil increases due to its use at 100°C, to what will the Reynolds number change? How is this flow classified?

ISO Grade	**SAE International Equivalent**	***Viscosity (Kinematic)***				***Density***	
		cSt		**10^{-6} Re (pound-seconds per square inch [$lb{\cdot}s/in^2$])**		**kg/m^3**	**lb/in^3**
		40°C	100°C	104 degrees Fahrenheit (°F)	212°F		
32	10W	32	5.4	4.0	0.6	857	0.0310
46	20	46	6.8	5.7	0.8	861	0.0311
68	20W	68	8.7	8.5	1.1	865	0.0313
100	30	100	11.4	12.6	1.4	869	0.0314
150	40	150	15.0	19.0	1.8	872	0.0315
220	50	220	19.4	27.7	2.4	875	0.0316

5. The kinematic viscosity of a hydraulic oil is 100 cSt. If the oil is flowing in a system with 0.75-in-diameter piping, find the head loss in feet if the fluid velocity is 12 ft/s.
6. The kinematic viscosity of a hydraulic oil is 0.0001 m^2/s. If the oil is flowing in a 20-mm-diameter pipe, find the head loss if the velocity of the fluid is 2 m/s.
7. **Find the friction factor for a given pipe.** The absolute roughness (pipe identification surface roughness) of the pipe is 0.003 ft. The diameter of the pipe is 8 in. The kinematic viscosity of the fluid is 14.1* 10^{-6} ft^2/s. The velocity of the fluid is 0.5 ft/s. State how you found this Moody diagram, app, or online calculator.
8. **Find the friction factor for a given pipe.** The absolute roughness (pipe identification surface roughness) of the pipe is 0.025 mm. The diameter of the pipe is 125 mm. The kinematic viscosity of the fluid is 50 cSt. The velocity of the fluid is 0.2 m/s. State how you found this Moody diagram, app, or online calculator.
9. **Find the friction loss for a given pipe in meters of head loss.** The length of the pipe is

600 m, and the absolute roughness (pipe identification surface roughness) of the pipe is 0.0005 m. The diameter of the pipe is 0.2 m. The kinematic viscosity of the fluid is $1.306 * 10^{-6}$ m^2/s. The volume flow rate of the fluid is 0.04 m^3/s.

10. **Find the friction loss for a given pipe in feet of head loss.** The length of the pipe is 1,000 ft, and the absolute roughness (pipe identification surface roughness) of the pipe is 0.003 ft. The diameter of the pipe is 0.5 ft. The kinematic viscosity of the fluid is $14.1 * 10^{-6}$ ft^2/s. The volume flow rate of the fluid is 0.5 ft^3/s.
11. Find the friction loss in kilopascals (3) and pounds per square inch (4) for problems 3 and 4 above using a specific gravity of 0.9.

Reference

The Engineering ToolBox. 2008. "Colebrook Equation." https://www.engineeringtoolbox.com/colebrook-equation-d_1031.html.

Figure Credits

Fig. 11-1: R. Nave / Georgia State University, http://en.wikipedia.org/wiki/File:Laminar_flow.gif, 2005.
Fig. 11-2: R. Nave / Georgia State University, http://en.wikipedia.org/wiki/File:Laminar_flow_profile.gif, 2005.
Fig. 11-3: Dubaj, http://commons.wikimedia.org/wiki/File:Laminar_and_turbulent_flows.svg, 2008.
Fig. 11-4: Anassagora, http://commons.wikimedia.org/wiki/File:Turbulence_in_Abdominal_Aorta_Aneurisma.svg, 2009.
Fig. 11-5: Osborne Reynolds, http://en.wikipedia.org/wiki/File:Reynolds_observations_turbulence_1883.svg, 1883.
Fig. 11-6: Antonysigma, http://en.wikipedia.org/wiki/File:Poiseuille_abstraction.svg, 2011.
Fig. 11-8: Copyright © by Donebythesecondlaw (CC BY-SA 3.0) at http://en.wikipedia.org/wiki/File:Moody_diagram.jpg.
Fig. 11-9: Copyright © by Sthidraulica (CC BY-SA 3.0) at https://commons.wikimedia.org/wiki/File:Alloy_valves.JPG.
Fig. 11-10: Sthidraulica, "Sectioned Ball Valve," http://www.engineer-and-technician.com/ball-valves/. Copyright © 2007 by engineer-and-technician.com.
Fig. 11-11: Copyright © by Sunspeanzler (CC BY-SA 4.0) at http://commons.wikimedia.org/wiki/File:Esquema_valvula_de_bola_1-6.svg.
Fig. 11-12: Copyright © by Ruben Castelnuovo (CC BY-SA 3.0) at http://en.wikipedia.org/wiki/File:Ball.PNG.
Fig. 11-13: Copyright © by Qurren (CC BY-SA 3.0) at http://en.wikipedia.org/wiki/File:Yagisawa_power_station_inlet_valve.jpg.
Fig. 11-14: Copyright © by Heather Smith (CC BY 3.0) at http://en.wikipedia.org/wiki/File:Butterfly-valve--The-Alloy-Valve-Stockist.JPG.
Fig. 11-15: Copyright © by Heather Smith (CC BY 3.0) at http://commons.wikimedia.org/wiki/File:6-Moly-Gate_Valve--The-Alloy-Valve-Stockist.jpg.
Fig. 11-16: U.S. National Archives and Records Administration, http://commons.wikimedia.org/wiki/File:%225_ft._and_6_ft._gate_valves,_220_ft._head.%22_-_NARA_-_294545.tif.
Fig. 11-17: Copyright © by Heather Smith (CC BY 3.0) at http://commons.wikimedia.org/wiki/File:Globe-valve.JPG.
Fig. 11-18: Petteri Aimonen, http://commons.wikimedia.org/wiki/File:Globe_valve_diagram.svg, 2009.
Fig. 11-19: Scs, https://commons.wikimedia.org/wiki/File:Malefemalepipe.jpg, 2007.
Fig. 11-20: Copyright © by frozen-tundra (CC BY 2.0) at http://commons.wikimedia.org/wiki/File:VictaulicPiping.jpg.
Fig. 11-21: Copyright © by Torsten Bätge (CC BY-SA 3.0) at http://commons.wikimedia.org/wiki/File:Kupferfittings_4062.jpg

Appendix

Fluid Power Symbols

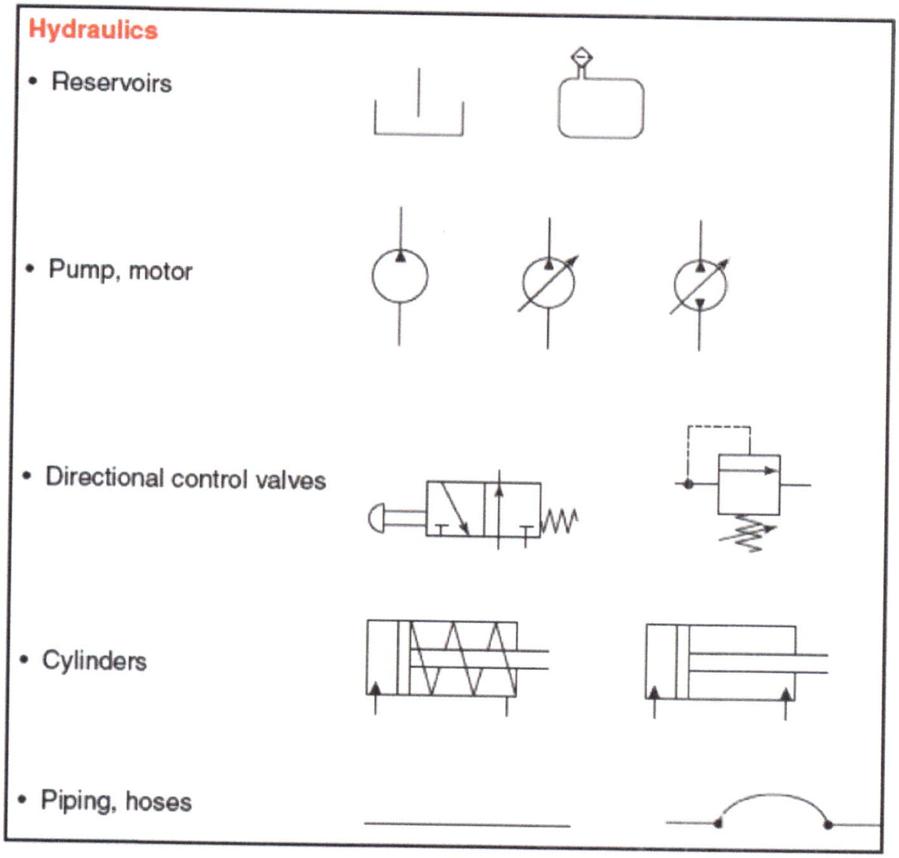

Image A.1

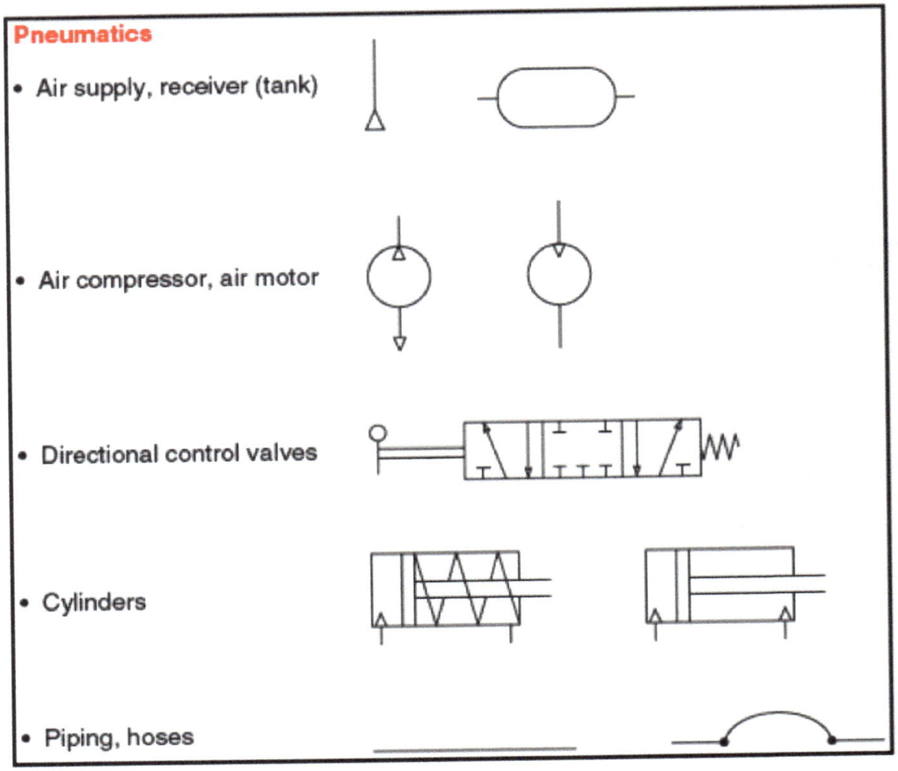
Pneumatics
• Air supply, receiver (tank)
• Air compressor, air motor
• Directional control valves
• Cylinders
• Piping, hoses

Image A.2

Properties of Water

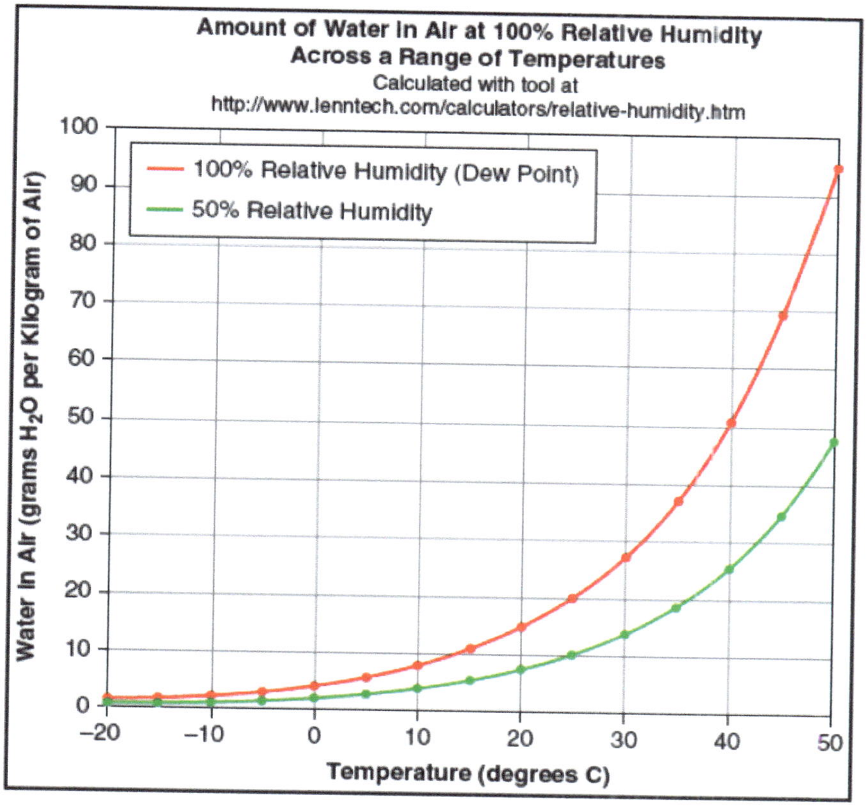

Image A.3

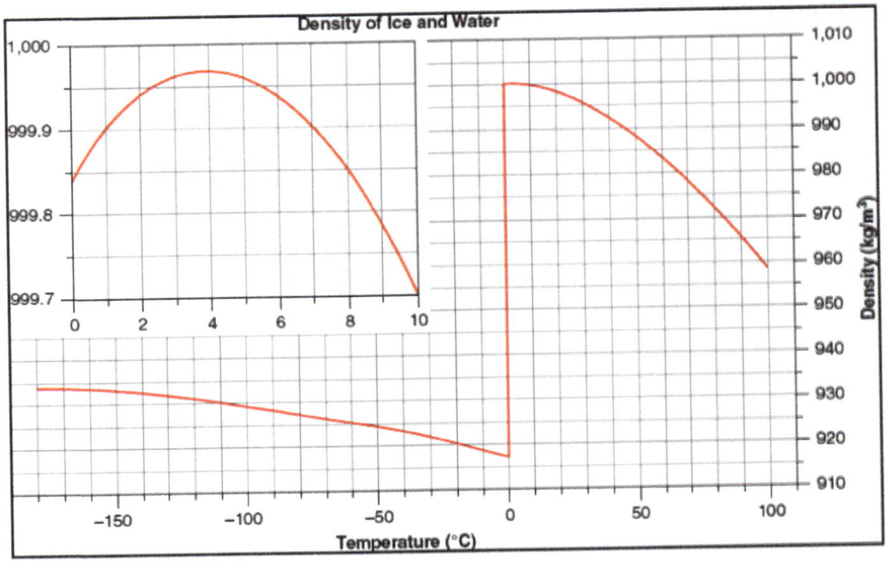

Image A.4

Density of water at 4°C	999.9720 kilogram (kg)/cubic meter (m^3)
Bulk modulus	2.2 gigapascal (GPa)

Variation of Viscosity With Temperature

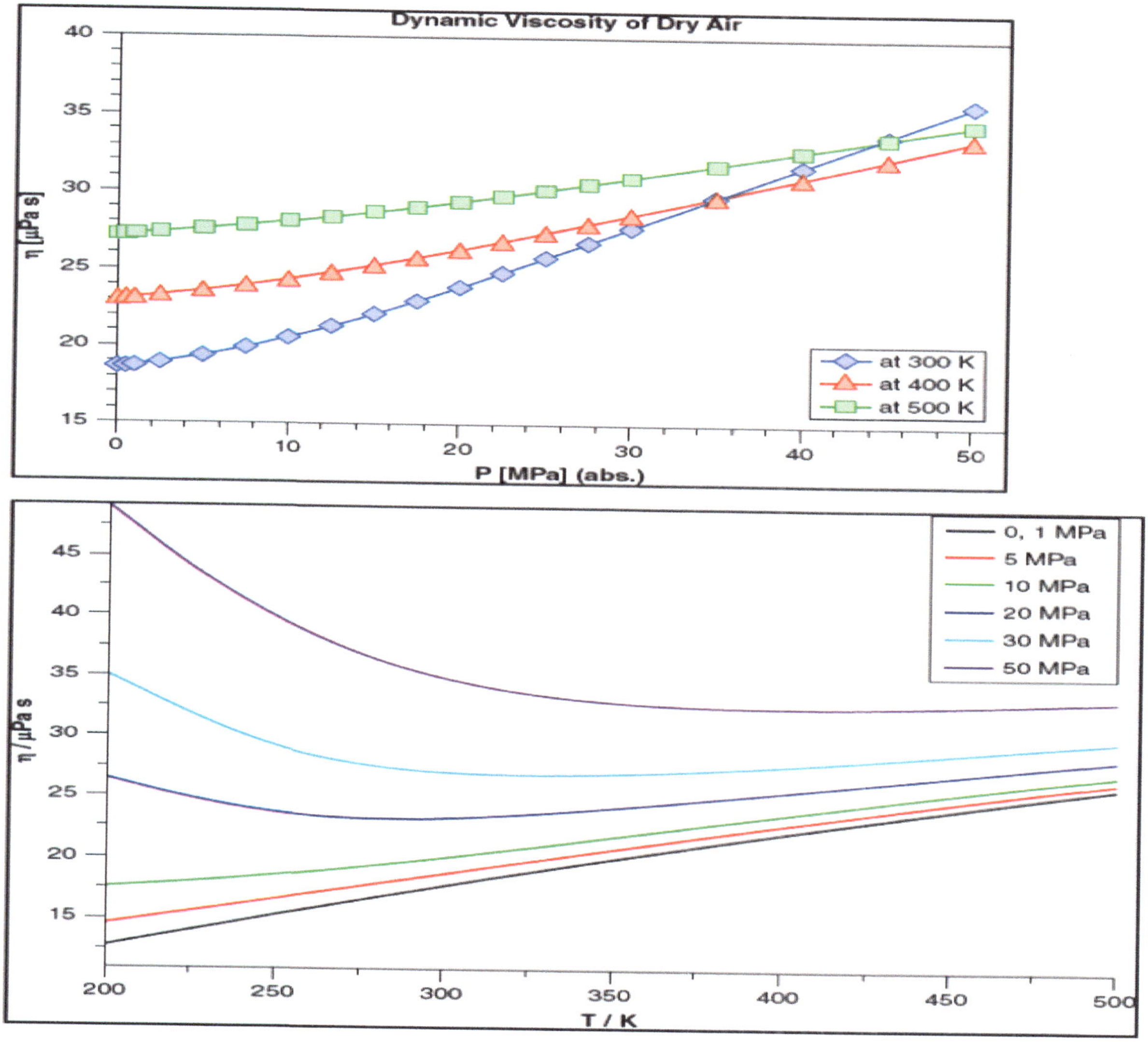

Image A.5

Conversion Factors

ATMOSPHERE (atm):

The pressure exerted by 76 centimeters (cm) mercury with a density of 13.5951 gram (gm)/cubic centimeter (cm^3) at 1 G (g; the standard barometric pressure at sea level).

1 atm = 760 millimeters of mercury (mm Hg)

1 atm = 14.696 pounds per square inch (psi)

1 atm = 101.329 kilopascals (kPa)

CUBIC CENTIMETER (cc or cm^3):

1 cc = 3.531 × 10-5 cubic foot (ft^3)

1 cc = 0.061023 cubic inch (in^3)

1 c = 1 × 10-6 cubic meter (m^3)

1 cc = 1000 cubic millimeter (mm^3)

1 cc = 2.6417 × 10-4 gallons (gal; U.S. fluid)

1 cc = 0.0338 ounces (oz; U.S. fluid)

1 cc = 2.113 × 10-3 pints (U.S. fluid)

1 cc/second (sec) = 0.0021186 ft^3/minute (min)

CUBIC FOOT:

1 ft^3 = 1728 in^3

1 ft^3 = 28.32 liters (L)

1 ft^3 = 0.02832 m^3

1 ft^3/min = 472.0 cc/sec

1 ft^3/min = 0.4720 L/sec

1 ft^3/min = 62.43 pounds (lb) H_2O/min

1 ft^3/sec = 1699.3 L/min

CUBIC INCH:

1 in^3 = 5.787 × 10-4 ft^3

1 in^3 = 1.639 × 102 L

1 in^3 = 1.639 × 10-5 m^3

CUBIC METER:

1 m^3 = 35.3144 ft^3

1 m^3 = 6.1023 × 104 in^3

1 m^3 = 999.973 L

DEGREE (ANGULAR):

1 degree (deg) = 60 min

1 deg = 0.01745 radian (rad)

1 deg = 3600 sec

DEGREES TO RADIANS:

1 deg = Pi (π)/180 rad

$60^\circ \times \pi/180 = 1.0472 = 1$ rad

DEGREES CENTIGRADE (°C):

°C = 5/9 × (degrees Fahrenheit [°F] – 32)

DEGREES FAHRENHEIT:

°F = (9/5 · °C) + 32

FOOT (ft):

1 ft = 30.48 centimeter (cm)

1 ft = 12 inches (in)

1 ft = 0.3048 meters (m)

FEET PER MINUTE:

1 ft/min = 0.3048 m/min

1 ft/min = 0.005080 m/sec

1 ft/min = 0.011364 miles per hour (mph)

FEET PER SECOND:

1 ft/sec = 1.0973 kilometer (km)/hour (hr)

1 ft/sec = 0.5921 knot

1 ft/sec = 0.6818 mph

FOOT-POUND (ft-lb):

1 ft-lb = 0.001285 British thermal unit (Btu)

1 ft-lb = 1.3558 × 107 ERG

1 ft-lb = 3.2389 × 10-4 kilocalories (kcal)

1 ft-lb/min = 3.0303 × 10-5 horsepower (hp)

1 ft-lb/min = 0.01667 ft-lb/sec

1 ft-lb/min = 0.022597 watts

1 ft-lb/sec = 0.001818 hp

1 ft-lb/sec = 0.01943 kcal/min

1 ft-lb/sec = 1.3558 watts

G (g):

The acceleration of gravity (also the acceleration of a vehicle):

1 g = 9.81 m/s^2

1 g = 32.2 ft/s^2

GRAM (gm):

1 gm = 0.001 kilograms (kg)

1 gm = 1000 milligrams (mg)

1 gm = 0.03527 oz

1 gm = 0.0022046 lb

1 gm/cm^3 = 62.428 lb/ft^3

1 gm/hr = 0.540 lb/day

1 gm/hr = 0.0003757 lb/min

1 gm/liter = 0.062427 lb/ft^3

1 gm/cm^2 = 9.6784 × 10-4 atm

1 gm/cm^2 = 980.665 dynes/square centimeter (cm^{2})

1 gm/cm^2 = 0.9356 mm Hg

1 gm/cm^2 = 0.014223 psi

HORSEPOWER (hp):

1 hp = 3.300 × 104 ft-lb/min

1 hp = 550 ft-lb/sec

1 hp = 10.688 kcal/min

1 hp = 745.7 watts

INCH (in):

1 in = 2.540 cm

1 in = 0.0833 ft

1 in = 25.40 mm

INCH OF WATER (in H_2O):

1 in H_2O (at 4°C) = 0.002458 atm

1 in H_2O (at 4°C) = 2490.82 $dynes/cm^2$

1 in H_2O (at 4°C) = 0.0361 psi

1 in H_2O (at 4°C) = 1.868 mm Hg

JOULE (joule):

1 joule = 1 watt-sec

KILOGRAM (kg):

1 kg = 1000 gm

1 kg = 2.205 lb

1 kg = 35.28 oz

KILOMETER (km):

1 km = 105 cm

1 km = 3281 ft

1 km = 103 m

1 km = 0.6214 statute miles (mi)
1 km = 0.54 nautical miles (nm)

KILOPASCAL (kPa):
1kPa = 6.895 psi

1kPa = 4.754 × 105 dynes/cm^2
1 psi = 0.145 kPa

KILOMETERS PER HOUR:
1 km/hr = 0.9113 ft/sec
1 km/hr = 0.5396 knot
1 km/hr = 0.6214 mph

LITER (L):

1 L = 0.03531 ft^3

1 L = 61.02 in^3
1 L = 1000 milliliters (mL)

1 L/min = 5.886 × 10-4 ft^3/sec

1 L/sec = 2.12 ft^3/min

METER (m):
1m = 100 cm
1m = 3.281 ft
1m = 39.37 in

METER PER SECOND (m/sec):
1 m/sec = 3.281 ft/sec
1 m/sec = 3.600 km/hr
1 m/sec = 2.2369 mph

MICRON (μ or mu):
A unit of length equal to one-millionth of a meter or one-thousandth of a millimeter, usually called a micrometer.
1 μ = 10-6 meters
1 μ = 3.937 × 10-5 in
1 μ = 0.001 mm

MIL (mil):
1 mil = 0.001 in
1 mil = 0.0254 mm
1 mil = 25.40μ (microns)

MILES (statute; mi):

1 mi = 1,609 × 105 cm

1 mi = 5,280 ft

1 mi = 1.609 km

MILES PER HOUR (mph):

1 mph = 88 ft/min

1 mph = 1.4667 ft/sec

1 mph = 1.6093 km/hr

1 mph = 0.8684 knot

MILLIGRAM (mg):

1 mg = 0.001 gm

1 mg = 3.5274 × 10-5 oz

1 mg = 2.2046 × 10-6 lb

1 mg/m^3 = 6.243 × 10-4 lb/ft^3

MILLILITER (mL):

1 mL = 1.000028 cc

1 mL = 0.061025 in^3

1 mL = 0.001 L

1 mL = 0.0338 oz (U.S. fluid)

MILLILITERS PER HOUR:

1 mL/hr = 0.06102 in^3/hr

MILLIMETER (mm):

1 mm = 0.10 cm

1 mm = 0.03937 in

1 mm = 1000 μm

MILLIMETER OF MERCURY (mm Hg):

1 mm Hg (at 0°C) = 0.0013158 atm

1 mm Hg (at 0°C) = 1333.22 dyne/cm^2

1 mm Hg (at 0°C) = 1.3595 gm/cm^2

1 mm Hg (at 0°C) = 0.019337 psi

1 mm Hg (at 0°C) = 0.535 in H_2O

1 mm Hg (at 0°C) = 0.1333 kPa

MILLISECONDS (msec):

1 msec = 0.001 sec

NEWTONS TO POUNDS:
1 N = 0.22507 lbf

NEWTON CENTIMETERS TO INCH-OUNCES (in-oz):
1 N-cm = 1/.706 in-oz
70N-cm × 1/.706 = 99.15 = 100 in-oz

NEWTON METERS TO FOOT-POUNDS:
1 Nm = 1/1.356 ft-lb
15 Nm × 1/1.356 = 11.06 = 11 ft-lb

NEWTON METERS TO INCH-POUNDS (in-lb):
1 Nm = 1/.11298 in-lb
4 Nm × 1/.11298 = 35.4 = 3 5 in-lb
Physical Constants:

Acceleration of gravity (g) = 32.17 ft/sec^2 = 980.6 cm/sec^2

Density of air at 0°C and 760 mm Hg = 0.0012929 gm/cm^3(0.0807 lb/ft^3)

Density of water at 0°C = .099984 gm/cm^3
Pi (π) = 3.14159265
Specific heat of air = Cp = 0.238 calories (cal)/gm (°C)
Velocity of light (c) = 2.99792458 × 1010 cm/sec
Velocity of sound in dry air at 0°C, 1 atm = 33,136 cm/sec = 1,089 ft/sec

POISE: Unit of viscosity

1 poise = 1 dyne/sec, cm^2
1 poise = 1 gm/cm, sec
1 poise = 0.067196 lb/ft, sec

POUND (lb):
1 lb = 453.5924 gm
1 lb = 0.45359 kg
1 lb = 16 oz
1 lb = 4.448 N
1 lb/day = 18.89 gm/hr
1 lb/hr = 0.7559 gm/min
1 lb/hr = 10.886 kg/day

POUNDS PER SQUARE INCH (psi):
1 psi = 0.06805 atm

1 psi = 6.8947 × 104 dynes/cm^2

1 psi = 70.307 gm/cm^2
1 psi = 51.715 mm Hg
1 psi = 27.7 in H_2O
1 psi = 145.03 Pa

RADIAN (rad):
1 radian = ½π circumference revolution (0.15915)
1 radian = 57.296 deg
1 radian/sec = 57.296 deg/sec
1 rad/sec = 9.549 revolutions per minute (rpm)
1 rad/sec^2 = 572.96 rpm^2

REVOLUTIONS PER MINUTE (rpm):
1 rpm = 6 deg/sec
1 rpm = 0.10472 rad/sec

SQUARE CENTIMETER (cm^2):
1 cm^2 = 1.076 × 10-3 ft^2
1 cm^2 = 0.1550 in^2
1 cm^2 = 100 mm^2

SQUARE FOOT (ft^2):
1 ft^2 = 929.0304 cm^2
1 ft^2 = 144 in^2

SQUARE INCH (in^2):
1 in^2 = 6.4516 cm^2
1 in^2 = 0.006944 ft^2
1 in^2 = 645.1626 mm^2

SQUARE MILLIMETER (mm^2):
1 mm^2 = 0.01 cm^2
1 mm^2 = 0.001550 in^2

Source: National Aeronautics and Space Administration. n.d. "Man-Systems Integration Standards: Volume II—Appendices." http://msis.jsc.nasa.gov/volume2/Appx_e_units.htm.

Printed and bound by CPI Group (UK) Ltd, Croydon, CR0 4YY
13/09/2026
14961053-0002